AF581743

Wood Modification: Characterization, Modelling and Applications

Wood Modification: Characterization, Modelling and Applications

Editors

Anna Sandak
Jakub Sandak

MDPI • Basel • Beijing • Wuhan • Barcelona • Belgrade • Manchester • Tokyo • Cluj • Tianjin

Editors

Anna Sandak
InnoRenew CoE
University of Primorska
Slovenia

Jakub Sandak
InnoRenew CoE
University of Primorska
Slovenia

Editorial Office
MDPI
St. Alban-Anlage 66
4052 Basel, Switzerland

This is a reprint of articles from the Special Issue published online in the open access journal *Coatings* (ISSN 2079-6412) (available at: https://www.mdpi.com/journal/coatings/special_issues/wood_modific_character_modle_appl).

For citation purposes, cite each article independently as indicated on the article page online and as indicated below:

LastName, A.A.; LastName, B.B.; LastName, C.C. Article Title. *Journal Name* **Year**, *Volume Number*, Page Range.

ISBN 978-3-0365-2382-8 (Hbk)
ISBN 978-3-0365-2383-5 (PDF)

Cover image courtesy of Anna Sandak and Jakub Sandak.

Contents

About the Editors

Anna Sandak Research Group Leader of the Wood Modification group at InnoRenew Centre of Excellence and assistant professor and research associate at the Faculty of Mathematics, Natural Science and Information Technology at University of Primorska. She was previously employed at the Trees and Timber Institute of Italian National Research Council, where she coordinated the Laboratory of Surface Characterization. She has a PhD in Wood Science and an M.Sc. in Biology. Her research activities include the investigation of the multi-scale relationship and performance of modified and functionalized bio-based materials and implementing them as new architectural elements. Her passion is to search for biomimetic solutions for design of new materials and to promote knowledge-based use of bio-inspired materials in modern sustainable buildings. Her recent book "Bio-based building skin" was published by Springer Nature as part of their Environmental Footprints and Eco-design of Products and Processes book series.

Jakub Sandak A third-generation carpenter currently working as a researcher at InnoRenew Centre of Excellence and assistant professor and research associate at the Faculty of Mathematics, Natural Science and Information Technology at the University of Primorska. He was previously employed at the Trees and Timber Institute of the Italian National Research Council. He received his PhD in Agricultural Sciences from Tottori University (Japan), a Master of Science in Natural Resources Process Engineering from Shimane University (Japan), and is an engineer of Wood Science and Technology from University of Life Sciences in Poznan (Poland). His research interests include multi-sensor evaluation of wood properties, expert systems dedicated for wood machining process control and monitoring, organization of manufacturing processes in furniture industries, vision systems in the wood industry, characterization of the wood surface, physics of the wood cutting, wood mechanics and fracture, spectroscopy and chemometry, hyperspectral imaging, numerical modelling of wood at different scales, and mechatronics. His passion is designing, assembling, testing, and implementing prototype scientific instrumentation.

Editorial

Special Issue "Wood Modification: Characterization, Modelling, and Applications"

Anna Sandak [1,2,*] and Jakub Sandak [1,3,*]

1 InnoRenew CoE, Livade 6, 6310 Izola, Slovenia
2 Faculty of Mathematics, Natural Sciences and Information Technologies, University of Primorska, Glagoljaška 8, 6000 Koper, Slovenia
3 Andrej Marušič Institute, University of Primorska, Titov trg 4, 6000 Koper, Slovenia
* Correspondence: anna.sandak@innorenew.eu or anna.sandak@famnit.upr.si (A.S.); jakub.sandak@innorenew.eu or jakub.sandak@upr.si (J.S.); Tel.: +386-40282949 (A.S.); +386-40282959 (J.S.)

Citation: Sandak, A.; Sandak, J. Special Issue "Wood Modification: Characterization, Modelling, and Applications". *Coatings* **2021**, *11*, 869. https://doi.org/10.3390/coatings11070869

Received: 12 July 2021
Accepted: 16 July 2021
Published: 20 July 2021

Publisher's Note: MDPI stays neutral with regard to jurisdictional claims in published maps and institutional affiliations.

Wood has been recognized as an attractive alternative to several other traditional construction solutions, and it is often called the "building material of the 21st century". However, compared with other traditional materials, wood possesses some technical limitations and properties that, because they are less understood, remain difficult to control. The most problematic are low-dimensional stability, thermal steadiness, limited fire resistance, biotic and abiotic degradation resistance, and varying mechanical properties highly affected by the morphological structure of wood. These properties need to be improved to further enlarge wood application fields as well as reinforcing the confidence of architects, engineers, and consumers when using wood. New developments in the field of wood modification resulted in the discovery of highly innovative materials with enhanced properties for natural timber. Several of these reached the highest readiness level and are now mass-produced in high volumes. Further intensive research is in progress to discover new wood modification solutions to ensure improvement of wood properties and functionality, allowing elongated service life and reducing the risk of unexpected product failure. This Special Issue presents the newest research outcome in the field of the enhancement of native wood properties through a wide range of chemical, biological, and physical agents. It contains two reviews and ten research reports authored by researchers from four continents and 13 countries, namely, Australia, Finland, Greece, Iceland, Iran, New Zealand, Norway, Poland, Romania, Slovenia, Switzerland, the UK, and the USA.

An overview of functional treatments for modified wood is provided in a comprehensive review paper [1]. The manuscript presents two parallel but closely connected aspects of timber modification: the material functionalization strategies combined with the expected (multi)-functionality of timber products. The wide range of covered modification processes and the scope of applications proved that wood modification is not only the sole concern of a few wood scientists but it became a truly inter-disciplinary research area. Novel formulations, characterization methods, and serviceability are presented and deeply discussed in several publications contributing to the Special Issue [2–7]. A broad portfolio of characterization methods can be successfully implemented for the evaluation of complex mechanisms of deterioration process occurring during service life. Artificial [2,3], natural [4,5], and combined [6] weathering methods are used for the assessment of the performance of different substrates and coatings. The performance of wood finished with a broad range of coatings systems is highly improved when compared to that of unmodified wood. This confirms the great advantage of wood surface protection, especially when used in outdoor applications. Investigated coatings included transparent, semi-transparent, and opaque solutions based on different chemical compositions [3,4,6]. The performance of an innovative, bioinspired, and fully biobased coating system containing living fungal cells provides testimony for technological development. [5]. In this case, the desired occurrence of fungi as a part of the coating formulation provides self-healing properties. However, evidence of unwanted microbial cells is considered a major cause of the failure of timber products in use. Therefore, the fundamentals of fungal decay, staining, and mold growth processes, as well as analysis of the fungal attack mechanisms in the context of the state-of-the-art knowledge, is reviewed [8]. Several suggestions for the future research directions in the new biocidal and non-biocidal coatings treatments are provided. These are essential for creating integrated coatings systems capable of limiting fungal attack while providing long-term protection against environmental factors.

A critical limitation of the natural weathering procedures is the fact that the test is very long-lasting. The novel methodology for the acceleration of the natural weathering test is presented [9]. It can be used for the generation of the weather dose–response models that are essential to estimate the future service life performance of timber elements. Machinability, mechanical properties, and other engineering aspects of the modified wood are addressed with consideration given to the specificity of arctic driftwood [10] as well as laminated wood composites [11]. The diverse positive effects of wood modification may be associated with the high environmental costs of the transformation processes. For this reason, a systematic comparison of the impacts associated with the most relevant modification technologies is presented in a dedicated review [12]. The impact of the expected service life extension, highly imparted by the proper use of modified wood in buildings, is simulated and profoundly discussed.

The broad spectrum of topics presented in this Special Issue provides a comprehensive update regarding ongoing research in this field. We do sincerely believe that such a compilation can be an inspiration for the further development of multifunctional and sustainable coatings, revolutionizing the wood sector of the future.

Acknowledgments: We would like to thank all the authors for their valuable contributions to this Special Issue, the reviewers for their reviews and useful comments allowing the improvement of the submitted papers, and the journal editors for their kind support throughout the production of this Special Issue.

Conflicts of Interest: The authors declare no conflict of interest.

References

1. Spear, M.J.; Curling, S.F.; Dimitriou, A.; Ormondroyd, G.A. Review of Functional Treatments for Modified Wood. *Coatings* **2021**, *11*, 327. [CrossRef]
2. Jankowska, A.; Rybak, K.; Nowacka, M.; Boruszewski, P. Insight of Weathering Processes Based on Monitoring Surface Characteristic of Tropical Wood Species. *Coatings* **2020**, *10*, 877. [CrossRef]
3. Salca, E.-A.; Krystofiak, T.; Lis, B.; Hiziroglu, S. Glossiness Evaluation of Coated Wood Surfaces as Function of Varnish Type and Exposure to Different Conditions. *Coatings* **2021**, *11*, 558. [CrossRef]
4. Sandak, A.; Földvári-Nagy, E.; Poohphajai, F.; Diaz, R.H.; Gordobil, O.; Sajinčič, N.; Ponnuchamy, V.; Sandak, J. Hybrid Approach for Wood Modification: Characterization and Evaluation of Weathering Resistance of Coatings on Acetylated Wood. *Coatings* **2021**, *11*, 658. [CrossRef]
5. Poohphajai, F.; Sandak, J.; Sailer, M.; Rautkari, L.; Belt, T.; Sandak, A. Bioinspired Living Coating System in Service: Evaluation of the Wood Protected with Biofinish during One-Year Natural Weathering. *Coatings* **2021**, *11*, 701. [CrossRef]
6. Bobadilha, G.d.S.; Stokes, C.E.; Ohno, K.M.; Kirker, G.; Lopes, D.J.V.; Nejad, M. Physical, Optical, and Visual Performance of Coated Cross-Laminated Timber during Natural and Artificial Weathering. *Coatings* **2021**, *11*, 252. [CrossRef]
7. Rammou, E.; Mitani, A.; Ntalos, G.; Koutsianitis, D.; Taghiyari, H.R.; Papadopoulos, A.N. The Potential Use of Seaweed (Posidonia oceanica) as an Alternative Lignocellulosic Raw Material for Wood Composites Manufacture. *Coatings* **2021**, *11*, 69. [CrossRef]
8. Goodell, B.; Winandy, J.E.; Morrell, J.J. Fungal Degradation of Wood: Emerging Data, New Insights and Changing Perceptions. *Coatings* **2020**, *10*, 1210. [CrossRef]
9. Sandak, A.; Sandak, J.; Noël, M.; Dimitriou, A. A Method for Accelerated Natural Weathering of Wood Subsurface and Its Multilevel Characterization. *Coatings* **2021**, *11*, 126. [CrossRef]
10. Chuchala, D.; Sandak, A.; Orlowski, K.A.; Sandak, J.; Eggertsson, O.; Landowski, M. Characterization of Arctic Driftwood as Naturally Modified Material. Part 1: Machinability. *Coatings* **2021**, *11*, 278. [CrossRef]

11. Koutsianitis, D.; Ninikas, K.; Mitani, A.; Ntalos, G.; Miltiadis, N.; Vasilios, A.; Taghiyari, H.R.; Papadopoulos, A.N. Thermal Transmittance, Dimensional Stability, and Mechanical Properties of a Three-Layer Laminated Wood Made from Fir and Meranti and Its Potential Application for Wood-Frame Windows. *Coatings* **2021**, *11*, 304. [CrossRef]
12. Hill, C.; Hughes, M.; Gudsell, D. Environmental Impact of Wood Modification. *Coatings* **2021**, *11*, 366. [CrossRef]

Review

Review of Functional Treatments for Modified Wood

Morwenna J. Spear *, Simon F. Curling, Athanasios Dimitriou and Graham A. Ormondroyd

The BioComposites Centre, Bangor University, Deiniol Road, Bangor LL57 2UW, UK;
s.curling@bangor.ac.uk (S.F.C.); a.dimitriou@bangor.ac.uk (A.D.); g.ormondroyd@bangor.ac.uk (G.A.O.)
* Correspondence: m.j.spear@bangor.ac.uk

Abstract: Wood modification is now widely recognized as offering enhanced properties of wood and overcoming issues such as dimensional instability and biodegradability which affect natural wood. Typical wood modification systems use chemical modification, impregnation modification or thermal modification, and these vary in the properties achieved. As control and understanding of the wood modification systems has progressed, further opportunities have arisen to add extra functionalities to the modified wood. These include UV stabilisation, fire retardancy, or enhanced suitability for paints and coatings. Thus, wood may become a multi-functional material through a series of modifications, treatments or reactions, to create a high-performance material with previously impossible properties. In this paper we review systems that combine the well-established wood modification procedures with secondary techniques or modifications to deliver emerging technologies with multi-functionality. The new applications targeted using this additional functionality are diverse and range from increased electrical conductivity, creation of sensors or responsive materials, improvement of wellbeing in the built environment, and enhanced fire and flame protection. We identified two parallel and connected themes: (1) the functionalisation of modified timber and (2) the modification of timber to provide (multi)-functionality. A wide range of nanotechnology concepts have been harnessed by this new generation of wood modifications and wood treatments. As this field is rapidly expanding, we also include within the review trends from current research in order to gauge the state of the art, and likely direction of travel of the industry.

Keywords: modified wood; functional wood; multifunctional wood; nanotechnology; transparent wood; durability; weathering; fire performance

Citation: Spear, M.J.; Curling, S.F.; Dimitriou, A.; Ormondroyd, G.A. Review of Functional Treatments for Modified Wood. *Coatings* **2021**, *11*, 327. https://doi.org/10.3390/coatings11030327

Academic Editor: Mariaenrica Frigione

Received: 15 February 2021
Accepted: 9 March 2021
Published: 12 March 2021

1. Introduction

Wood modification is a group of technologies that have shown steady growth in the past two decades. While timber is a ubiquitous material that has been known to humankind for millennia, and used in a wide range of forms, including structures, furniture and many household objects; it has also been applied in high technology industries such as engineered components for automotive, energy and aerospace industries [1,2]. Wood modification is a relatively new field, which has contributed to this development, and is likely to allow continued growth. Some of the limitations of wood, including its dimensional change in changing moisture environments, and susceptibility to insect attack or decay, have traditionally been addressed through good design, chemical treatments and strategic choice of wood species for the intended applications. However, wood modification has offered an alternative approach. The majority of dominant wood modification systems (chemical modification, thermal modification and polymer or resin impregnation) have sought to alter the relationship between wood and moisture, and as a result restrict dimensional change, and reduce susceptibility to decay. These attributes are well reported by Hill [3] and many recent reviews of the technology [4–6].

However, if we cast the net slightly wider, some wood modifications offer benefits over and above the primary gains in dimensional stability and durability. Some seek specifically to enhance hardness or wear-resistance, while others seek to further gain UV weathering resistance. Other novel treatments seek to utilise the stability of the modified wood substrate to develop new technologies or ideas, for example as printable surfaces for electrical circuits or sensors and new bio-based electronics. These diverse options introduce multi-functionality to modified wood, so it is worth considering, do they themselves count as emerging multi-functional wood modification systems, or are they simply post-modification treatment options? This paper will firstly review some of the options for creating (multi)functional wood using modified wood as a substrate, then newer options for creating multifunctional wood using emerging systems and then consider and discuss this question further.

2. The Main Modification Systems

Wood modification has been defined by Hill [3] as a process that "involves the action of a chemical, biological or physical agent upon the material, resulting in a desired property enhancement during the service life of the modified wood." This is frequently considered to be a separate technology to the established wood preservative treatments using biocides [7,8]. Several systems are well developed and commercially available, including:

- Acetylation—the process of esterifying wood using acetic anhydride, including Accoya® [9,10]
- Thermal modification—the process of altering hygroscopicity through use of elevated temperatures (160 °C to approx. 230 °C) in an inert atmosphere, steam or vacuum, including Thermowood®, VacWood®, LunaWood®, VAP Holzsystemr® [11]
- Resin or polymer modification—the process of impregnating wood with thermosetting resins, monomers or oligomers capable of cross-linking in situ in the wood cell wall and/or cell lumena to form a non-leachable polymeric system, including furfurylation (Kebony®, NobelWood® [10]), resin modification (Impreg™, Indurite™, Lignia® [12]) and polymer impregnations (Permagrain, Permali®, Jabroc®).

It will be assumed that readers can refer to detailed reviews elsewhere [9–12] for the principles and properties of each of these modifications, or gain specific details from product technical information via the relevant companies. The main modifications and estimated commercial volumes per year by region are presented in Table 1, based on data from Jones and Sandberg [6].

Table 1. Estimated global production of modified wood (Source: [6]).

Modification	Estimated Volumes (m^3)				
	Europe	China	N America	Oceania/Japan	Other
Thermally modified timber	695,000	250,000	140,000	15,000	10,000
Densified wood	2000	<1000	-	<1000	-
Acetylation	120,000	-	-	-	-
Furfurylation	45,000	-	-	-	-
Other methods	35,000	290,000 *	-	5000	TBD

* Figures are a combination of furfurylation processes other than Kebony® and NobelWood® as well as DMDHEU and other resin treatments. Empty fields indicate that no data is available or that no modified wood of this type is known in this region to the authors of reference [6].

Other chemical modification systems exist, and some have been commercially available for intervals of time, e.g., the modification using DMDHEU (known as Belmadur®, or its new name HartHolz®), or have been well investigated but currently left at a medium level of advancement, not yet developed through the scale-up and commercialisation process stages towards market availability. Such options include reaction of wood with other acid anhydrides including succinic or maleic anhydride; reaction of wood with citric acid and various polyols (glycerol, sorbitol, etc.); reaction of wood with silanes; reaction of wood with chitosan; impregnation and/or reaction of wood with natural oils, waxes or paraffins, impregnation and/or reaction of wood with silicates or water glass. Some of these systems may be already in use for specific small applications, such as conservation of archaeological woods, and others are still in active process of investment or development to the commodity market. Densification of wood, through the use of steam, heat or reagents under pressure has also been investigated, and a small quantity of densified or surface densified wood is reportedly available in Europe [6].

The main enhanced properties advertised for the wood modification systems are summarised in Table 2, with a note of the main technical limitations, such as corrosiveness to metal fixings, or loss of strength or toughness [6].

Table 2. Commonly achieved properties of modified wood from various systems (based on [6]).

Process	Typical Species Used Commercially	Dimensional stability	Durability	Hardness	Electrical Resistance	High Density	Low Mould Resistance	Corrosive to Fasteners	Reduced Mechanical Properties	Oil Exudation	Notes
Accoya®	Radiata pine	yes	yes	-	-	-	yes	yes	-	-	-
Thermal modification	Norway spruce, Scots pine, various hardwoods	yes	mod.	-	-	-	-	-	yes	-	mod. = moderate
Kebony® NobelWood®	Radiata pine, Scots pine	yes	yes	yes	-	-	-	yes	yes	-	-
Compreg™	European beech veneer	-	yes	yes	yes	yes	-	-	-	-	-
Impreg™	European beech, Scots pine	yes	yes	-	-	-	-	-	yes	-	-
Lignia®	Radiata pine	yes	yes	-	-	-	-	-	-	-	-
HartHolz® (DMDHEU)	European beech, Scots pine	-	yes	-	-	-	-	-	-	-	-
Organowood® (Silicate modification)	Norway spruce, Scots pine	-	yes	-	-	-	-	-	-	-	Stable silver grey surface
Linotech (Linseed oil derrivatives)	Scots pine, (Norway spruce)	-	yes	-	-	-	-	-	-	yes	-

Empty fields indicate that no benefit is claimed by the manufacturers, to the best of the authors' knowledge.

There are differences between modification systems in which mechanical properties are affected, for example a reduction of bending strength, hardness and toughness is reported for thermally modified wood, depending on the level of modification applied. For the furfurylated products, the loss of mechanical properties relates only to toughness, and for Indurite the reported losses are for hardness and toughness. An interesting challenge is the poor mould resistance of thermally modified wood, and several of the investigations to add functionality to thermally modified wood relate to this issue—as we will discover in this review.

In his definition, Hill [3] indicates that wood modification is distinct from the more general band of wood treatments, such as traditional preservatives or fire retardants, which are frequently applied as aqueous solutions of emulsions, without any chemical grafting or reaction in situ [8]. However, the limitations of some modifications provide the impetus for adding functionality through secondary processes. This gives rise to combination treatments, with partial modification and partially traditional methods. For example, a commonly identified trait of thermally modified wood is its colour change on weathering. The thermally modified timber is generally a darker colour than the original untreated timber, however exposure to sunlight and rainfall leads to a relatively rapid change of colour towards a silver-grey form [13–15]. This has led to many research teams seeking to add a second treatment, either painted on, or impregnated into the wood, or using more innovative approaches, that would increase UV stability of the thermally modified timber [16,17].

Other combined wood treatments are developed using the initial chemical modification as a starting reaction, which then permits a second new chemical reaction, e.g., using functional groups introduced during step 1, or capitalising on changes achieved in step 1 (e.g., delignification). Many of these combined modification methods achieve stunning additional functionality in the wood, e.g., photoluminescence, electrical conductivity, sensors, transparency, etc. However, this may be at the expense of original material properties. In many cases the dimensional stabilisation that is achieved by the wood modification contributes to a more uniform and predictable starting material for the application of nanotechnology approaches. These systems will be introduced and discussed in greater detail to highlight future possibilities and limitations.

The sections which follow will consider each of the main modification systems, and the ways in which secondary treatments have been used to add functionality or overcome challenges identified in Table 2. We will also consider methods where wood modification methods have been used innovatively as part of a several step process. In some cases hybrid systems emerge, for example where a pre-treatment rather than post-treatment is used, but a wood modification technique remains as the central enabling step to permit new properties to be gained. The fields of application vary widely, as will be seen, and a large amount of potential exists for ongoing exploration and development in this field. Thirdly, we have included in the second part of the text a number of examples where the same active ingredients for functionalisation of wood have been used without modification, but where we see that there is significant potential for them to be included in modification approaches within the near future. This is true for many of the nanotechnology approaches, where a rapid growth of activity is showing interchangability between modified and unmodified wood substrates and combinations of active ingredients. We return to this consideration in the discussion section.

3. Adding Functionality to Modified Wood

3.1. Thermally Modified Wood with Functionality

3.1.1. TiO_2 for Weathering Resistance

To improve weathering performance of thermally modified wood, the titania sol (TiO_2) in paraffin was impregnated into pine wood that had been previously thermally treated at 212 °C. The paraffin assisted in retaining the titania on the wood surface (water stability), and UV resistance of the treated impregnated wood was observed [17]. Others have considered titanium dioxide or other oxides such as zinc or cerium in acrylic binders, as is common for UV protection of unmodified wood [18,19]. Both TiO_2 and ZnO nanoparticles were considered by Miklecic et al. [16] in a polyacrylate system, as a coating for thermally modified beech wood. Both nanoparticle types improved colour stability of the thermally modified beech, but the zinc oxide system was prone to more crack formation in the surface, and peeling of the coating. A TiO_2/Ce xerogel system was demonstrated for protecting wood by Guo et al. [20], which could also contribute to future development in this area for modified woods. The TiO_2/Ce xerogel blocked virtually all light with wavelength below 370 nm in their study. When dip coated onto wood a microsheet film was formed with thickness 7 μm (five dip cycles). Colour change during UV weathering test was greatly reduced for spruce wood treated with the xerogel.

Shen et al. [21] measured colour and water contact angle during an ASTM 154 weathering test on thermally modified timber, and thermally modified timber with TiO_2 nanoparticle coatings, and thermally modified timber with polydimethyoxysiloxane (PDMS) treatment, or combined TiO_2 and PDMS treatments. The TiO_2 treatment greatly reduced the development of lightness over weathering exposure period, whereas PDMS allowed bleaching to occur. The combined TiO_2 and PDMS treatments both gave low total colour change (ΔE) values compared to the control samples and the treatments when applied as stand-alone.

The water contact angle of thermally modified timber was initially high (over 100°) but rapidly decreased [21]. The PDMS treatment increased the initial contact angle to approx. 110°, and the TiO_2 treatments with PDMS second step further increased the contact angle to 125–135°. Both the PDMS, and the TiO_2/PDMS systems retained their contact angles during the weathering exposure, whereas the untreated thermally modified timber lost hydrophobicity by the 168 h measurement time, and contact angle continued to decrease as weathering time increased up to 1176 h exposure.

3.1.2. Silica Nanosols

Nanosol treatments are stable dispersions of inorganic nanoparticles, such as silica, that can be painted or sprayed onto a surface [22]. A number of studies have looked at using this technology with wood, and differing nanoparticles affect wood in different ways, e.g., silica- alkoxysilane sols can provide water repellence [23,24], silicon oxide/boron oxide complexes can give flame retardancy, and silica oxide or boric acid sols can be used for biocidal properties [23]. With modified wood, and in particular thermally treated wood, iron salt sols and silica/iron oxide have been applied to improve weathering stability and water resistance Incorporation of pigments such as iron oxide red with the sol not only enhanced weathering protection and increased hydrophobicity but also enhanced aesthetic qualities [23,25].

3.1.3. Suppressing Mould Growth

Another aspect of exterior use of thermally modified wood can be mould growth. A few different approaches have been tried to improve upon this. Kwaśniewska-Sip et al. [26] used caffeine, impregnated into wood prior to thermal modification, to alter the susceptibility to mould growth. Others have used methacrylate as a secondary treatment on thermally modified wood [27,28]. The methacrylate suppressed mould growth; however, it did not reduce the colour change due to UV.

3.1.4. Other Improvements through Concurrent Treatments

Other approaches for altering the properties of thermally modified timber include addition of reagents to the wood prior to heat treatment. One example is the use of a zwitterionic buffer during thermal modification to alter the reactions which occur. In a study by Duarte et al. [29], two zwitterionic buffers were used (bicine and tricine) alone or in combination with thermal modification of wood. It is proposed that the amine component of these zwitterionic buffers is able to contribute to Maillard-type reaction with the polysaccharide components of the wood at elevated temperatures [30] and influence the characteristics of the thermally modified wood. In the study by Duarte et al. [29] the presence of bicene and tricene had an effect on termite survival, but further investigations of the interactions of these molecules with wood during thermal modification was recommended.

Another attempt to modify the properties of thermally modified wood used vinylic monomers, impregnated into the wood before thermal treatment, the monomer solution also contained boron derivatives, to alter the termite resistance of the wood [31]. The polyglycerolmethacrylate and maleic anhydride-polyglycerol adducts improved the boron fixation, even after leaching. In addition, samples without boron treated at 220 °C showed benefit in termite resistance, indicating a synergistic effect between the thermal modification and reaction with monomers.

3.2. Resin Modified Wood with Functionality

UV Resistance Promotion for Resin Modified Timber

The UV stability of resin modified timbers was investigated by Kielmann and Mai [32]. Two types of resin modified timber were studied—based on N-methylol melamine (NMM) and phenol formaldehyde resin (PF), in addition, both resins were also studied when a dye had been incorporated into the resin during wood treatment. The timbers were coated with a translucent waterborne acrylic binder, with and without UV protective agents. The UV protective system (Tinuvin 533-DW) was a light stabilizer blend containing UV absorbers and hindered amine light stabiliser (HALS). While this acrylic treatment is a typical paint or varnish-type treatment, not a surface functionalisation or modification study, it is included in this review to reflect the interest in achieving UV stability enhancement for modified wood. The NMM modified wood, and NMM wood with dye both became whitish during weathering, relating to loss of polyphenolics from lignin, and leaching of the dye if present. The PF modified woods became a darker brown colour, while the PF wood with dye only became a slightly lighter shade of black, but with some lightish regions on the sample face. The darkening of PF modified woods has been previously reported in other studies [33,34] and is related to formation of new chromophores within the PF resin during UV exposure, which offsets the degradation of lignin and loss of polyphenolics from the wood itself.

In the coating study by Kielmann and Mai [32] the NMM, PF and PF with dye samples that had been coated with acrylic without UV agents gained a darker colour after weathering, which was attributed to the lack of removal of polyphenolic fragments, due to the coating inhibiting the action of water to mobilise these fragments. The samples with acrylic binder and light stabiliser blend turned only somewhat darker on UV weathering exposure, relating to the action of the UV stabilisers. The greatest benefit from the light stabiliser formation was seen for the NMM samples containing the dye.

Evans et al. [33] used hindered amine light stabilisers (HALS) in the PF resin formulation that they used to resin modify wood veneers. Resin at 10, 20 and 30% resin solids content were tested, and the presence of 1% or 2% HALS additives. The PF resin and the HALS additive had a beneficial effect, reducing strength loss resulting from natural weathering. SEM images of the UV exposed PF treated wood containing HALS (150 h exposure) revealed that middle lamellae (ML) were intact, and very limited damage to cell walls, whereas the untreated wood tracheids had clearly separated due to loss of lignin from the ML. The PF treated wood without HALS showed an intermediate state of damage, with some small fissures beginning to develop between cells along the line of ML. After 1070 h of UV exposure clear differences remained between the samples, and macro scale observations indicated greater structural integrity of the PF modified wood with the HALS.

Resin modification has been used as a second step on succinic anhydride modified wood by Wang et al. [35], with the claim that this led to enhanced dimensional stability, water repellency (determined by contact angle) and flame retardancy (determined by LOI, limiting oxygen index). The combination of treatments showed ASE values than the individual succinylation or melamine formaldehyde impregnation treatments. Both the resin impregnation and the succinylation are known to have a beneficial effect on dimensional stability, however it was suggested that the succinic anhydride promoted cross linking between the resin and the wood.

3.3. Acetylation and Esterification of Wood with Added Functionality

3.3.1. Chemical Modification with UV Stability

Esterification is a popular method for reducing the colour change of wood when exposed to weathering conditions. For example, acetylated timber shows only minor changes on weathering, if sufficient weight percent gain (WPG) has been achieved [36,37]. Acetylation shows good colour stability, while succinic and maleic anhydride performed acceptably, but phthalic anhydride modified wood tended to gain more chromophores during UV exposure test. Esterification of fir with benzoyl chloride was also shown to reduce colour change, in tests by Pandey and Chandrashekar [38]. For untreated wood bleaching is common as lignin is degraded by UV light, and water removes fragments of polyphenol decomposition products from the wood surface [7,39]. Rosu et al. [40] investigated further enhancement of UV stability of esterified wood (prepared with succinic acid anhydride modification) by an epoxidized soybean oil (ESO) treatment. The ring opening of the succinic anhydride during the wood modification step presented carboxylic acid functional groups on the wood surface for further grafting of the ESO, thus this is a two-step modification process, with the ESO being grafted on the surface of the wood. The colour change of SA modified wood was reduced compared to the untreated wood, while the ESO grafted SA modified wood had very low colour change, which on the first observation interval (20 h) was below the threshold for detection by the human eye. An ESO treatment for unmodified wood was previously reported by Olsson et al. [41], in an experiment with additional UV absorbing agent, 2-hydroxy-4(2,3-epoxy propoxy)-benzophenone (HEPBP), where it showed reduction in colour change on artificial and natural weathering.

3.3.2. Esterification as a Pre-Treatment for Other Purposes

Esterification with maleic anhydride was used by Li et al. [42] to pre-swell wood prior to a second step—grafting of a mix of glycidyl methacrylate and methyl methacrylate monomers. They reported better interfacial compatibility between the phases of wood and polymer, and a high level of dimensional stability was reported. Pre-treatment of wood with succinic anhydride (another cyclic anhydride molecule) was reported by Wang et al. [35] for resin modifications, as noted above.

3.4. *Polymer Impregnation with Functionality*

Polymer impregnation of wood has been recognised for many years. Initial interest in impregnating the cellular structure of wood with monomers or polymers such as methyl methacrylate or epoxy or microscopy led to advances in larger scale pieces of polymer impregnated wood for high performance applications. Schneider and Phillips [43] reported investigations on methyl methacrylate impregnated timber giving it the term wood polymer composite (which has since been assigned to other composites with a wood flour filler in any polymer matrix). Several commercial systems were developed in the UK, North America and elsewhere, and found niche applications such as neutron shielding, or machinable wood blanks for manufacturing high wear components for motorsport and engineering, or hardwearing flooring. These included products which are still available today: Permali, Jabroc and Insulam. Work in this area continues today, considering polystyrene [44], polymerised glycidyl methacrylate [45] and polymethylmethacrylate (PMMA) [46].

3.4.1. Biodegradability in Impregnated Wood

Some of the new generation of modification systems have sought to address dimensional stability while retaining the biodegradability of wood. A good example is the polymer modification proposed by Ermeyden et al. [47], who generated poly-ε-caprolactone (PCL) in situ by ring opening polymerisation. PCL is a synthetic polymer with strongly biodegradable character, when subjected to the correct conditions. Depending on the solvent used during modification, the experiments indicated that good distribution of the PCL within the cell wall is possible, although toluene as a solvent gave a lumen-filling and cell wall coating, rather than bulking [47]. Consequently PCL synthesised in dimethyl formamide solution showed greater anti-swelling efficiency (ASE) than the toluene synthesised PCL modified wood.

Studies by Noel et al. [48,49] used four biopolymers, polylactic acid (PLA), polyglycolic acid (PGA), polybutylene adipate (PBA) and polybutylene succinate (PBS). A mixture of cell wall bulking and cell lumen filling properties were observed. The PLA and PBS showed very interesting affinity for the wood cell wall material, leading to migration into the cell wall structure during a heating stage, which enhanced dimensional stability [50,51]. Both

PBS and PLA show biodegradability, however PLA requires careful control of conditions to facilitate degradation, and may result in a less biodegradable modified wood product.

3.4.2. Pretreatments and Grafting for Polymer Impregnation Modifications

Other systems from the new generation of polymer functionalisation reactions make use of wood modification, prior to full polymer impregnation This can be to control the location of the polymer within the wood, or to ensure maximum grafting between the wood cell wall and the polymer phase. Berglund and Burgert [52] summarised the pretreatments and polymerisations, with typical applications, in their review. Atom transfer radical polymerisation (ATRP), reversible addition-fragmentation chain transfer polymerisation (RAFT) and ring opening polymerisation (ROP) can be used to graft the intended functionality onto the wood surface for subsequent reactions. Various surface initiated polymerisations can also be used.

In one example a pH responsive wood system was created, using two different grafted polymers (PMMA and PDMAEMA, poly(2-dimethylamino)ethyl methacrylate) to favour water uptake under acidic or basic conditions by Cabane et al. [53]. In a later work by Cabane et al. [54] they used a bromoisobutyryl bromide (BiBB) initiator and ATRP to graft polystyrene and poly-N-(isopropyl acrylamide) (PNIPAM) into wood, demonstrating different functionalities. The polystyrene modified wood was intended as a hydrophobic treatment, while PNIPAM is known to change its hygroscopicity with temperature-induced change in microstructure, offering potential to give control over the hygroscopicity of PNIPAM modified wood. The contact angle at room temperature was low for PNIPAM wood, but after heating to above the critical temperature (32 °C) the contact angle significantly increased, demonstrating very low wettability [54]. A wealth of other grafting-based reactions are possible. For example, Trey et al. [55] polymerised aniline within the cell lumena of Southern yellow pine veneers, creating semi-electrically conductive wood, with potential application in anti-static or charge-dispersing materials [56].

3.4.3. Delignification Combined with Polymer Impregnation—Transparent Wood

Polymer impregnation is a key step in the manufacture of transparent wood, however this is done following a prior delignification step, so is a two-step, or hybrid, modification [57–60]. However, once the delignification has been accomplished, various options exist for the polymer impregnated wood, as well reviewed by [59,60]. Polymethyl methacrylate is a commonly chosen polymer for this [57], although other researchers have considered epoxy resins, PVA or other systems [58,59,61–64]. In order to make a two-component material (such as composite or impregnated wood) transparent, the difference between the refracted indices of the two phases must be minimised to reduce diffraction at interfaces between the two phases [57].

Uses for this wood include optically transparent or opaque materials to permit light to enter structures [57,58,62,65,66], transparent materials and light diffusing layers for solar cells [67,68], and diffused luminescence wood structures [69]. Further functionality can be added by incorporating particles, dyes or other additives into the polymer during impregnation of the delignified wood. Examples of extended functionality include: wood based lasers (using a dye in the transparent wood to give laser activity) [70,71], heat-shielding transparent wood windows (incorporating nanoparticles, Cs_xWO_3) [72], thermo-reversible optical properties—using phase change materials [73], electrochromic properties [74], energy storage through encapsulation of phase change materials such as polyethylene glycol [75], luminescent properties by incorporating nanoparticles (Fe_2O_3@YVO_4:Eu^{3+}) [76], luminescence [69], magnetic properties by incorporating Fe_3O_4 nanoparticles [77].

Materials can be produced with high optical transmittance (over 80%) and haze (over 70%) [57]. It is also reported that thermal conductivity is low [66], and load bearing capacity is good—as reflected in tough, non-catastrophic failure behaviour, or the absence of a shattering behaviour [57,69]. In some applications the optical haze is seen as beneficial, to create a uniform and consistent distribution of daylight without any glare effect [66], while in other applications a greater level of clarity is sought through adjustment of refractive indices of the wood and the polymer matrix [64]. The process has now also been applied to bamboo [64,78].

The actual mechanical properties of the transparent wood have been tested and reported by Li et al. [57] and Zhu et al. [58] and show that the delignification has a large reduction in strength, which is then overcome by the polymer impregnation. This provides a stiffness of the transparent wood (MOE = 2.05 GPa) that is higher than the PMMA polymer used for impregnation (MOE = 1.8 GPa) [54]. For an epoxy matrix based transparent wood studied by Zhu et al. [58], the MOE value was 1.22 GPa when loaded in the radial direction (i.e., higher than the unmodified Basswood in this test orientation (0.19 GPa). Meanwhile, in the longitudinal direction the transparent wood MOE was 2.37 GPa, which was approximately half the strength of the natural wood prior to delignification and polymer resin impregnation (5.78 GPa).

A compression step was added to the transparent wood preparation by Li et al. [57], which further increased the stiffness value to 3.59 GPa, the tensile stress for this compressed transparent wood was higher than PMMA (ca. 90 MPa and ca. 40 MPa, respectively) when the wood was loaded longitudinally, whereas the delignified wood had a maximum stress of below 5 MPa [57]. The level of delignification in the study was high, reducing lignin content of the balsa wood used from 24.9% to 2.9% [57]. However, there is also interest in development of a method for creating optically transparent wood which has not been subjected to such intense delignification, to reduce the loss of strength, and to reduce the time-consuming delignification step [61]. A method where 80% of lignin remains in the wood has been proposed and demonstrated, using a hydrogen peroxide system [61]. The method focused on bleaching by removal of the chromophoric structures, while leaving bulk lignin in situ within the wood cell wall, preserving the structure of the wood. Reaction time was significantly reduced for four species studied by Li et al. [61] and the wet strength of the wood parallel to the grain was significantly higher than the full delignification method.

In a further development of the transparent wood concept, an intermediate step has been demonstrated by Montanari et al. [79]. Following delignification, the wood substrate was grafted with either maleic, itaconic or succinic anhydride from renewable sources. The esterification reaction onto the wood substrate permits interface tailoring in subsequent reaction with active agents and controls the wood-polymer interaction necessary for optical transparency. Additionally, the esterification reduces moisture sorption of the wood, as would be expected based on the well-known esterification family of chemical modification systems.

4. Functionality and Multi-Functionality through Other Wood Modifications

4.1. New Approaches for Functionality

Looking outside the literature relating to the primary, or fully commercialised wood modification systems, we find reviews which consider how wood can be given multi-functionality through any combination of cell wall modification, cell lumen surface modifications, and lumen filling. These can be considered to be 'new generation wood modifications', although there is some overlap with existing wood modifications mentioned in Section 2, as shown in Figure 1. Li et al. [59] comment that this can use "chemical reactions, physical adsorption or absorption, thermal treatment for phase separation effects, inorganic or metal particle precipitation from salt solution, and combined approaches." Many of these have been reviewed and compared in Burgert et al. [80], whose review also extended beyond wood modification to also consider using the hierarchical porous structure of the wood as a template for inorganic non-metallic materials, and applying biomimicry to cell wall structures, i.e., reassembling cellulose nanofibers to design new structures. These elements extend far outside the realms of wood modification, however many interesting principles can be considered based on the modification elements of Burgert et al.'s review [80]. The first insight is that the well-established wood modifications typically focus on making wood more inert (e.g., blocking or removing hydroxyl groups), whereas many of the nanotechnology wood modifications have focused on increasing functionality, or chemical reactivity. A second insight is that the commercialised wood modifications are typically carried out with relatively little nanostructural control, i.e., they applied to the bulk of the wood. They also highlighted the emerging field of wood nanotechnology, which includes advanced wood functionalisation, where molecular features of the wood, and pore spaces within the wood cell wall are targeted specifically [80]. The pore spaces at specific length scales can be deliberately utilised to fine-tune the synthesis processes or to facilitate novel properties.

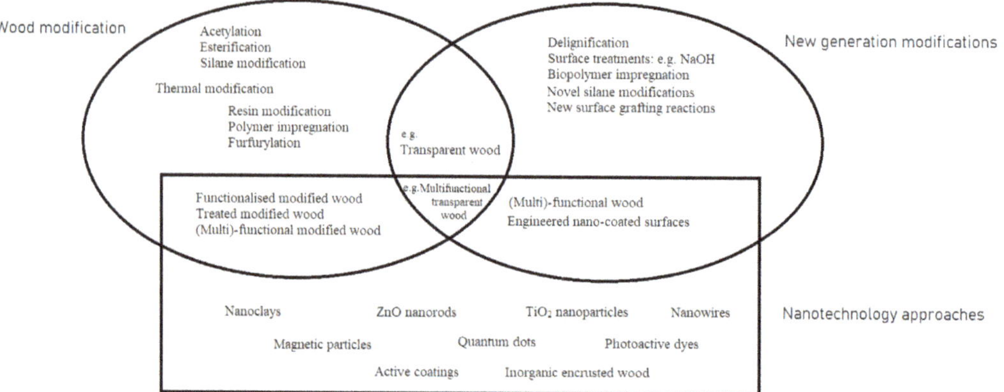

Figure 1. Interaction of wood modification and nanotechnology to create functional and multi-functional wood.

Through these impregnation or mineralisation approaches we find examples of fire-retardant wood [81], magnetic wood [76,77,82,83], UV stabilised wood [76,82,84], and conductive wood [55,56]. However, some references for conductive wood refer to systems where the wood is so highly carbonised that is should no longer be called modified wood, for example many studies have used carbonised wood as a substrate for lithium or other conductive materials for battery anodes [85]. Other studies [56] have used conductive polymers to impregnate the wood without causing this severe degradation, permitting the conducting or super-conducting behaviour. In some of the above studies, the functionalising agent was incorporated within the polymer during impregnation, while for others a pre-treatment with dye or functionalising agent was completed prior to polymer impregnation which sealed the treatment agent in situ [59]. Thus, not only is wood modification occurring, but in some systems nanoparticles are being utilised to extend the types of functionality given to the wood. These nanoparticle-based modifications will be considered in following sections, but are represented by the rectangular field in Figure 1, intersecting with both the traditional and new generation modifications.

4.2. Delignified Wood with Functionality

In the section on polymer modified woods (above), we identified the use of delignified wood with a polymer impregnation process to form transparent wood. Other work has considered the use of delignified wood without polymer, as a scaffold for further modifications. One example is the luminescent wood films proposed by Fu et al. [86], where the delignified wood also had a reduced hemicellulose content, enhancing hydrophobicity, and was infused with quantum dots before densification. Light scattering by the delignified wood was supplemented by the presence of quantum dots to generate different colours of light. After densification, the wood film was protected by a further modification using hexadecyl trimethoxy silane, so we observe a three-step modification, combining well known techniques to generate the desired functionality. The use of a silane to protect the wood film was selected to avoid debonding of a more rigid polymer such as the PMMA or epoxy that are often used in transparent wood systems. As a result, a high level of flexibility was formed in this product.

Frey et al. [87] used delignification followed by densification to produce a bulk composite with specific fibre folding patters that had excellent elastic (circa 35 GPa) and tensile (270 MPa) properties and could be easily shaped. They also reported the potential for further functionalisation of the material. Delignification of wood followed by freeze drying and compression has been used [88] to create highly compressible aerogels with very low thermal conductivity values (0.028 W/mK) that could have other applications such as noise adsorption or soaking up pollution like a sponge. Extreme modification has performed using delignified wood as a template for precision casting with a nanosol of cerium zirconium oxide to form a non-structural ceramic like material with potential applications in fuel cells or catalyst systems [89].

4.3. Sodium Hydroxide Treatment

A less intense chemical pre-treatment has been used for creation of wood scaffolds for metal organic frameworks (MOFs) by Tu et al. [90]. Sodium hydroxide pretreatment does not substantially delignify wood when conducted at room temperature, but does open up the wood structure to facilitate the necessary rough fibrillar structure for MOF formation. The sodium hydroxide solution also leads to substitution of sodium cations into the wood structure replacing carboxyl groups, and these sodium ions are believed to act as nucleation sites for the MOF deposition process [90]. Two MOFs were formed, including a zeolitic imidazole framework (ZIF-8) derived from zinc nitrate and 2-methyl imidazole, demonstrated in beech, spruce and basswood; and a demonstration of MOF-199 in beech only. The tensile and compressive strength of the beech timber was greatly increased by formation of the ZIF-8 composite, and the porous MOF deposited within the cell lumena showed good adsorption of gases, with potential suitability for environmental or energy applications.

4.4. Other Modifications with Functionality

Many of the emerging wood modifications for enhanced functionality rely on a set of additives that can be incorporated into the wood during polymer modification, or decorated onto the wood surface as a separate treatment step. Many are compiled in Table 3, although the list is far from exhaustive. Several of the inorganic or metallic particles listed are familiar ingredients for wood coatings, and indeed, there is some overlap between coatings and surface treatments for functionalised wood [91]. During the process of compiling this review the wide range of such additives has become apparent. It was also noted that there is considerable cross-over between the use of such treatments on or in modified wood, versus their use on unmodified wood to create similar functionalities. Therefore, some overlap between modified and unmodified woods is acknowledged, but as research continues, there is a high likelihood that treatments currently demonstrated on one substrate will be further developed on others, to generate modifications of the future. Other reviews which consider nanoparticles as treatments for wood or for surface coatings contain additional information [22,92–96].

Table 3. Typical or potential additives used to add functionality during or after a main wood modification step.

Additive	Used during Modification	Used after Modification	Properties Given or Enhanced	Refs.
Metal oxides nanoparticles: TiO_2, ZnO, CeO_2	-	Unmodified wood	Photocatalytic properties; UV-protection; Self-cleaning	[19]
Silicon nanoparticles; TiO_2 nanoparticles; ZnO nanorods	-	Unmodified wood	Superhydrophobic properties; Fire retardancy	[97,98]
TiO_2 (anatase form)	Polymer impregnated wood	Unmodified wood [99] Thermal mod. [17,21]	Hydrophobic surface treatment	[17,21,99,100]
TiO_2 (rutile form)	-	Unmodified wood [101]; Thermal mod. [16]	UV resistance	[16,101]
Nanoclay (layered aluminosilicates)	PF Resin modification, Polymer impregnation	-	Quicker reaction, Improved dimensional stability	[100,102]
Halloysite, Montmorillonite, Calcium carbonate, TiO_2, silica nanoparticles	-	-	As a carrier for biocides	[103]
Montmorillonite clay	-	Delignified wood	Fire retardancy; Readily splits into multiple platelets	[84,98]
Fe_3O_4 particles; $MnFe_2O_4$ particles; $CoFe_2O_4$ particles	-	-	Magnetic properties	[83]
Fe_3O_4 nanoparticles	Before PMMA impregnation	-	Magnetic properties	[78]
$CoFe_2O_4$ nanoparticles	Before silane treatment	-	Magnetic properties; hydrophobicity when combined with OTS	[77]
Metal organic frameworks (MOFs) ZIF-8 & MOF199	-	NaOH pretreated scaffold	Adsorption of gases	[89]
Metal oxides (atomic layer deposition)	-	Unmodified wood	Hygroscopicity and resistance to mould growth	[104]
MXene (transition metal carbides, carbonitrides and nitrides)	-	Unmodified wood	Elecrtomagnetic shielding	[105]
Silver nanoparticles	Before thermal modification	Unmodified wood	Decay resistance; electrical conduction	[106–108]
Quantum dots (Si and CdSe) (CdSe and ZnS)	PMMA impregnation (transparent)	Delignified scaffold	Luminescence—red & green excitation caused by blue or UV light	[68,109]
Electrochromic polymers (ECPs)	PMMA impregnation (transparent)	-	Colour change on electrical current	[74]

Table 3. *Cont.*

Additive	Used during Modification	Used after Modification	Properties Given or Enhanced	Refs.
Rh6G dye (Rhodamine 6G)	PMMA impregnation (transparent)	-	Red colour of transparent wood; laser output (565 to 570 nm) from the wood perpendicular to input laser (532 nm)	[70,71]
Carnauba wax and organoclay	-	Thermal modification	Reduction in water uptake; increase in strength properties; improved photostability	[110]

One common treatment seen in many of the nanotechnology based systems is the reaction of wood with organosilanes. Several have already been mentioned in this review for their ability to assist in grafting of the treatment [17,80,109], or to protect the treatment once in situ [111–113]. Silanes are also used in the functionalisation of nanoclays, to enhance their compatibility with the intended polymer matrix, or to promote binding of active substances. For example, Devi et al. [100] used a functionalised nanoclay in preparing resin modified wood, the nanoclay had previously modified with octadecylamine and aminopropyl triethoxy silane.

The grafting of different silanes onto wood to increase hydrophobicity were considered by Donath et al. [114] and depending on the pendant functional groups many different levels of effect may be achieved. Other organic silicon compounds can include chlorine or fluorine species, further altering the properties provided to the wood [115]. One frequent aim is to enhance hydrophobicity through alkyl or other low polarity functional groups. In performing these reactions, a sol-gel process is common, where a hydrolysis reaction of the silane creates silanols leading to sol formation, followed by gel formation through cross linking reactions. Some treatments utilise the moisture content of the wood to supply water for hydrolysis reactions to occur in situ, while other systems may generate the sol in a reactor prior to impregnation. Donath et al. [108] favoured the in situ reaction for good results in wood modification, with better distribution of the silane in the wood structure. The use of emulsions, plasma treatments and other methods for surface coating or for impregnation into the bulk of the wood were considered in the review by Mai and Militz [115].

5. Targeted Properties

5.1. Hydrophobicity

As introduced above, one issue for modified wood is the loss of hydrophobicity on weathering exposure, for example thermally modified wood has relatively high hydrophobicity immediately after treatment, but contact angle decreases with UV exposure [21]. Ways of improving the contact angle of unmodified wood have also been studied, using similar approaches, such as deposition of titanium dioxide on the wood surfaces to enhance hydrophobicity. Sun et al. [99] used a hydrothermal method to achieve the surface deposition of TiO_2, with certain deposition conditions generating a micromorphology on the surface that demonstrated hydrophobic (contact angle of over 90°) or superhydrophobic (contact angle of over 150 °C) behaviour. Such micromorphologies can result in the 'lotus effect' with either Cassie-Baxter or Wenzel mode and are found in various plant leaves, giving rise to a well-investigated field of biomimetic surfaces [116,117]. For example, a water contact angle of 154 °C was demonstrated for hydrothermal growing process at 130 °C, whereas lower temperatures gave contact angles of 85° or above. This growth process followed an earlier deposition stage in a treatment solution of tetrabutyl orthotita (TBOT) in ethyl alcohol with sodium dodecyl sulphate (SDS), conducted at 70 °C for 4 h. The pH of the starting solution also had a significant influence on deposition of the

TiO_2 and contact angle achieved, with lower pH favouring higher hydrophobicity [100]. TiO_2 occurs in several crystalline forms, or polymorphs, and the anatase form (tetragonal) shows photoactivity [118], this form was also achieved during the deposition method of Sun et al. [99].

Similar techniques have since been used with modified wood. Titanium dioxide sols have been applied to thermally modified Scots pine to enhance weathering resistance [17]. They demonstrated that TiO_2 sol was not resistant to water, but the addition of paraffin emulsion helped the retention of the TiO_2 by the wood surface. The samples impregnated with the combined TiO_2 and paraffin emulsion showed improved resistance to the effects of ultraviolet light and water on the wood surface, and degradation of the wood components was slowed by the presence of the treatment. In a second study, using a two-step process, Shen et al. [21] applied titanium dioxide nanoparticles, followed by a silane coating. The titanium nanoparticles were grown from acidic solution of ammonium fluorotitanate and boric acid, into which wood samples were immersed for three days. The TiO_2 treated wood was then washed and dried before silane treatment with hydroxyl-terminated polydimethyl siloxane (PDMS) by brush application. A weathering test (ASTM 154) was used to evaluate UV resistance. Water contact angle was measured at intervals over the weathering experiment, and TiO_2 treatments, with and without the PDMS retained the high contact angle that indicates hydrophobicity (values between 110 to 145°). Colour change was also inhibited by the combined treatment.

Zinc oxide nanorods have also shown enhanced hydrophobicity when grown on the wood surface [83,97]. Control of the nanorod morphology and coating of the nanorod assembly with PTES (perfluorotriethoxy silane) led to contact angles of 130° (without PTES) and 160° (with PTES). Thus, the treated wood surfaces demonstrated superhydrophobic behaviour, with water droplets sliding off the surface. In a study by Kong et al. [119] high-density zinc nanorods were demonstrated to give not only water-repellency but also photostability and flame retardancy to wood.

Silica nanoparticles were used by Jia et al. [120] to prepare superhydrophobic wood that also had high wear resistance. The nanoparticles were added to the wood in a hydrothermal process with vinyl triethoxy silane (VTES) and ethanol in sealed containers that were placed in the oven at 100 °C for 90 min. This process was repeated three times to obtain treated samples. Similar samples were also prepared on wood that had received a mild sodium hydroxide pretreatment, and this gave rise to very high contact angles, whereas the nanoparticle and VTES treatment alone did not. A further treatment includes sodium hydroxide in the single step hydrothermal treatment with the SiO_2 and VTES, giving contact angle values that were almost as high as the pretreated SiO_2 and VTES samples. The wear resistance of the pretreated SiO_2 and VTES treated wood was tested using abrasion tests (with sandpaper and knife-scratches), and a boiling procedure, in all cases the high contact angle was retained [120].

Other surface modifications have been attempted using bio-derived extracts. Filgueira et al. [121] used water insoluble condensed tannins and hydroxypropylated condensed tannins extracted from *Pinus radiata* bark to increase hydrophobicity on the surface of beech wood. The process was enabled using laccase at pH 10 to produce a heterogenous modification. The hydrophobicity of the surface of beech wood (*Fagus sylvatica* L.) was improved by this treatment.

5.2. Self-Healing

The development of new nanotechnology-based coatings brings the need for robust and wear resistant surfaces. Some researchers have begun to address this, either considering the abrasion resistance [120], or through developing self-healing surfaces [122]. A silane modification was used by Tu et al. [122] as a pretreatment for the wood, followed by spray coating with TiO_2 nanoparticles dispersed in a waterborne perfluoroalkyl methacrylic copolymer. The coating performed well in abrasion tests, retaining a high contact angle and only small increase in sliding angle (where the water droplet rolls off the surface). The silane pretreatment was shown to assist in long-term UV exposure tests, whereas samples prepared with only the nanoparticles in copolymer suspension showed decreasing contact angle with duration of UV.

5.3. Indoor Environment Improvement

Comfort of the interior environment is an issue that is becoming more of a design feature. There is a perception that natural materials create more comfortable environments, and wood can contribute significantly to this [123]. Some of this perception is psychological but there is also a physical effect based on temperature and humidity. Hygroscopic materials such as wood have been shown to affect humidity in the indoor environment via moisture buffering—i.e., absorbing moisture in high humidity environments and conversely releasing moisture in low humidity environments [124,125]. This can help keep room humidity in the comfortable zone for most people. This moisture buffering also affects temperature via differences in latent heat of sorption [126,127]. Different wood types have different moisture buffering effects and therefore different effects on room comfort [128]. There is a challenge to maintain moisture buffering (which relies on hygroscopicity) whilst also protecting wood against liquid water. Lozhechnikova et al. [129] have used Carnauba wax in a non-continuous film to treat Spruce wood and showed that good moisture buffering and good water repellence was obtained. In comparison to full wax films and lacquers reduced moisture buffering significantly whilst linseed oil did not provide enough hydrophobicity. Adding zinc oxide to the Carnauba wax [130] and using a layer-by-layer deposition increased both the hydrophobicity and the UV resistance of the wood whilst maintaining the moisture buffering capability.

Other functionalised woods have been developed with indoor environmental comfort in mind. Montanari et al. [75] developed a transparent wood which incorporated a phase change material within the PMMA main polymer. Phase change materials have been an area of development within building design in recent years. Transparent wood has already demonstrated good thermal insulation properties, with potential to outperform traditional glazing [65,66]. The intention of the study was to allow thermal regulation of the indoor environment, while also permitting light to be transmitted into the building. It was found that the transparent wood with thermal energy storage was superior to normal glass, with its combination of thermal energy storage and thermal insulation properties. A thermal conductivity of 0.30 $Wm^{-1}{\cdot}K^{-1}$ was reported for the transparent wood, compared with 1.36 $Wm^{-1}{\cdot}K^{-1}$ for glass [75]. The low thermal conductivity of transparent wood for glazing applications was also demonstrated by Mi et al. [62], with an even lower value of thermal conductivity, 0.19 $Wm^{-1}{\cdot}K^{-1}$, however, did incorporate phase change materials.

5.4. UV Resistance

As mentioned above, TiO_2 can be used to promote UV resistance, for example during weathering [21,101]. The preparation of TiO_2 particles of various sizes on wood surface has been demonstrated. Sun et al. [101] reported a one-pot method for hydrothermal treatment of wood creating sub-micron sized particles of TiO_2. Yang et al. [131] showed that anatase TiO_2 has greatest use for photocatalysis, whereas rutile form shows greater UV blocking capability. Sun et al. [101] created anatase and rutile TiO_2 on wood surfaces using different reaction conditions, and showed that the rutile form had greater UV hindering capacity. Colour change analysis after QUV weathering exposure revealed the lowest change for the rutile TiO_2 treated wood, with anatase TiO_2 showing intermediate colour change, and untreated wood the greatest change.

Copper micro-particle deposition on wood was studied by Gascon-Garrido et al. [132] and demonstrated a reduction in UV induced strength loss. A pulsed-arc discharge plasma was used to deposit the copper microparticles on the wood surface. Micro-veneers with and without the copper surface treatment were exposed to accelerated weathering (EN 927-6), then analysed by various techniques including FTIR, SEM-EDAX, zero span tensile test and blue stain fungal test (EN 152). The FTIR results indicated that lignin degradation was not inhibited by the presence of copper, whereas the strength results indicated that UV-initiated degradation of the polysaccharides was inhibited. The stain fungi were unable to colonise the copper coated surface of the wood. The results are therefore promising for exterior weathering conditions, where mould and stain fungi frequently contribute to discoloration of wood surfaces.

Ultra-violet protection was also reported by Gan et al. [82] in a study that generated magnetic wood by deposition of cobalt ferrite ($CoFe_2O_4$) nanoparticles in the wood, followed by an octadecyltrichlorosilane (OTS) modification to provide hydrophobicity and UV resistance. The OTS treatment used a room temperature immersion followed by 50 °C oven drying step. Colour change on UV exposure was substantially reduced for the combined treatment. The combination of $CoFe_2O_4$ particles with the OTS increased the contact angle from 100° for OTS only, to 150°, although the use of nanoparticles on their own gave a very low contact angle of 7°, which was lower than the wood itself.

5.5. Photochromic Behaviour

The potential to give wood photochromic properties has been demonstrated by Hui et al. [133] who used spironaphoxazine, an organic photochromic compound with good fatigue resistance. This compound alternates between a colourless (closed ring structure) and coloured (open ringed structure) on the presence of UV light. In the study the photochromic material was incorporated in a surface coating of polyvinyl alcohol (PVA) and dextrin [133]. A second formulation incorporated octadecyltrichlorosilane (OTS) to increase hydrophobicity of the coating. The presence of OTS also slowed the colour change seen during weathering test, but did not hinder the photochromic effect.

5.6. Acoustic Properties

One interesting additional use of wood modification is to change the acoustic properties of the wood either for sound adsorption or sound production. Noise pollution is a factor of modern life and yet it can have serious impacts on wellbeing. Modification to increase the sound adsorption of wood could therefore be a highly useful procedure to increase wellbeing of occupants of buildings. Kang et al. [134] used steam explosion to increase the permeability of yellow poplar, resulting in 15 to 50% increase in sound adsorption in the fibre direction, (dependent on the treatment). Wang et al. [135] used microwave treatment to significantly increase sound absorption of Scots pine, especially in mid-range frequencies, via increased permeability due to micro-cracks and voids formed in the wood.

Thermal treatments have been used with higher temperatures appearing to be the most effective in increasing acoustic properties of the wood. Thermal treatment of *Paulownia tomentosa* between 100 °C and 200 °C had an insignificant effect on permeability and therefore sound absorption [136] although heat treatment of Mulberry at 145 °C to 165 °C was shown to increase absorption coefficients in the 250 Hz range [137]. In contrast, high temperature treatment (190 °C) of Malas wood increased permeability in the longitudinal direction with a resulting increase in sound adsorption [138]. Even higher temperature treatments (200–240 °C) have been used to increase the sound absorption coefficient of *Larix kaempferi*, with a particular increase in the absorption of high frequencies [139]. At lower temperatures, using a hygrothermal treatment on *Paulownia coreana* at 110 °C did produce good results, with an 81.8% improvement of the sound absorption coefficient [140].

Noise, at least in the form of music, also has a mostly beneficial impact on wellbeing and wood is used in a myriad of musical instruments. Of course, the influence of humidity and dimensional movement can have serious effects on these instruments both in terms of tone and tuning, and in strain or catastrophic failures when joints and timber are placed under high tension. Wood modification has therefore been used to alleviate some of these issues. Of the chemical methods acetylation, has been tested [141] to solve the issues of dimensional changes on wood in instruments. It was found that whilst stabilising the wood, both the sound velocity and sound adsorption were reduced, although no conclusions on sound quality were made. Succinylation, however, did not improve acoustic properties although in the same study crosslinking with glyoxal and carboxymethyl cellulose did [142]. Vapour formaldehyde treatment of a violin [143] was shown to improve the tone of a violin whilst also protecting it against humidity effects. Further treatments [143,144] using low molecular phenolic resin or vapour formaldehyde combined with saligenin impregnation were found to affect tone and timbre of wooden instruments. Ahmed et al. [145] compared acetylation, phenol formaldehyde, and furfurylation treatments to thermal treatments of a number of woods. They identified a range of differing modulus of elasticity to damping coefficient (tanδ) ratios that would be beneficial for differing instruments. Most modifications produced better acoustic properties compared to the original wood types though thermally modified wood and acetylated wood showed the most potential. Conversely phenol formaldehyde and furfurylation treatments showed the least potential.

Seeing that thermal modification seems to give good results Mania et al. [146] report that the most resonant wood has a high modulus of elasticity and low density. Thermal modification decreases both density and hygroscopicity. Modification at 160 °C increased the modulus of elasticity and acoustic parameters such as sound propagation and acoustic impedance, (determined by the modulus). Pfriem [147] found that thermal modification at 160 °C to 180 °C in low oxygen gave good improvements of acoustic properties. Hydrothermal treatment at temperatures between 100 °C and 140 °C has also been used [148] to decrease density and increase Young's modulus of mulberry wood to improve acoustic properties. However, if intermediate relative humidities, e.g., 29–64%, are used for the hygrothermal treatment then gains are actually reversible and so provide no long term solution [149]. Mania and Gasiorek [150] found hot oil treatment to have different effects based on the size of the element being treated. Small pieces absorbed comparatively high levels of oil which increased density and decreased acoustic performance. In contrast larger pieces showed more of an increase in MOE which overcame the increase in density to result in a small increase in acoustic parameters.

5.7. Electrical Conductivity

Several different approaches have been taken to introduce electrical conductivity to wood in various forms. Polymer impregnation or polymer grafting have been demonstrated by Lv et al. [151] with polypyrrole, and Trey et al. [55] with polyaniline. Proposed applications were for use in supercapacitors and in anti-static applications, respectively. Others have looked at different approaches, such as creation of a fine mesh of conductive nanowires on the wood surface. Guo et al. [152] used copper nanowires, and developed a fusion welding process, based on photonic curing, to ensure good mesh formation and avoid degradation of the thermally sensitive wood substrate during this process. In other transparent materials both copper and silver nanowires have been used, however, the greater abundance of copper leads to greater potential for widespread use in flexible electronics. The use of silver nanoparticles for conductive wood was investigated by Gao et al. [108]. In their study, a post-treatment with a fluoroalkyl silane led to a superhydrophobic surface, using the texture of the nanoparticles combined with the hydrophobicity of the silane. Another method is the impregnation of the wood scaffold with metals [153], however this requires high temperatures, to melt the metal, and could lead to wood thermal degradation, so it is most frequently performed on charred wood, rather than untreated wood, and falls outside the scope of this review. Unlike metals, many of the semiconducting metal oxides (such as TiO_2) show potential for adaptation to introducing electrical conductivity to material substrates, including wood, and could be further tailored to offer sensing properties [95]

There is interest in using wood as a scaffold for electrically conductive or semiconducting materials for various reasons. One is the potential of the wood grain to contribute to different levels of conductivity in different orientations along or across the wood, relating to the distribution of conductive polymer or metal within cell lumena, or as a coating on the cell wall of each lumen space. The flexibility of wood veneers was also listed as an advantage by Lv et al. [151], the flexible wood-polypyrrole material had high capacitance, and showed potential for use in energy storage.

Piezoelectric energy generation using wood-based materials has also been considered. Sun et al. [154] used delignified balsa wood as a 'wood sponge' with greater deformability than untreated wood, and a higher piezoelectric yield on deformation. A large-scale demonstrator used 30 wood sponges connected in to power a commercial LCD screen through compression in a flooring element. This holds considerable potential for wearable electronics and other motion-based energy supply.

5.8. Fire-Retardant Properties

A desirable area for development in modified wood is fire retardancy, either as an indirect result of deposition of the types of inorganic materials used in some of the emerging nanotechnology-based modifications listed above, or as a deliberate effort to move commercially modified timbers into higher performance applications within buildings. As a result, the tests used to claim that new modified woods demonstrate fire retardancy can vary widely, from small lab experiments such as thermogravimetric analysis up to the more thorough fire test standards used on construction sized timber elements or assemblies. Both have their role to play in understanding the many mechanisms through which fire retardancy may be achieved, and an excellent overview can be found in Lowden and Hull [155]. This understanding is especially important as new generation fire-retardant treatments are developed. In this review we made effort to clarify how the reported levels of fire retardancy were tested, to help differentiate between these different scales.

Martinka et al. [156] has shown that thermal treatment can change the fire properties of timber. Thermally treated spruce had a lower total heat release and heat rate comparing to untreated timber in a cone calorimeter study. During thermal treatment, hemicellulose and some of the extractives which have the lowest combustion temperature are decomposed. The absence of these components can explain the thermally modified wood better reaction to fire. Others have shown the opposite, indicating a higher burn rate for thermally modified wood from higher temperature processes, for example Čekovská et al. [157], who used spruce wood in their study, which used a novel 45° ignition method. The treatment conditions chosen (180 °C and 210 °C) were representative for the Thermowood S and Thermowood D product categories, respectively. However, according to Luptakova et al. [158], in order to have any beneficial effect on fire performance of thermally modified timber, the treated temperature should be over 200 °C. Working with larch timber in a cone calorimeter tests, Xing and Li ([159] observed that the heat release rate and mass loss rate curves for thermally treated larch (180 °C, 190 °C and 210 °C) were suppressed, which concords with other cone calorimetry studies, such as Martinke et al. [156]. It should be pointed out that no thermally modified timber, regardless of the temperature conditions used, has sufficient gain in fire retardancy in the full-scale tests required to meet the building regulations in Europe. Therefore, it is highly likely that additional treatments will continue to be required to protect thermally modified timber. Gasparik et al. [160] showed that the application of a two-substance fire-retardant coating can improve the fire performance of thermally treated oak. This was a commercially available product, using three layers of a foaming formulation containing oxalic acid, ammonium phosphates and fire-retardant additives, which was followed by an acrylic cover lacquer.

Acetylated wood also shows a small change in flammability, as a result of the chemical modification, however this is insufficient to provide fire resistance. Mohebby et al. [161] tested acetylated beech plywood in an ignitability test on small samples. They reported a delay in ignition for the acetylated beech, and a reduction in observed glowing, which they attributed to the weight percent gain from the acetylation treatment.

In 2007, an Australian collaborative project was funded by the Commonwealth Scientific and Industrial Research Organization (CSIRO) to develop a single step treatment that could provide fire protection and preservation to wood. According to Marney and

Russell [162], to produce a multifunctional system to act both as fire retardant and wood preservative there are four different approaches. The first approach would be to modify an existing preservative with the addition of fire-retardant chemical. Another approach would be to treat or chemically modify the wood with fire retardants that have proven to have good biocidal performance. A similar approach would be the usage of preservatives that have shown good fire resistance when fixed into the wood structure. The fourth approach would be the wood modification with inorganic substances forming a wood-inorganic composites.

Sweet et al. [163] tested a variety of fire retardants in combination with wood preservatives. This follows point two of Marney and Russel's suggested four potential approaches [162], in this case the examples are based on treatments rather than wood modifications, in the strict sense, however the list is included here to show the range of substances currently under investigation. Sweet et al. [163] tested the compatibility of the different products in leaching, indicating their ability to fulfil the requirements for external usage. According to their findings urea dicyandiamide phosphoric acid and formaldehyde (UDPF), melamine dicyandiamide phosphoric acid and formaldehyde (MDPF), dicyandiamide phosphoric formaldehyde (DPF) and dicyandiamide phosphoric acid (DP) were the fire retardant candidates to show resistance to leaching and be compatible with the chromated copper arsenate (CCA), didecyl dimethyl ammonium chloride (DDAC), DDCA with 3-iodo-2-propynyl butyl carbamate (NP-1), ammoniacal copper quaternary (ACQ), zinc naphthenate, waterborne emulsion (ZnNaph-wb) and ammoniacal copper arsenate (ACA) preservatives. Their study concluded that the combinations UDPF with DDAC and MDPF with NP-1 were the most effective system to provide both fire resistance and durability before and after weathering. Similarly, many researchers have used different combinations of fire retardants with wood preservatives such as zirconium and borate salts and urea phosphoric acid and ethanol with a mixture of boric acid and borax pentahydrate [162]. An example of wood modification combined with fire retardant is seen in the work of Baysal [164], who showed that the melamine formaldehyde with boric acid and borax improved the fire performance of wood. However, the efficiency of the system as preservative was not investigated in this study as it was assumed that boric acid would provide sufficient performance.

Lewin [165] tried the approach of chemically modify wood with proven fire-retardant substances. Yellow pine and spruce were subjected to bromination of lignin with bromate-bromide resulting to improved fire and decay resistance while being resistant to leaching. Moreover, the brominated timber did not show any significant reduction in mechanical properties while improving hydrophobicity and swelling. Moreover, in situ production of phosphoramides within Loblolly Pine and Sweetgum timber using phosphorus pentoxide and amines has been reported to provide good decay and fire resistance [162,166].

In an attempt to fix a substance into wood in order to provide fire and decay resistance properties, Tsunoda [167] proposed a boron vapor treatment and showed that there was no difference between fire performances of the suggested application with the application of boron in a liquid state. However, this study did not include any data about leaching after weathering. To fix the leaching problem of boron substances in a fire-decay resistance modified wood, Kartal et al. [168] introduced a chemical modification using allyl glycidyl ether (AGE) in combination with methyl methacrylate (MMA), grafted in situ into timber that had been treated with disodium octoborate tetrahydrate (DOT). Their findings were that the modified Japanese cedar timber showed improvement in dimensional stability, leaching, decay and termite resistance, however they did not include fire performance testing. Another approach to fix the boron into the treated wood was proposed by Baysal et al. [169] with the application of various water repellent agents, including styrene, isocyanate, paraffin wax and methyl methacrylate, as a post borate impregnation treatments. A vacuum impregnation system and thermally induced cure step (90 °C, 4 h) was used to cure the monomers in situ. They found that styrene application provided the most effective results in reducing borate leaching, yet they did not present fire performance data.

Jiang et al. [170] used a N-methylol resin to fix a guanylurea phosphate (GUP) and boric acid (BA) fire-retardant system into wood. The resin was DMDHEU (which is a known wood modification agent) was connected to an increase in flammability, when used alone, but with the GUP/BA treatment suppressed combustion. The treated wood was tested in full-scale tests and demonstrated class $B_{fl\text{-}s1}$ performance, suitable for certain building applications.

Yue et al. [171] used boric acid that was incorporated into phenol formaldehyde resin during synthesis to enhance the fire retardancy of PF modified wood. The treated wood was assessed in cone calorimetry tests, and showed a suppression of HRR, which became more prominent as the percentage of boric acid PF resin within the wood increased. Smoke formation was also reduced. Interestingly, the mechanical properties of the boric acid PF treated wood were greater than those of the PF treated wood. In a development of this work, Yue et al. [172] showed that compression of the boric acid PF resin impregnated timber led to further gains in combustion resistance, and increased mechanical properties.

Different silane compounds and different treatment methods have been investigated as systems to provide inorganic fire resistance and durability to timber and timber-based products with some promising results. However, some of the treatments could not minimize the leaching issue seen with boron compounds. Other approaches with combination of zinc and boron compounds have been also investigated with interesting results, however many of those studies do not include fire test data and they are focusing on the leachability of their substances [162]. Mineralisation of timber has also been investigated as a fire resistance and hydrophobicity improvement method. $CaCO_3$ can be found in many exoskeleton species such as corals, egg-shells and crustacean exoskeletons and it provides durability, increase of hardness and water uptake reduction. The incorporation of $CaCO_3$ into the timber structure as a bio-inspired mineralisation showed a high improvement in fire resistance. This could be an alternative approach to improve fire resistance of timber, however there was no leaching and water absorbing data in this study [79].

The modification of timber in order to provide durability has an effect on the fire performance. Rabe et al. [173] investigated the fire performance of acetylated (Accoya), thermally treated and N-methylol crosslinked (Belmadur) timber, and compared them to untreated timbers of the equivalent species. In addition, Scots pine and a tropical hardwood (Meranti) were included to investigate species effects. They used four test systems: Thermogravimetric analysis, bomb calorimetry, heat release capacity and dual cone calorimetry to conduct tests, in a comprehensive study. All of the treatments except for the acetylated timber showed a partial flame retardancy. The introduced acetyl groups into acetylated wood are producing volatile compounds with low ignition temperatures during pyrolysis in the event of fire which they negatively affect their fire resistance [173,174]. However, it seems that for the timbers investigated by Rabe et al. [173], the timber species had a greater effect on fire performance than the wood modification treatment itself. When species with higher lignin content (which is associated with higher char yield) were used, the improvement in fire performance became more prominent.

6. Discussion

This review has considered the functionalisation of modified timber and the modification of timber to provide (multi)-functionality as two overlapping fields (Figure 1). Additionally, we have identified rapid expansion in the field of nanotechnology applied to wood, whether this is in combination with wood modification; or as a wood modification process in its own right; or as a process with potential for application to modified wood through future research. Our intention has been to highlight the many parallel research fields, and some of their inter-connectivity, which leads to multifunctionality in new generation wood. Throughout the review we have encountered similar active substances, combined in different sequences, or different combinations.

It is clear from the review that for some properties, the commercially available modified timbers may be easily subjected to a secondary treatment, to deliver additional functionality. A good example is the addition of hydrophobicity to thermally modified wood, often at the same time as introducing UV stability and weathering resistance. Another example is where thermally modified timber is routinely available with a secondary commercially applied fire-retardant formulation of the same type as is used on unmodified timber, to deliver fire retardancy. In this particular example, we have not presented formulation data, or scientific journal citations, as this is typically proprietary information available from the timber merchant or fire-retardant treatment company. Similarly, Accoya have stated that the fire class of their timber for European cladding applications is class D-s2, but note that fire-retardant treatment using commercially available systems can be used to achieve higher requirements, up to class B, with appropriate systems [175].

As a contrast, at a laboratory research level, various groups have looked to alter the wood modification processes to incorporate fire-retardant agents, or to trap active chemicals that were previous applied (e.g., using resin impregnation systems), or to modify the wood so that it is more receptive to secondary fire-retardant treatments, and better able to retain these active components during leaching or other weathering and exposure tests. This is an essential process, and can provide good results, suitable for future incorporation in commercial systems or development into new stand-alone products.

A similar mix of strategies can be seen in other fields of adding functionality to modified wood, or adding functionality through the modification process. Examples of adding functionality to modified wood that we have seen include: resistance to mould growth (by adding biocides, or by adding secondary modifications); termite resistance (by adding active agents before or during the thermal modification); promotion of UV resistance on resin modified woods (by incorporating dye in resin, or including HALS in the resin, or the use of surface coatings, or use of pre-modification with succinic anhydride); improvement of biodegradability (through the use of biopolymers rather than the traditionally chosen polymers and resins); and adding novel functionality through new polymer impregnants (leading to pH response, electrical conductivity, or energy storage by inclusion of phase change materials within the resin).

There have been examples of combining two modification systems to successful effect. Notably this is the basis for forming transparent wood by using delignified wood with polymer impregnation; but is also well demonstrated in the various chemical pretreatments (silanes, anhydrides) to enhance resin adhesion in resin modified timber. Some pretreatments have also been developed that could essentially be considered to be modifications, for example the use of NaOH to increase cell wall texture prior to grafting or deposition of other components, or the delignification of wood to alter nanostructure and chemical functionality.

This research has led to some new modification systems: delignification with compression (for flexible functionalised films); and the impregnation or encrustation with inorganic particles. This latter modification is not always considered to be a modification in the truest sense, as it overlaps with the impregnation of micronized copper or other traditional preservative systems. However, in this review we believe it is justified in some cases, where for example the deposition is highly controlled to provide closely spaced nanorods of ZnO as a surface coating, or TiO_2 of the correct crystalline form on the wood cell wall. This is clearly more engineered than simple impregnation with preservatives or FR systems; there is tailoring of crystalline forms and particle sizes. These systems are most closely associated with the treatments giving hydrophobicity and superhydrophobicity.

Many examples of simultaneously adding several functionalities while modifying wood can be found in the transparent wood family of products—where different systems have demonstrated combinations of secondary or tertiary functionalities through added nanoparticles, chemicals, dyes or active substances. Traits that have been achieved include: heat shielding, lasing, luminescence, heat storage and magnetic properties. It is clear from the many publications in this field that there is considerable scope to achieve new combinations through continued research.

Many of these multi-functional materials are not targeting traditional sectors where wood is used as a commodity material. The use of veneers or thin slices of wood for transparent materials in glazing or solar cells is a good example; it is likely to displace glass, rather than replace structural timber. However, good mechanical properties will still be required, to sustain the wear and tear of the domestic environment—weather, cleaning, human activity. Thus, the reported advances in mechanical properties [57,58,61] are a necessary part of the whole product. When transparent wood with energy regulating properties induced by phase change materials is fully developed, will it be used in posts and beams for routine construction? This is unlikely, a more strategic location for these wood materials will be chosen where their light diffusing qualities can be combined with greatest capacity for thermal gain and thermal release. It will be necessary to not embed the functionalised timber into standard wall elements as structural units, but leave them exposed within the wall to ensure a real contribution to the regulation of the interior environment.

A big question is: can multi-functional wood still be defined as wood or modified wood? Looking back at the definition proposed by Hill [3], many of the cases described in this review have retained the necessary traits:

- The action of chemical, physical or biological agents on the cell wall—in the case of this review, primarily chemical processes have been described, but also compression which is a physical process.
- Achievement of property enhancement has been achieved, in a very diverse set of properties.

In his definition, Hill continues to say that "The modified wood should itself be nontoxic under service conditions, and furthermore, there should be no release of any toxic substances during service, or at end of life, following disposal or recycling of the modified wood. If the modification is intended for improved resistance to biological attack, then the mode of action should be non-biocidal" [3]. This is probably the more challenging part of the definition, and the answers to different functionalised modified woods described in this review will vary significantly in their toxicity in service, or their likelihood of releasing toxic substances during service life or after disposal or recycling. To consider these questions would take detailed individual study, and during the route to scale-up and commercialisation of novel modified wood products, researchers must continue to ask these questions, to ensure that new products are correctly declaring their sustainable or environmentally benign credentials.

In order to do this, various green chemistry approaches, ecotoxicity studies, life cycle assessment (LCA) studies and other research will be required to answer pertinent questions about end-of-life options for modified wood or functionalised wood. There is clearly much to be done.

Certainly, this review has revealed that the range of modifications covered, and the scope of applications, show that wood modification may no longer be the sole remit of wood scientists but a true inter-disciplinary area. Ongoing work will involve metallurgy, ceramic, electronics, nanotechnology and polymer science on one hand and will involve engineers, designers, and even health professionals on the other.

7. Conclusions

Wood modification, by chemical, impregnation or thermal methods, has already been used to address issues such as dimensional instability with changes in moisture content, or by degradation and deterioration due to biotic factors or weathering with varying success. Ongoing research has looked to new chemical modifications and other fields of inspiration such as biomimetics and nanotechnology to further augment the properties of wood. This review has considered two parallel, and closely connected themes: (1) the functionalisation of modified timber and (2) the modification of timber to provide (multi)-functionality. In several examples we have observed new wood modification processes, which are emerging through the pursuit of functionalised wood for high technology applications such as energy and electronics.

A large body of work is also emerging through nanotechnology approaches that have been tested on wood (but not necessarily applied to modified wood yet). These were noted for their potential to radically change the way that wood is used in future. Wood, and especially delignified wood, is increasingly seen as a scaffold for enhancement through decoration with functional particles. In time, some of these nanotechnology techniques will be applied to modified wood as part of a multi-step process, where the benefits complement or offset the properties of the modified wood.

In the shorter term, a good example that is well advanced is the development of weathering resistant surfaces for thermally modified timber, which boosts the performance of the thermally modified timber in an area where it previously had a weakness. It is clear there are many other treatments and modifications in the pipeline. Certainly, as wood modification moves to render dimensional change negligible for wood, new aspirations for using this stable substrate in precision applications such as electronics or responsive materials have become tangible.

Author Contributions: Conceptualization, M.J.S., S.F.C., and G.A.O.; investigation, M.J.S., S.F.C., and A.D.; resources, G.A.O.; writing—original draft preparation, M.J.S., S.F.C., and A.D.; writing—review and editing, M.J.S., S.F.C., G.A.O., and A.D.; visualization, M.J.S.; supervision, M.J.S. and G.A.O. All authors have read and agreed to the published version of the manuscript.

Funding: The authors gratefully acknowledge the support of SEEC (Smart Efficient Energy Centre) at Bangor University, funded by European Regional Development Fund (ERDF), administered through the Welsh Government.

Institutional Review Board Statement: Not applicable.

Informed Consent Statement: Not applicable.

Data Availability Statement: Data sharing not applicable. No new data were created or analyzed in this study. Data sharing is not applicable to this article.

Conflicts of Interest: The authors declare no conflict of interest.

References

1. Ross, P.; Downes, G.; Lawrence, A. *Timber in Contemporary Architecture*; TRADA Technology Ltd.: London, UK, 2009.
2. Ansell, M.P. Hybrid wood composites–integration of wood with other engineering materials. In *Wood Composites*; Ansell, M.P., Ed.; Chapman & Hall: London, UK, 2015; pp. 411–426.
3. Hill, C.A.S. *Wood Modification—Chemical, Thermal and Other Processes*; Wiley Series in Renewable Resources; John Wiley & Sons: Chichester, UK, 2006; p. 260.
4. Ormondroyd, G.A.; Spear, M.J.; Curling, S.C. Modified wood: Review of efficacy and service life testing. *Constr. Mater.* **2015**, *168*, 187–203. [CrossRef]
5. Sandberg, D.; Kutnar, A.; Mantanis, G. Wood modification technologies—A review. *iForest* **2017**, *10*, 895–908. [CrossRef]
6. Jones, D.; Sandberg, D. A review of wood modification globally–updated findings from COST FP1407. *Interdiscip. Perspect. Built Environ.* **2020**, *1*, 1. [CrossRef]
7. Spear, M.J. Preservation, protection and modification of wood composites. In *Wood Composites*; Ansell, M.P., Ed.; Woodhead Publishing: Cambridge, UK, 2015; pp. 253–310.
8. Eaton, R.A.; Hale, M.D.C. *Wood: Decay Pests and Protection*; Chapman and Hall: London, UK, 1993.
9. Rowell, R.M. Acetylation of wood—A Review. *Int. J. Lignocellul. Prod.* **2014**, *1*, 1–27.
10. Mantanis, G.I. Chemical reaction of wood by acetylation and furfurylation: A review of the present scaled up technologies. *BioResources* **2017**, *12*, 4478–4489. [CrossRef]
11. Esteves, B.M.; Pereira, H.M. Wood modification by heat treatment: A review. *BioResources* **2009**, *3*, 142–154. [CrossRef]
12. Stefanowski, B.; Spear, M.; Pitman, A. Review of the use of PF and related resins for modification of solid wood. In *Timber 2018*; Spear, M.J., Ed.; IOM3: London, UK, 2018; pp. 165–179.
13. Srinivas, K.; Pandey, K. Photodegradation of thermally modified wood. *J. Photochem. Photobiol.* **2012**, *117*, 140–145. [CrossRef]
14. Huang, X.; Kocaefe, D.; Kocaefe, Y.; Boluk, Y.; Krause, C. Structural analysis of heat-treated birch (*Betula papyrifera*) surface during artificial weathering. *Appl. Surf. Sci.* **2013**, *264*, 117–127. [CrossRef]
15. Ayadi, N.; Lejeune, F.; Charrier, F.; Charrier, B.; Merlin, A. Color stability of heat-treated wood during artificial weathering. *Holz Als Roh-Und Werkst.* **2003**, *61*, 221–226. [CrossRef]

16. Miklecic, J.; Turkulin, H.; Jirous-Rajkovic, V. Weathering performance of surface of thermally modified wood finished with nanoparticles-modified waterborne polyacrylate coatings. *Appl. Surf. Sci.* **2017**, *408*, 103–109. [CrossRef]
17. Shen, H.; Zhang, S.; Cao, J.; Jiang, J.; Wang, W. Improving anti-weathering performance of thermally modified wood by TiO_2 or/and paraffin emulsion. *Constr. Build. Mater.* **2018**, *169*, 372–378. [CrossRef]
18. Veronovski, N.; Verhofsek, D.; Godmjavec, J. The influence of surface treated nano-TiO_2 (rutile) incorporation in water-based acrylic coatings on wood protection. *Wood Sci. Technol.* **2013**, *47*, 317–328. [CrossRef]
19. Faure, B.; Salazar-Alvarez, G.; Ahniyaz, A.; Villaluenga, L.; Berriozabal, G.; De Miguel, R.D.M.; Bergstrom, L. Dispersion and surface functionalisation of oxide nanoparticles for transparent photocatalytic and sunscreens. *Sci. Technol. Adv. Mater.* **2013**, *14*, 023001. [CrossRef]
20. Guo, H.; Klose, D.; Hou, Y.; Jesche, G.; Burgert, I. Highly efficient UV protection of the biomaterial wood by a transparent TiO_2/Ce xerogel. *ACS Appl. Mater. Interfaces* **2017**, *9*, 39040–39047. [CrossRef]
21. Shen, H.; Cao, J.; Jiang, J.; Xu, J. Antiweathering properties of a thermally treated wood surface by two-step treatment with titanium dioxide nanoparticle growth and polydimethylsiloxane. *Prog. Org. Coat.* **2018**, *125*, 1–7. [CrossRef]
22. Papadopoulos, A.; Taghiyari, H.R. Innovative wood surface treatments based on nanotechnology. *Coatings* **2019**, *9*, 866. [CrossRef]
23. Mahltig, B.; Swaboda, C.; Roessler, A.; Böttcher, H. Functionalising wood by nanosol application. *J. Mater. Chem.* **2008**, *18*, 3180. [CrossRef]
24. Wang, X.; Chai, Y.; Liu, J. Formation of highly hydrophobic wood surfaces using silica nanoparticles modified with long-chain alkylsilane. *Holzforschung* **2013**, *67*, 667–672. [CrossRef]
25. Mahltig, B.; Arnold, M.; Löthman, P. Surface properties of sol–gel treated thermally modified wood. *J. Sol-Gel Sci. Technol.* **2010**, *55*, 221–227. [CrossRef]
26. Kwaśniewska-Sip, P.; Bartkowiak, M.; Cofta, G.; Bolesław Nowak, P. Resistance of Scots Pine (*Pinus sylvestris* L.) after Treatment with Caffeine and Thermal Modification against *Aspergillus niger*. *BioResources* **2019**, *14*, 1890–1898.
27. Karlsson, O.; Myronycheva, O.; Jones, D.; Sandberg, D. Thermally modified wood treated with methacrylate. In Proceedings of the IRG Annual Meeting IRG49 Scientific Conference on Wood Protection, Sandton, Johannesburg, South Africa, 29 April–3 May 2018.
28. Kim, I.; Karlsson, O.; Myronycheva, O.; Jones, D.; Sandberg, D. Methacrylic Resin for Protection of Wood from Discoloration by Mould Growth and Weathering. *BioResources* **2020**, *15*, 7018–7033.
29. Duarte, S.; Nunes, L.; Kržišnik, D.; Humar, M.; Jones, D. Influence of zwitterionic buffer effects with thermal modification treatments of wood on symbiotic protists in *Reticulotermes grassei* Clement. *Insects* **2021**, *12*, 139. [CrossRef]
30. Popescu, C.-M.; Jones, D.; Kržišnik, D.; Humar, M. Determination of the effectiveness of a combined thermal/chemical wood modification by the use of FT–IR spectroscopy and chemometric methods. *J. Mol. Struct.* **2020**, *1200*, 12713. [CrossRef]
31. Salman, S.; Thevenon, M.-F.; Petrissans, A.; Dumarcay, S.; Candelier, K.; Gerardin, P. Improvement of the durability of heat-treated wood against termites. *Maderas Cienc. Y Tecnol.* **2017**, *19*, 317–328. [CrossRef]
32. Kielmann, B.C.; Mai, C. Application and artificial weathering performance of translucent coatings on resin-treated and dye-stained beech-wood. *Prog. Org. Coat.* **2016**, *95*, 54–63. [CrossRef]
33. Evans, P.D.; Gibson, S.K.; Cullis, I.; Liu, C.; Sebe, G. Photostabilisation of wood using low molecular weight phenol formaldehyde resin and hindered amine light stabilisers. *Polym. Degrad. Stab.* **2013**, *98*, 158–168. [CrossRef]
34. Stefanowski, B.K.; Spear, M.J.; Curling, S.F.; Pitman, A.J.; Bailey, R.J.D.; Mathias, J.; Evans, N.; Ayala, L.; Theobald, T.; Ormondroyd, G.A. Properties of LIGNIA modified wood. In *Timber 2020*; Spear, M., Ed.; IOM3: London, UK, 2020; pp. 177–188.
35. Wang, Z.; Han, X.; Wang, S.; Lv, Y.; Pu, J. Enhancing the thermal stability, water repellency, and flame retardancy of wood treated with succinic anhydride and melamine-urea formaldehyde resins. *Holzforschung* **2020**, *74*, 957–965. [CrossRef]
36. Evans, P.D.; Wallis, A.F.A.; Owen, N.L. Weathering of chemically modified wood surfaces: Natural weathering of Scots pine acetylated to different weight gains. *Wood. Sci. Technol.* **2000**, *32*, 151–165. [CrossRef]
37. Chang, S.-T.; Chang, H.-T. Comparisons of the photostability of esterified wood. *Polym. Degrad. Stab.* **2001**, *71*, 261–266. [CrossRef]
38. Pandey, K.K.; Chandrashekar, N. Photostability of wood surfaces esterified by benzoyl chloride. *J. Appl. Polym. Sci.* **2005**, *99*, 2367–2374. [CrossRef]
39. Feist, W.C.; Hon, D.N.S. Chemistry of weathering and protection. In *The Chemistry of Solid Wood*; Rowell, R.M., Ed.; Advances in Chemistry Series; American Chemistry Society: Washington, DC, USA, 1984; Volume 207.
40. Rosu, D.; Bodirlau, R.; Teaca, C.-A.; Rosu, L.; Varganici, C.D. Epoxy and succinic anhydride functionalized soybean oil for wood protection against UV light action. *J. Clean. Prod.* **2016**, *112*, 1175–1183. [CrossRef]
41. Olsson, S.K.; Johansson, M.; Westin, M.; Östmark, E. Reactive UV-absorber and epoxy functionalized soybean oil for enhanced UV-protection of clear coated wood. *Polym. Degrad. Stab.* **2014**, *110*, 405–414. [CrossRef]
42. Li, Y.; Wu, Q.; Li, J.; Liu, Y.; Wang, X.-M.; Liu, Z. Improvement of dimensional stability of wood via combination treatment: Swelling with maleic anhydride and grafting with glycidyl methacrylate and methyl methacrylate. *Holzforschung* **2012**, *66*, 59–66. [CrossRef]
43. Schneider, M.H.; Phillips, J.G. Elasticity of wood and wood polymer composites in tension and compression loading. *Wood Sci. Technol.* **1991**, *25*, 361–364. [CrossRef]
44. Ermeyden, M.A.; Babacan, M.; Tomak, E.D. Evaluation of dimensional stability, weathering and decay resistance of modified pine wood by in-situ polymerisation of styrene. *J. Wood Chem. Technol.* **2020**, *40*, 294–305. [CrossRef]

45. Devi, R.R.; Maji, T.K. Effect of glycidyl methacrylate on the physical properties of wood-polymer composites. *Polym. Compos.* **2007**, *28*, 1–5. [CrossRef]
46. Curling, S.F.; Spear, M.J.; Ormondroyd, G. A Gibson R Physical properties and durability of methacrylate impregnated timber. In Proceedings of the 13th Annual Meeting of the Northern European Network for Wood Science and Engineering, Copenhagen, Denmark, 28–29 September 2017; pp. 32–37.
47. Ermeyden, M.A.; Cabane, E.; Hass, P.; Koetz, J.; Burgert, I. Fully biodegradable modification of wood for improvement of dimensional stability and water absorption properties by poly(ε-caprolactone) grafting into the cell walls. *Green Chem.* **2014**, *16*, 3313–3321. [CrossRef]
48. Noel, M.; Grigsby, W.J.; Vitkeviciute, I.; Volkmer, T. Modifying wood with biopolyesters: Analysis and performance. *Int. Wood Prod. J.* **2014**, *6*, 14–20. [CrossRef]
49. Noel, M.; Grigsby, W.J.; Volkmer, T. Evaluating the extent of bio-polyester polymerization in solid wood by thermogravimetric analysis. *J. Wood Chem. Technol.* **2015**, *35*, 325–336. [CrossRef]
50. Noel, M.; Grigsby, W.J.; Ormondroyd, G.A.; Spear, M.J. Influence of water and humidity on chemically modifying wood with polybutylene succinate bio-polyester. *Int. Wood Prod. J.* **2016**, *7*, 80–88. [CrossRef]
51. Grosse, C.; Noel, M.; Thevenon, M.-F.; Rautkari, L.; Gerardin, P. Influence of water and humidity on wood modification with lactic acid. *J. Renew. Mater.* **2017**, *6*, 259–269. [CrossRef]
52. Berglund, L.; Burgert, I. Bioinspired wood nanotechnology for functional materials. *Adv. Mater.* **2018**, *30*, 1704285. [CrossRef] [PubMed]
53. Cabane, E.; Keplinger, T.; Merk, V.; Hass, P.; Burgert, I. Renewable and functional wood materials by grafting polymerization within cell walls. *ChemSusChem* **2014**, *7*, 1020–1025. [CrossRef]
54. Cabane, E.; Keplinger, T.; Künniger, T.; Merk, V.; Burgert, I. Functional lignocellulosics prepared by ATRP from a wood scaffold. *Sci. Rep.* **2016**, *6*, 31287. [CrossRef] [PubMed]
55. Trey, S.; Jafarzadeh, S.; Johansson, M. In-situ polymerization of polyaniline in wood veneers. *ACS Appl. Mater. Interfaces* **2012**, *4*, 1760–1769. [CrossRef] [PubMed]
56. Hassel, B.I.; Trey, S.; Leijonmarck, S.; Johansson, M. A study on the morphology mechanical and electrical performance of polyaniline-modified wood—A semiconducting material. *BioResources* **2014**, *9*, 5007–5023. [CrossRef]
57. Li, Y.; Fu, Q.; Yu, S.; Yan, M.; Berglund, L. Optically transparent wood from nanoporous cellulosic template combining functional and structural performance. *Biomacromolecules* **2016**, *17*, 1358–1364. [CrossRef] [PubMed]
58. Zhu, M.; Song, J.; Li, T.; Gong, A.; Wang, Y.; Dai, J.; Yao, Y.; Luo, W.; Henderson, D.; Hu, L. Highly anisotropic, highly transparent wood composites. *Adv. Mater.* **2016**, *28*, 5181–5187. [CrossRef] [PubMed]
59. Li, Y.; Fu, Q.; Yang, X.; Berglund, L. Transparent wood for functional and structural applications. *Phil. Trans. R. Soc. A* **2017**, *376*, 20170182. [CrossRef]
60. Wu, J.M.; Wu, Y.; Yang, F.; Tang, C.Y.; Huang, Q.T.; Zhang, J.L. Impact of delignification on morphological, optical and mechanical properties of transparent wood. *Compos. Part A* **2019**, *117*, 324–331. [CrossRef]
61. Li, Y.; Fu, Q.; Rojas, R.; Yan, M.; Lawoko, M.; Berglund, L. Lignin-retaining transparent wood. *ChemSusChem* **2017**, *10*, 3445–3451. [CrossRef]
62. Mi, R.; Li, T.; Dalgo, D.; Chen, C.; Kuang, Y.; He, S.; Zhao, X.; Xie, W.; Gan, W.; Zhu, J.; et al. A clear strong and thermally insulated transparent wood for energy efficient windows. *Adv. Funct. Mater.* **2020**, *30*, 1907511. [CrossRef]
63. Wu, Y.; Wang, Y.; Yang, F.; Wang, J.; Wang, X. Study on the properties of transparent bamboo prepared by epoxy resin impregnation. *Polymers* **2020**, *12*, 863. [CrossRef] [PubMed]
64. Chen, H.; Montanari, C.; Yan, M.; Popov, S.; Li, Y.; Sychugov, I.; Berglund, L.A. Refractive index of delignified wood for transparent biocomposites. *RSC Adv.* **2020**, *67*, 40719. [CrossRef]
65. Wang, X.; Zhan, T.; Liu, Y.; Shi, J.; Pan, B.; Zhang, Y.; Cai, L.; Shi, S.Q. Large-size transparent wood for energy-saving building applications. *ChemSusChem* **2018**, *11*, 4086–4093. [CrossRef] [PubMed]
66. Li, T.; Zhu, M.; Yang, Z.; Song, J.; Dai, J.; Yao, Y.; Luo, W.; Pastel, G.; Hu, L. Wood composite as an energy efficient building material: Guided sunlight transmittance and effective thermal insulation. *Adv. Energy Mater.* **2016**, *6*, 1601122. [CrossRef]
67. Zhu, M.; Li, T.; Davis, C.S.; Yao, Y.; Dai, J.; Wang, Y.; AlQatari, F.; Gilman, J.F.; Hu, L. Transparent and haze wood composites for highly efficient broadband light management in solar cells. *Nano Energy* **2016**, *26*, 332–339. [CrossRef]
68. Chen, L.; Xu, Z.; Wand, F.; Duan, G.; Xu, W.; Zhang, G.; Yang, H.; Liu j Jiang, S. A flame-retardant and Transparent wood/polyamide composite with excellent mechanical strength. *Compos. Commun.* **2020**, *20*, 100355. [CrossRef]
69. Li, Y.; Yu, S.; Veinot, J.G.C.; Linnros, J.; Berglund, L.; Sychugov, I. Luminescent transparent wood. *Adv. Opt. Mater.* **2017**, *5*, 1600834. [CrossRef]
70. Vasileva, E.; Li, Y.; Sychugov, I.; Mensi, M.; Berglund, L.; Popov, S. Lasing from organic dye molecules embedded in transparent wood. *Adv. Opt. Mater.* **2017**, *5*, 1700057. [CrossRef]
71. Koivurova, M.; Vasileva, E.; Li, Y.; Berglund, L.; Popov, S. Complete spatial coherence characterization of quasi-random laser emission from dye doped transparent wood. *Opt. Express* **2018**, *26*, 13474. [CrossRef]
72. Yu, Z.; Yao, Y.; Yao, J.; Zhang, L.; Chen, Z.; Gao, Y.; Luo, H. Transparent wood containing Cs_xWO_3 nanoparticles for heat-shielding windows. *J. Mater. Chem. Part A* **2017**, *5*, 6019–6024. [CrossRef]

73. Qiu, Z.; Wang, S.; Wang, Y.; Li, J.; Xiao, Z.; Wang, H.; Lian, D. Transparent wood with thermo-reversible optical properties based on phase-change material. *Compos. Sci. Technol.* **2020**, *200*, 108407. [CrossRef]
74. Lang, A.W.; Li, Y.; De Keersmaecker, M.; Shen, D.E.; Österholm, A.M.; Berglund, L.; Reynolds, J. Transparent wood smart windows: Polymer electrochromic devices based on poly(3,4-ethylenedioxythiophene):poly(styrene sulphonate) electrodes. *ChemSusChem* **2018**, *11*, 854–863. [CrossRef]
75. Montanari, C.; Li, Y.; Chen, H.; Yan, M.; Berglund, L.A. Transparent wood for thermal energy storage and reversible optical transmittance. *Appl. Mater. Interfaces* **2019**, *11*, 20465–20472. [CrossRef] [PubMed]
76. Gan, W.; Xiao, S.; Gao, L.; Gao, R.; Li, J.; Zhan, X. Luminescent and Transparent Wood Composites Fabricated by Poly(methyl methacrylate) and γ-Fe_2O_3@YVO_4,Eu^{3+} Nanoparticle Impregnation. *ACS Sustain. Chem. Eng.* **2017**, *5*, 3855–3862. [CrossRef]
77. Gan, W.; Gao, L.; Xiao, S.; Zhang, W.; Zhan, X.; Li, J. Transparent magnetic wood composites based on immobilising Fe_3O_4 nanoparticles into a delignified wood template. *J. Mater. Sci.* **2017**, *52*, 3321–3329. [CrossRef]
78. Wang, X.; Shan, S.; Shi, S.Q.; Zhang, Y.; Cai, L.; Smith, L.M. Optically transparent bamboo with high strength and low thermal conductivity. *ACS Appl. Mater. Interfaces* **2021**, *13*, 1662–1669. [CrossRef] [PubMed]
79. Montanari, C.; Olsen, P.; Berglund, L.A. Interface tailoring by a versatile functionalisation platform for nanostructured wood biocomposites. *Green Chem.* **2020**, *22*, 8012. [CrossRef]
80. Burgert, I.; Cabane, E.; Zollfrank, C.; Berglund, L. Bio-inspired functional wood-based materials–hybrids and replicates. *Int. Mater. Rev.* **2015**, *60*, 431–450. [CrossRef]
81. Merk, V.; Chanana, M.; Keplinger, T.; Gaand, S.; Burgert, I. Hybrid wood materials with improved fire retardance by bio-inspired mineralisation on the nano- and submicron level. *Green Chem.* **2015**, *17*, 1423–1428. [CrossRef]
82. Gan, W.; Gao, L.; Sun, Q.; Jin, C.; Lu, Y.; Li, J. Multifunctional wood with magnetic, superhydrophobic and anti-ultraviolet properties. *Appl. Surf. Sci.* **2015**, *332*, 565–572. [CrossRef]
83. Trey, S.; Olsson, R.T.; Ström, V.; Berglund, L.; Johnasson, M. Controlled deposition of magnetic particles within the 3D template of wood: Making use of the natural hierarchical structure of wood. *RSC Adv.* **2014**, *4*, 35678. [CrossRef]
84. Guo, H.; Fuchs, P.; Cabane, E.; Michen, B.; Hagendorfer, H.; Romanyuk, Y.E.; Burgert, I. UV-protection of wood surfaces by controlled morphology fine-tuning of ZnO nanostructures. *Holzforschung* **2016**, *70*, 699–708. [CrossRef]
85. Zhang, Y.; Luo, W.; Wang, C.; Li, Y.; Chen, C.; Song, J.; Dai, J.; Hitz, E.M.; Xu, S.; Yang, C.; et al. High capacity, low-tortuosity and channel-guided lithium metal anode. *Proc. Natl. Acad. Sci. USA* **2017**, *114*, 3584–3589. [CrossRef]
86. Fu, Q.L.; Medina, L.L.; Li, Y.Y.; Carosio, F.; Alireza, H.J.; Berglund, L.A. Nanostructured wood hybrids for fire-retardancy prepared by clay impregnation into the cell wall. *ACS Appl. Mater. Interfaces* **2017**, *9*, 36154–36163. [CrossRef] [PubMed]
87. Frey, M.; Widner, D.; Segmehl, J.S.; Casdorff, K.; Keplinger, T.; IBurgert, I. Delignified and Densified Cellulose Bulk Materials with Excellent Tensile Properties for Sustainable Engineering. *ACS Appl. Mater. Interfaces* **2018**, *10*, 5030–5037. [CrossRef] [PubMed]
88. Song, J.; Chen, C.; Yang, Z.; Kuang, Y.; Li, T.; Li, Y.; Huang, H.; Kierzewski, I.; Liu, B.; He, S.; et al. Highly Compressible, Anisotropic Aerogel with Aligned Cellulose Nanofibers. *ACS Nano* **2018**, *12*, 140–147. [CrossRef]
89. Deshpande, A.; Burgert, I.; Paris, O. Hierarchically Structured Ceramics by High-Precision Nanoparticle Casting of Wood. *Small* **2006**, *2*, 994–998. [CrossRef]
90. Tu, K.; Puertolas, B.; Adobes-Vidal, M.; Wang, Y.; Sun, J.; Traber, J.; Burgert, I.; Perez-Ramirez, J.; Keplinger, T. Green synthesis of hierarchical metal-organic framework/wood functional composites with superior mechanical properties. *Adv. Sci.* **2020**, *7*, 1902897. [CrossRef]
91. Blanchet, P.; Landry, V. Nanocomposite coatings and plasma-treatments for wood-based products. In *Wood Composites*; Ansell, M.P., Ed.; Woodhead Publishing: Cambridge, UK, 2015; pp. 335–355.
92. Petric, M. Surface modification of wood: A critical review. *Rev. Adhes. Adhes.* **2013**, *1*, 216–247. [CrossRef]
93. Nikolic, M.; Lawther, J.M.; Sanadi, A.R. Use of nanofillers in wood coatings: A scientific review. *J. Coat. Technol. Res.* **2015**, *12*, 445–461. [CrossRef]
94. Mishra, P.K.; Giagli, K.; Tsalagas, D.; Mishra, H.; Talegaonkar, S.; Gyrc, V.; Wimmer, R. Changing face of wood science in modern era: Contribution of nanotechnology. *Recent Pat. Nanotechnol.* **2018**, *12*, 13–21. [CrossRef]
95. Ramanavicius, S.; Ramanavicius, A. Insights in the application of stoichiometric and non-stoichiometric titanium oxides for the design of sensors for the determination of gases and VOCs (TiO_{2-x} and Ti_nO_{2n-1} vs. TiO_2). *Sensors* **2020**, *20*, 6833. [CrossRef]
96. Ramanavicius, S.; Ramanavicius, A. Progress and insights in the application of MXenes as new 2D nano-materials suitable for biosensors and biofuel cell design. *Int. J. Mol. Sci.* **2020**, *21*, 9224. [CrossRef] [PubMed]
97. Guo, H.; Fuchs, P.; Casdorff, K.; Michen, B.; Chanana, M.; Hagendorfer, H.; Romanyuk, Y.E.; Burgert, I. Bio-inspired superhydrophobic and omniphobic wood surfaces. *Adv. Mater. Interfaces* **2017**, *4*, 1600289. [CrossRef]
98. Vakhitova, L.N. Fire retardant nanocoating for wood protection. In *Nanotechnology in Eco-Efficient Construction*; Pacheco-Torgal, F., Diamanti, M.V., Nazari, A., Goran-Granqvist, C., Pruna, A., Amirkhanian, S., Eds.; Woodhead: Cambridge, UK, 2019; pp. 361–383.
99. Sun, Q.; Lu, Y.; Liu, Y. Growth of hydrophobic TiO_2 on wood surface using a hydrothermal method. *J. Mater. Sci* **2011**, *46*, 7706–7712. [CrossRef]
100. Devi, R.R.; Maji, T.K. Study on properties of simul wood (*Bombax ceiba* L.) impregnated with styrene acrylonitrile polymer, TiO_2 and nanoclay. *Polym. Bull.* **2012**, *69*, 105–123. [CrossRef]

101. Sun, Q.; Lu, Y.; Zhang, H.; Zhao, H.; Yu, H.; Xu, J.; Fu, U.; Yang, D.; Liu, Y. Hydrothermal fabrication of rutile TiO_2 submicrospheres on wood surface: An efficient method to prepare UV-protective wood. *Mater. Chem. Phys.* **2012**, *133*, 253–258. [CrossRef]
102. Leemon, N.F.; Ashaari, Z.; Uyup, M.K.A.; Bakar, E.S.; Tahir, P.M.; Saliman, M.A.R.; Ghani, M.A.; Lee, S.H. Characterisation of phenolic resin and nanoclay admixture and its effect on *impreg* wood. *Wood Sci. Technol.* **2015**, *49*, 1209–1224. [CrossRef]
103. Teng, T.-J.; Arip, M.N.M.; Sudesh, K.; Nemoikina, A.; Jalaludin, Z.; Ng, E.-P.; Lee, N.-L. Conventional technology and nanotechnology in wood preservation: A review. *BioResources* **2018**, *13*, 9220–9252. [CrossRef]
104. Gregory, S.A.; McGettigan, C.P.; McGuinness, E.K.; Rodin, D.M.; Yee, S.K.; Losego, M.D. Single-cycle atomic layer deposition on bulk wood lumber for managing moisture content, mold growth and thermal conductivity. *Langmuir* **2020**, *36*, 1633–1641. [CrossRef] [PubMed]
105. Wang, Z.; Han, X.; Wang, S.; Han, X.; Pu, J. MXene/wood-based composite materials with electromagnetic shielding properties. *Holzforschung* **2020**. [CrossRef]
106. Bayani, S.; Taghiyari, H.R.; Papodopoulos, A.N. Physical and mechanical properties of thermally modified beech wood impregnated with silver nano-suspension and their relationship with the crystallinity of cellulose. *Polymers* **2019**, *11*, 1538. [CrossRef]
107. Akhtari, M.; Ganjipour, M. Effect of nano-silver and nano-copper and nano zinc oxide on Paulownia wood exposed to white rot fungus. *Agric. Sci. Dev.* **2013**, *2*, 116–119.
108. Gao, L.; Lu, Y.; Li, J.; Sun, S. Superhydrophobic conductive wood with oil repellency obtained by coating with silver nanoparticles modified with fluoroalkyl silane. *Holzforschung* **2016**, *70*, 63–68. [CrossRef]
109. Fu, Q.; Tu, K.; Goldhahn, C.; Keplinger, T.; Adobes-Vidal, M.; Sorieul, M.; Burgert, I. Luminescent and hydrophobic wood films as optical lighting materials. *ACS Nano* **2020**, *14*, 13775–13783. [CrossRef] [PubMed]
110. Wang, W.; Ran, Y.; Wang, J. Improved performance of thermally modified wood via impregnation with carnauba wax/organoclay emulsion. *Constr. Build. Mater.* **2020**, *247*, 118586. [CrossRef]
111. Tshabalala, M.A.; Libert, R.; Schaller, C.M. Photostability and moisture uptake properties of wood veneers coated with a combination of thin sol-gel films and light stabilizers. *Holzforschung* **2011**, *65*, 215–220. [CrossRef]
112. Gholamlyan, H.; Tarmiam, A.; Ranjbar, Z.; Abdulkhani, A.; Asadfallah, M.; Mai, C. Silane nanofilm formation by sol-gel processes for promoting adhesion of waterborne and solventborne coatings to wood surface. *Holzforschung* **2016**, *70*, 429–437. [CrossRef]
113. Mahr, M.S.; Hubert, T.; Stephan, I.; Bucker, M.; Militz, H. Reducing copper leaching from treated wood by sol-gel derived TiO_2 and SiO_2 depositions. *Holzforschung* **2016**, *67*, 429–435. [CrossRef]
114. Donath, S.; Militz, H.; Mai, C. Creating Water-Repellent Effects on Wood by Treatment with Silanes. *Holzforschung* **2006**, *60*, 40–46. [CrossRef]
115. Mai, C.; Militz, H. Modification of wood with silicon compounds. Treatment systems based on organic silicon compounds—A review. *Wood Sci. Technol.* **2004**, *37*, 453–461. [CrossRef]
116. Bhusan, B.; Nosonovski, M. The rose petal effect and the modes of superhydrophobicity. *Philos. Trans. R. Soc. A* **2010**, *368*, 4713–4728. [CrossRef]
117. Spear, M. Emerging nature-based materials and their use in new products. In *Designing with Natural Materials*; Ormondroyd, G.A., Morris, A.F., Eds.; Taylor Francis: Boca Raton, FL, USA, 2019; pp. 217–320.
118. Yang, D.; Liu, H.; Zheng, Z.; Yuan, Y.; Zhao, J.-C.; Waclawik, E.R.; Ke, X.; Zhu, H. An efficient photocatalyst structure: TiO_2(B) nanofibers with a shell of anatase nanocrystals. *J. Am. Chem. Soc.* **2009**, *131*, 17885–17893. [CrossRef]
119. Kong, L.; Tu, K.; Guan, H.; Wang, X. Growth of high-density ZnO nanorods on wood with enhanced photostability, flame retardancy and water repellency. *Appl. Surf. Sci.* **2017**, *407*, 479–484. [CrossRef]
120. Jia, S.; Liu, M.; Wu, Y.; Luo, S.; Qing, Y.; Chen, H. Facile and scalable preparation of highly wear-resistance superhydrophobic surface on wood substrates using silica nanoparticles modified by VTES. *Appl. Surf. Sci.* **2016**, *386*, 115–124. [CrossRef]
121. Filgueira, D.; Moldes, D.; Fuentealba, C.; García, D.E. Condensed tannins from pine bark: A novel wood surface modifier assisted by laccase. *Ind. Crops Prod.* **2017**, *103*, 185–194. [CrossRef]
122. Tu, K.; Wang, X.; Kong, L.; Guan, H. Facile preparation of mechanically durable, self-healing and multifunctional hydrophobic surfaces on solid wood. *Mater. Des.* **2018**, *140*, 30–36. [CrossRef]
123. Nyrud, A.Q.; Bringslimark, T. Is interior wood use psychologically beneficial? A Review of psychological responses toward wood. *Wood Fiber Sci.* **2010**, *42*, 202–218.
124. Simonson, C.; Salonvaara, M.; Ojanen, T. Humidity, comfort and air quality in a bedroom with hygroscopic wooden structures. In Proceedings of the 6th Symposium on Building Physics in the Nordic Countries; SINTEF: Oslo, Norway, 2002; Volume 2, pp. 743–750.
125. Rode, C.; Peuhkuri, R.; Lone, H.; Time, B.; Gustavsen, A.; Ojanen, T.; Ahonen, J.; Svennberg, K.; Harderup, L.-E.; Arfvidsson, J. (Eds.) *Moisture Buffering of Building Materials*; DTU BYG-Rapporter No. R-126; Technical University of Denmark DTU: Kongens Lyngby, Denmark, 2006.
126. Bhatta, S.R.; Tiippana, K.; Vahtikari, K.; Kiviluoma, P.; Hughes, M.; Kyttä, M. Quantifying the sensation of temperature: A new method for evaluating the thermal behaviour of building materials. *Energy Build.* **2019**, *195*, 26–32. [CrossRef]
127. Salonvaara, M.; Ojanen, T.; Holm, A.; Kunzel, H.; Kuragiozis, A. Moisture buffering effects on indoor air quality-Experimental and simulation results. In *Performance of Exterior Envelopes of Whole Buildings: Proceedings of the IX International Conference ASHRAE*; Oak Ridge National Laboratory: Oak Ridge, TN, USA, 2004.

128. Legros, C.; Piot, A.; Woloszyn, M.; Pailha, M. Effect of moisture buffering on surface temperature variation: Study of different indoor cladding materials. In *12th Nordic Symposium on Building Physics, E3S Web of Conferences*; EDP Sciences: Les Ulis, France, 2020; p. 06002.
129. Lozhechnikova, A.; Vahtikari, K.; Hughes, M.; Österberg, M. Toward energy efficiency through an optimized use of wood: The development of natural hydrophobic coatings that retain moisture-buffering ability. *Energy Build.* **2015**, *105*, 37–42. [CrossRef]
130. Lozhechnikova, A.; Bellanger, H.; Michen, B.; Burgert, I.; Österberg, M. Surfactant-free carnauba wax dispersion and its use for layer-by-layer assembled protective surface coatings on wood. *Appl. Surf. Sci.* **2017**, *396*, 1273–1281. [CrossRef]
131. Yang, H.; Zhu, S.; Pan, N. Studying the mechanisms of titanium dioxide as ultraviolet-blocking additive for films and fabrics by an improved scheme. *J. Appl. Polym. Sci.* **2004**, *92*, 3201–3210. [CrossRef]
132. Gascon-Garrido, P.; Mainusch, N.; Militz, H.; Viol Mai, C. Effects of copper-plasma deposition on weathering properties of wood surfaces. *Appl. Surf. Sci.* **2016**, *366*, 112–119. [CrossRef]
133. Hui, B.; Li, Y.; Huang, Q.; Li, G.; Li, J.; Cai, L.; Yu, H. Fabrication of smart coatings based on wood substrates with photoresponsive behaviour and hydrophobic performance. *Mater. Des.* **2015**, *84*, 277–284. [CrossRef]
134. Kang, C.-W.; Kim, G.-C.; Park, H.-J.; Lee, N.-H.; Kang, W.; Matsumura, J. Changes in Permeability and Sound Absorption Capability of Yellow Poplar Wood by Steam Explosion Treatment. *J. Fac. Agric.* **2010**, *55*, 327–332.
135. Wang, D.; Peng, L.; Zhu, G.; Fu, F.; Zhou, Y.; Song, B. Improving the Sound Absorption Capacity of Wood by Microwave Treatment. *BioResources* **2014**, *9*, 7504–7518. [CrossRef]
136. Kang, C.-W.; Jang, E.-S.; Jang, S.S.; Cho, J.-I.; Kim, N.-H. Effect of Heat Treatment on the Gas Permeability, Sound Absorption Coefficient, and Sound Transmission Loss of *Paulownia tomentosa* Wood. *J. Korean Wood Sci. Technol.* **2019**, *47*, 644–654.
137. Esmailpour, A.; Norton, J.; Taghiyari, H.; Zolfaghari, H.; Asadi, S. Effects of heat treatment on sound absorption coefficients in nanosilver-impregnated and normal solid woods. *IET Nanobiotechnology* **2017**, *11*, 365–369. [CrossRef]
138. Kang, C.-W.; Li, C.; Jang, E.-S.; Jang, S.-S.; Kang, H.-Y. Changes in Sound Absorption Capability and Air Permeability of Malas (*Homalium foetidum*) Specimens after High Temperature Heat Treatment. *J. Korean Wood Sci. Technol.* **2018**, *46*, 149–154.
139. Chung, H.; Park, Y.; Yang, S.Y.; Kim, H.; Han, Y.; Chang, Y.-S.; Yeo, H. Effect of heat treatment temperature and time on sound absorption coefficient of *Larix kaempferi* wood. *J. Wood Sci.* **2017**, *63*, 575–579. [CrossRef]
140. Kolya, H.; Kang, C.-W. Hygrothermal treated paulownia hardwood reveals enhanced sound absorption coefficient: An effective and facile approach. *Appl. Acoust.* **2021**, *174*, 107758. [CrossRef]
141. Yano, H.; Minato, K. Improvement of the acoustic and hygroscopic properties of wood by a chemical treatment and application to the violin parts. *J. Acoust. Soc. Am.* **1992**, *92*, 1222–1227. [CrossRef]
142. Chang, S.-T.; Chang, H.-T.; Huang, Y.-S.; Hsu, F.-L. Effects of Chemical Modification Reagents on Acoustic Properties of Wood. *Holzforschung* **2000**, *54*, 669–675. [CrossRef]
143. Yano, H.; Minato, K. Controlling the timbre of wooden musical instruments by chemical modification. *Wood Sci. Technol.* **1993**, *27*, 287–293. [CrossRef]
144. Yano, H.; Kajita, H.; Minato, K. Chemical treatment of wood for musical instruments. *J. Acoust. Soc. Am.* **1994**, *96*, 3380–3391. [CrossRef]
145. Ahmed, S.A.; Adamopoulos, S. Acoustic properties of modified wood under different humid conditions and their relevance for musical instruments. *Appl. Acoust.* **2018**, *140*, 92–99. [CrossRef]
146. Mania, P.; Moliński, W.; Roszyk, E.; Górska, M. Optimization of Spruce (*Picea abies* L.) Wood Thermal Treatment Temperature to Improve Its Acoustic Properties. *BioResources* **2020**, *15*, 505–516.
147. Pfriem, A. Thermally Modified Wood for Use in Musical Instruments. *Drvna Ind.* **2015**, *66*, 251–253. [CrossRef]
148. Mohebby, B.; Yaghoubi, K.; Roohnia, M. Acoustical properties of hydrothermally modified mulberry (*Morus alba* L.) wood. In Proceedings of the Third European Conference on Wood Modification, Cardiff, UK, 15–16 October 2007; pp. 283–286.
149. Obataya, E. Effects of natural and artificial ageing on the physical and acoustic properties of wood in musical instruments. *J. Cult. Herit.* **2017**, *27*, S63–S69. [CrossRef]
150. Mania, P.; Gasiorek, M. Acoustic Properties of Resonant Spruce Wood Modified Using Oil-Heat Treatment (OHT). *Materials* **2020**, *13*, 1962. [CrossRef]
151. Lv, S.; Fu, F.; Wang, S.; Huang, J.; Hu, L. Novel wood-based all-solid-state flexible supercapacitors fabricated with a natural porous wood slice and polypyrrole. *RSC Adv.* **2015**, *5*, 2813–2818. [CrossRef]
152. Guo, H.; Büchel, M.; Li, X.; Wäckerlin, A.; Chen, Q.; Burgert, I. Dictating anisotropic electric conductivity of a transparent copper nanowire coating by the surface structure of wood. *J. R. Soc. Interface* **2018**, *15*, 20170864. [CrossRef] [PubMed]
153. Ansell, M.P. Carbonised and mineralised wood composites. In *Wood Composites*; Ansell, M.P., Ed.; Woodhead Publishing: Cambridge, UK, 2015; pp. 395–409.
154. Sun, J.; Guo, H.; Ribera, J.; Wu, C.; Tu, K.; Binelli, M.; Panzarasa, G.; Schwarze, F.W.M.R.; Wang, Z.L.; Burgert, I. Sustainable and biodegradable wood sponge nanogenerator for sensing and energy harvesting applications. *ACS Nano* **2020**, *14*, 14665–14674. [CrossRef]
155. Lowden, L.A.; Hull, T.R. Flammability behaviour of wood and a review of the methods for its reduction. *Fire Sci. Rev.* **2013**, *2*, 4. [CrossRef]
156. Martinka, J.; Chrebet, T.; Kral, J.; Balog, K. An examination of the behavior of thermally treated spruce wood under fire conditions. *Wood Res.* **2013**, *58*, 599–606.

157. Čekovská, H.; Gaff, M.; Osvald, A.; Kačik, F.; Kubš, J.; Kaplan, L. Fire resistance of thermally modified spruce wood. *BioResources* **2017**, *12*, 947–959. [CrossRef]
158. Luptakova, J.; Kacik, F.; Mitterova, I.; Zachar, M. Influence of temperature of thermal modification on the fire-technical characteristics of spruce wood. *BioResources* **2019**, *14*, 3795–3807.
159. Xing, D.; Li, J. Effect of heat treatment on thermal decomposition and combustion performance of *Larix* spp. Wood. *BioResources* **2014**, *9*, 4274–4287. [CrossRef]
160. Gasparik, M.; Osvaldova, M.L.; Cekovska, H.; Potucek, D. Flammability characteristics of thermally modified oak wood treated with fire retardant. *BioResources* **2017**, *12*, 8451–8467.
161. Mohebby, B.; Talaii, A.; Najafi, S.K. Influence of acetylation on fire resistance of beech plywood. *Mater. Lett.* **2007**, *61*, 359–362. [CrossRef]
162. Marney, D.C.O.; Russell, L.J. Combined fire retardant and wood preservative treatments for outdoor wood applications- A review of the literature. *Fire Technol.* **2008**, *44*, 1–14. [CrossRef]
163. Sweet, M.S.; LeVan, S.L.; Tran, H.C.; DeGroot, R. *Fire Performance of Wood Treated with Combined Fire-Retardant and Preservative Systems*; FPL-RP-545; USDA Forest Products Laboratory: Madison, WI, USA, 1996; 12p.
164. Baysal, E. Determination of oxygen index levels and thermal analysis of Scots Pine (*Pinus sylvestris* L.) impregnated with melamine formaldehyde–Boron combinations. *J. Fire Sci.* **2002**, *20*, 373–389. [CrossRef]
165. Lewin, M. Flame retarding of wood by chemical modification with bromate-bromide solutions. *J. Fire Sci.* **1997**, *15*, 29–51. [CrossRef]
166. Lee, H.L.; Chen, G.C.; Rowell, R.M. Chemical modification of wood to improve decay and thermal resistance. In Proceedings of the 5th Pacific Rim Biobased Composites Symposium, Canberra, Australia, 10–13 December 2000.
167. Tsunoda, K. Preservative properties of vapor-boron-treated wood and wood-based composites. *J. Wood Sci.* **2001**, *47*, 149–153. [CrossRef]
168. Kartal, S.N.; Yoshimura, T.; Imamura, Y. Decay and termite resistance of boron-treated and chemically modified wood by in situ co-polymerization of allyl glycidyl ether (AGE) with methyl methacrylate (MMA). *Int. Biodeterior. Biodegrad.* **2004**, *53*, 111–117. [CrossRef]
169. Baysal, E.; Sonmez, A.; Colak, M.; Toker, H. Amount of leachant and water absorption levels of wood treated with borates and water repellents. *Bioresour. Technol.* **2006**, *97*, 2271–2279. [CrossRef]
170. Jiang, T.; Feng, X.; Wang, Q.; Xiao, Z.; Wang, F.; Xie, Y. Fire performance of oak wood modified with N-methylol resin and methylolated guanyl urea phosphate/boric acid-based fire retardant. *Constr. Build. Mater.* **2014**, *72*, 1–6. [CrossRef]
171. Yue, K.; Chen, Z.; Lu, W.; Liu, W.; Li, M.; Shao, Y.; Tang, L.; Wan, L. Evaluating the Mechanical and fire-resistance properties of modified fast-growing Chinese fir timber with boric-phenol-formaldehyde resin. *Constr. Build. Mater.* **2017**, *154*, 956–962. [CrossRef]
172. Yue, K.; Wu, J.; Xu, L.; Tang, Z.; Chen, Z.; Liu, W.; Wang, L. Use impregnation and densification to improve mechanical properties and combustion performance of Chinese fir. *Constr. Build. Mater.* **2020**, *241*, 118101. [CrossRef]
173. Rabe, S.; Klack, P.; Bahr, H.; Schartel, B. Assessing the fire behavior of woods modified by N-methylol crosslinking, thermal treatment, and acetylation. *Fire Mater.* **2020**, *44*, 530–539. [CrossRef]
174. Morozovs, A.; Buksans, E. Fire performance characteristics of acetylated ash (*Fraxinus excelsior* L.) wood. *Wood Mater. Sci. Eng.* **2009**, *4*, 76–79. [CrossRef]
175. Accsys. Accoya Wood Information Guide, v3.9. 2020. Available online: https://www.accoya.com/app/uploads/2020/04/Accoya_WoodInfoGuide-1.pdf (accessed on 13 February 2021).

Article

Insight of Weathering Processes Based on Monitoring Surface Characteristic of Tropical Wood Species

Agnieszka Jankowska [1], Katarzyna Rybak [2], Malgorzata Nowacka [2] and Piotr Boruszewski [1,*]

1 Institute of Wood Sciences and Furniture, Warsaw University of Life Sciences—SGGW, 159 Nowoursynowska St., 02-776 Warsaw, Poland; agnieszka_jankowska@sggw.edu.pl

2 Institute of Food Sciences, Warsaw University of Life Sciences—SGGW, 159C Nowoursynowska St., 02-776 Warsaw, Poland; katarzyna_rybak@sggw.edu.pl (K.R.); malgorzata_nowacka@sggw.edu.pl (M.N.)

* Correspondence: piotr_boruszewski@sggw.edu.pl; Tel.: +48-2-2593-8528

Received: 5 August 2020; Accepted: 9 September 2020; Published: 11 September 2020

Abstract: The main aim of the presented research was to compare the influence of selected ageing factors, such as UV radiation and complex artificial weathering methods, on the colour, wettability and roughness changes in garapa, tatajuba, courbaril and massaranduba from South America—tropical wood species that are popular for external usage in European countries. Both processes caused wood surfaces to become darker and turn to shades of brown. The highest total colour changes were shown in courbaril wood (wood with the highest extractives content). The wood surface roughness demonstrated variation, depending on the wood section and measurement direction, and increased after ageing treatments. Changes in surface contact angle were significant after the inclusion of water and drying in the weathering process (wettability decreased). Anatomical analyses of the tested tropical woods revealed structural changes after used artificial weathering treatments (distortion between cell elements, degradation of the middle lamella, micro-cracks in cell walls, thinning and degradation of parenchyma cells, cracks along pits within vessels). As a result of desorption tension, the changes caused by UV irradiation were much smaller than those caused by full artificial weathering. Fourier-transform infrared spectroscopy (FTIR) analysis indicated the occurrence of lignin and hemicelluloses oxidative changes after the weathering process which resulted in the formation of carbonyl and carboxyl compounds. The depolymerisation of cellulose was also identified. The results show that the observed changes may affect the long-term durability of finishes applied over wood subjected to weathering factors for a short period before finishing.

Keywords: tropical wood; surface preservation; wood weathering; FTIR; weathered wood roughness; wood degradation

---

1. Introduction

During the external exposure of wood there are some key procedures which should be followed to reduce the effects of weathering. The natural weathering was defined as a process of irreversible changes to the appearance and properties of wood effected by the long-term impact of outdoor factors, such as the solar radiation, air and oxygen contained within it, and changes in the temperature and humidity (no direct influence of biotic factors should be assumed) [1–7]. This complex phenomenon is caused by solar radiation, hydrolysis, and the leaching of wood components [8–10]. The harsh outside environment makes it necessary to consider wood durability, in accordance with EN 350-2 [11]. Next, a proper construction solution is required, such as proper usage conditions. Moreover, surface treatments can be used to prolong the service life of wood [12,13].

Despite the high durability of many tropical wood species against biological factors such as fungi or insects, it is always recommended to protect wood surface during outside exposure due to weathering. On the other hand, one of the most important factors in securing sustainable development is utilization of renewable natural materials, which undoubtedly include wood [14]. In order to reduce the environmental burden, the surfaces of wooden elements are not treated with any painting and varnishing products. In addition to the traditional interior design elements, the use of non-treated wood is expanding even further to external use. According to contemporary trends, it is recommended to use untreated wood and allowing it turn grey upon exposure to weather under aboveground conditions over using non-durable wood with applied surface coatings. Many tropical species are considered to be the most durable wood. However, the use of tropical wood has several negative consequences including illegal deforestation. Hence, impacting significant climate changes. Pronounced through net carbon emissions, deforestation leads to a global warming [15]. However, as wood of tropical species is utilized in Europe, knowledge of their weathering characteristic is important to optimize their performance. Materials should be recognizable in terms of their surface properties. The wettability of wood is one of the most significant parameters influencing the gluing as well as the coating processes [16–19]. One of the most critical factors for extending the durability of painted wood is freshness of the wood surface and only a fresh, high-energy surface guarantees optimum adhesion conditions. The loss of coating ability and glueability due to the increasing age of a wood surface was studied over time by several researchers [20–22], who arrived at the conclusion that the changes caused by the weathering are the effect of the migration of wood extractives to wood surfaces after their preparation, which causes a decrease in wood surfaces' wettability.

It is also convenient to determine the direction and degree of colour changes in individual tropical wood species caused by external factors. The aesthetic function of wooden products used externally can be extended by selecting the most resistant wood species in terms of its surface properties stability. Some knowledge about changes occurring in tropical wood species was gained, but mainly during colour testing wood subjected to different weathering treatments [23,24]. This paper is a part of an extensive study determining the influence of artificial weathering on the surface properties of several species of wood from tropical and subtropical zones. The main aim of the presented research is to compare the influence of selected ageing factors such as ultraviolet radiation and complex artificial weathering methods on the colour, wettability and surface roughness changes in garapa (*Apuleia leiocarpa* (Vogel) J.F. Macbr.), tatajuba (*Bagassa guianensis* Aubl.), courbaril (*Hymenea courbaril* L.) and massaranduba (*Manilkara bidendata* (A. DC.) A. Chev.) wood species, as these are popular for external usage in European countries.

2. Materials and Methods

The wood species used in this study are presented in Table 1. All test materials were heartwood, because the heartwood of tropical species is more commercially usable than sapwood. All species came from South America (Brazil) and wood was acquired from DLH Poland (Warsaw, Poland). The material was identified using macroscopic techniques and was deciduous diffuse-porous in all cases. Characterization of the tested wood species was supplemented with density determination, performed in accordance with the ASTM D2395 standard [25].

Wood samples were prepared for investigation and analyses by using standardized methods. To avoid differences in the tested properties caused by differences in wood anatomy, identical samples of each wood species were collected from one log. Each one was sawn to produce planks approximately 4 cm thick. The obtained planks were air-dried in a room with relative humidity up to 50% and a temperature of 21 °C for approximately 6 months before testing. Then, the defect-free planks were sized into samples for the tests. Forty samples of each wood species were prepared, each with a radial and tangential cross-section of 10 × 10 mm^2 and a length of 70 mm. Following Gardner [26] and Liptakova et al. [27], the radial-oriented or tangential-oriented surface of the wood block was planned. The aim was to make wood showing the roughness caused by the cellular structure of wood and only a negligibly small roughness caused by cutting. Moreover, the wood surface is chemically heterogeneous, and therefore does not comply with the requirements of the physicochemical theory of contact angle in a strict sense.

Table 1. Wood species used in tests.

Wood Name	Scientific Name	Plant Family	Special Features	Wood Density * (kg/m^3)
Garapa	*Apuleia leiocarpa* (Vogel) J.F.Macbr.	*Caesalpiniaceae*	Irregular fibre arrangement	739 (16)
Tatajuba	*Bagassa guianensis* Aubl.	*Moraceae*	Irregular fibre arrangement	884 (35)
Courbaril	*Hymenea courbaril* L.	*Fabaceae*	Irregular fibre arrangement, paratracheal parenchyma and in narrow bands	1038 (6)
Massaranduba	*Manilkara bidendata* (A. DC.) A. Chev.	*Sapotaceae*	Irregular fibre arrangement, parenchyma in narrow bands	1131 (18)

* means and standard deviations in parentheses.

2.1. Properties Measurements

The parameters of the colour of unmodified and modified wood were measured on the basis of the mathematical CIE (International Commission on Illumination known as the Commission Internationale de l'Elcairage) L^*C^*h colour space models. The parameter L^* represents lightness. The parameters C^* and h describe the saturation (colour intensity) and hue angle, respectively. The total colour change ΔE^* was determined in accordance with ISO 7724-3 [28]. The 3NH NH300 spectrophotometer made by X-Rite Europe GmbH (Regensdorf, Switzerland) was used to examine the colour parameters. The sensor head was 8 mm in diameter. Measurements were made using a D65 illuminate.

Surface properties can be characterized by the water contact angle (wettability). A single measurement of the water contact angle provides information on several important parameters, such as the surface free energy, contact angle and wetting coefficient or work of adhesion. To predict interactions with wetting materials such as lacquers or adhesives, the surface properties are characterized. The lower the contact angle value (θ), which is a measure of the wetting impact of the substrate by solution, the better the wettability of the material. As soon as the samples were placed in the contact angle measuring apparatus, measurements were started. The contact angle is defined as the angle between the solid surface and the tangent, drawn on the drop-surface, passing through the three-point liquid-solid atmosphere [29]. Using a contact angle analyzer, Haas Phoenix 300 Goniometer (Surface Electro Optics, Suwon City, South Korea), equipped with microscopic lenses, a digital camera connected to a computer with software—image analysis system (Image XP, Surface Electro Optics, version 5.8, Suwon City, South Korea) that provides an image of the drop on the examined wooden surfaces, the contact angles of the expanding droplets (advancing angels) were determined. To test the wettability, a re-distilled water was used as a liquid. All contact angles were measured along the grain direction followed by Gindl et al. [16]. Based on the research of Liptáková et al. [27], the measurements of the contact angle were taken 30 s after each drop of reference liquid.

The roughness was evaluated in accordance with the requirements of ISO 4287 [30]. As part of the conducted research, the arithmetic mean deviation of the assessed profile (R_a) was measured. The surface roughness was tested using the Surftest SJ-210 Series 178-Portable Surface Roughness Tester (Mitutoyo Corporation, Takatsu-ku, Japan). The parameter R_a was measured in parallel and perpendicular to the grain direction.

Properties were measured on a fresh wood surface after UV irradiation and complex artificial weathering treatment. Before and after the treatments, each of the samples were conditioned in a climatic chamber with a temperature of 21 °C and relative humidity of about 50%. Measurements were done for each tested variant 30 times. In case of roughness and wettability testing, measurements were made both on the tangential and radial cross-section.

2.2. UV Irradiation

Four fluorescent 100R's Lightech lamps of 100 W each and the spectrum 300–400 nm (90% of the radiation spectrum is a wavelength of 340–360 nm) were used for irradiating. The time of irradiation was 24 h. The source of radiation applied in this study imitated solar radiation and, in particular, the UVA component of solar radiation. This component causes the greatest changes in the appearance and structure of organic materials exposed in an outdoor environment. This is due to the fact that UVA radiation accounts for 90%–95% of the solar radiation reaching the Earth's surface [31].

2.3. Artificial Weathering

The artificial weathering method was based on the literature [10,32]. It took 30 h to complete one artificial weathering cycle. The first step of each cycle was soaking samples in water at 20 °C for 16 h. The conditions of the second step were 70 °C and 5%–10% RH for 8 h, and the third step was performed at 30 °C and 20%–25% RH with irradiation with UV rays (24 h). The tested wood species were subjected to one and four cycles of artificial weathering. That weathering method was previously used for tropical wood [33], but obtained results regarding the effect of long-term aging.

2.4. Scanning Electron Microscopy SEM

The microstructure of samples was examined using a scanning electron microscope HITACHI, model TM-3000 (Hitachi Ltd., Tokyo, Japan) with a digital image record. The samples were chosen randomly and put into the vacuum chamber. The photos at accelerating voltages equal to 5 kV were taken with 1000 magnification and the record was saved using SEM software (TM3000, Hitachi Ltd., Tokyo, Japan). The analysis was conducted at least in three repetitions for each sample.

2.5. FTIR

The spectra were recorded using an Agilent Cary 630 FTIR spectrometer (Agilent Technologies Inc., Santa Clara, CA, USA) equipped with a single bounce diamond ATR unit. Scans were performed in the range of 650–4000 cm^{-1} with a resolution of 4 cm^{-1}, and 64 scans for each sample. Background correction with 32 scans was performed after each measurement. The measurements were conducted in three repetitions.

2.6. Statistical Analysis

Statistical analysis of the test results was carried out using Statistica v. 10 software (TIBCO Software Inc., Palo Alto, CA, USA). Data were analysed and provided as the mean ± standard deviation, the scatter plot of results around the median, and minimum and maximum values. To compare and determine the significance of difference between data, *t*-test was used. Analysis of variance (ANOVA) was used to determine the influence of UV irradiation, full artificial weathering process and anatomical properties, such as the direction of measurements and cross-section (tangential or radial).

3. Results and Discussion

3.1. Colour Changes

The colour of wood exposed to external conditions can rapidly change in quite a short time [23]. Due to the photodegradation (photooxidation) of lignin and extractives, wood colour usually turns yellowish or brown. During UV radiation, the surface colour changed visibly. The deviation ΔE^* in all the wood species exceeds the value of 3 (Table 2), which is considered the limit for visibility to the naked eye [34]. The Figure 1 shows the clear contrast between wood surfaces when they were fresh and after different treatments. Distinctive differences in colour can be seen. In general, wood surfaces become darker and turn to shades of brown. The previous studies conducted on softwoods [35], and tropical wood species as well [36], confirmed that the most rapid changes in ΔE^* occurred during the first stages of UV radiation. The tested wood species are rich in extractives [37]. The rapid change in the first hours of exposure could have been caused by the reaction of extractives contained within the wood to UV radiation. Pandey [38] compared the behaviour of unextracted and extractive free wood of *Acacia auriculaeformis*. The unextracted wood surface showed a rapid colour change at the initial period of exposure, which decreased upon prolonged exposure. Tests on wood (coniferous, deciduous) from moderate climates during weathering in natural external conditions did not show a wide range of changes, which can be explained by the relatively low content of extractives. The maximum value of ΔE^* was for spruce—34.1 after 12 months of weathering. In the study here and in the case of courbaril wood, $\Delta E^* = 51.21$ was caused only after 24 h of UV radiation. Soaking wood samples in water meant that the wood dyes contained within were washed out, and accumulated on the top layers of wood, which caused wood darkening (lower values of L^*)—Table 3. The full artificial weathering process gave higher colour changes and also a response to changes in the appearance of wood surfaces in natural external conditions. In all cases of the tested wood species, wood became darker both after UV irradiation and after full weathering treatment. Conducting a higher number of artificial weathering cycles caused higher changes.

Table 2. Overall colour change (ΔE^*) during exposure.

Species	ΔE^*		
	After 24 h UV Irradiation	**After 1 Cycle of Artificial Weathering**	**After 4 Cycles of Artificial Weathering**
Garapa	4.4	6.3	8.4
Tatajuba	7.5	9.4	12.0
Courabril	52.2	58.0	58.4
Massaranduba	8.2	9.5	13.2

Table 3. Coordinates of CIE L^*C^*h colour system for individual woods (means and standard deviations in parentheses).

Species	Before Exposure			After 24 h UV Irradiation			After 1 Cycle of Artificial Weathering			After 4 Cycles of Artificial Weathering		
	L^*	C^*	h	L^*	C^*	h	L^*	C^*	h	L^*	C^*	h
Garapa	71.4 (3.1)	33.5 (2.9)	65.8 (1.3)	68.6 (0.7)	36.7 (1.3)	65.1 (2.7)	65.9 (2.5)	34.9 (1.4)	64.9 (3.4)	64.9 (2.1)	32.8 (1.5)	65.8 (2.1)
Tatajuba	58.6 (1.2)	32.8 (1.7)	69.3 (1.2)	51.8 (1.3)	35.1 (1.0)	65.6 (1.3)	49.9 (1.6)	34.0 (1.0)	63.7 (1.3)	46.9 (1.3)	32.2 (1.4)	60.9 (1.8)
Courabril	53.3 (3.7)	25.5 (1.9)	71.4 (1.3)	41.2 (1.8)	20.0 (2.1)	65.8 (2.3)	37.6 (2.2)	18.3 (1.9)	62.3 (1.5)	35.1 (1.7)	18.8 (2.0)	60.6 (1.3)
Massaranduba	50.0 (1.6)	23.9 (0.7)	42.5 (0.4)	43.8 (1.2)	21.2 (1.3)	48.1 (1.2)	43.0 (1.5)	21.9 (1.0)	49.2 (2.4)	40.3 (2.2)	18.1 (1.8)	48.8 (2.1)

In the case of wood species of a lighter colour—garapa and tatajuba—C^* and h values followed a similar trend: they increased after UV irradiation and decreased during the full artificial weathering process. In the case of courbaril and massaranduba wood (both of darker colour), C^* value was lower after each treatment. The different behaviours of individual wood species regarding chromatic parameters probably could have been caused by the presence of specific types of extractives in the wood. While studying the influence of extractives on discoloration, Pandey [38] found that this relation is, to a great extent, dependent on the nature and chemical composition. Both ΔC^* and Δh are determined mainly by the changes in the chromophore groups in extractives and change through the lignin degradation and later leaching. Thus, it can be assumed that a higher proportion of extractives within one wood species can cause a darker colour of the wood. The test results of tropical wood discoloration caused by simulated sunlight [23] confirmed that lighter samples manifest a larger colour change due to sunlight. In the research here, the results were different. The scope of change was similar for both garapa, tatajuba of light colour and massaranduba of dark colour. Massive colour changes (total colour change ΔE^*) showed courbaril wood, known for its high extractives content [37].

According to the above, the conclusion can be frawn that wood, as a complex structure, differing between species, requires a unique approach for each individual species. Especially in the case of tropical wood, which, despite many reports, remains little known. With a large number of different wood types, the formulation of dependencies is only indicative. Due to the huge variability of properties and structure details [39], deep knowledge is necessary, especially in the context of proper wood application.

Figure 1. Visual comparison of contrast among the exposed and non-exposed surface (from left: garapa, tatajuba, courbaril and massaranduba; from top: fresh wood, after 24 h of UV irradiation, after 1 cycle of artificial weathering, after 4 cycles of artificial weathering).

3.2. Roughness Changes

The results of the surface roughness R_a of tested wood species are shown on Figures 2 and 3. Tests were performed on both the tangential and radial wood section, parallel and perpendicular to the grain. The wood roughness is a complex phenomenon because wood is an anisotropic and heterogeneous material, and several factors such as anatomical differences and the machining properties should be considered in evaluating the surface roughness of wood [40]. The roughness of the tested wood species demonstrated variation, depending on the wood section and the measurement direction (parallel or perpendicular to the grain). As can be seen from Figures 2 and 3, mostly radial sections showed higher roughness. All tested wood species are characterized with interlocked fibres, which causes variable fibre orientation. As a result, they are cut in various ways on the radial section of the wood. In general, $R_{a\perp}$ values (measured perpendicular to the wood fibres) were twice as high as those measured parallel to the grain (R_{aII}). Wood, as a non-homogeneous material, shows differences in its properties depending on the direction. The roughness perpendicular to the fibres is mostly caused by irregularities in the structural element sizes, such as the vessels' diameter.

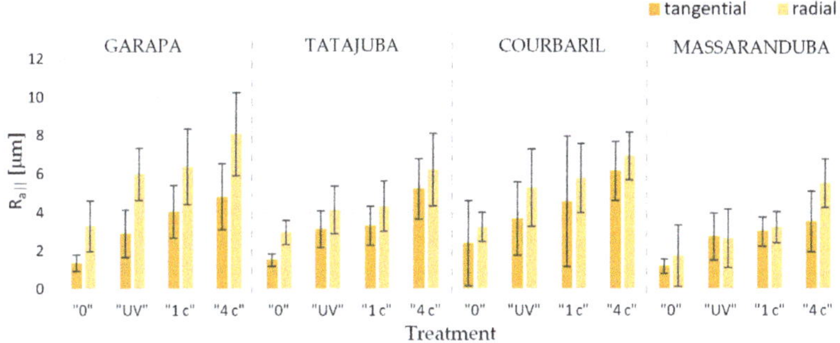

Figure 2. Roughness ($R_{a\parallel}$) along the fibres of tested wood species: "O"—fresh wood, "UV"—after 24 h of UV irradiation, "1 c"—after one cycle of artificial weathering, "4 c"—after four cycles of artificial weathering.

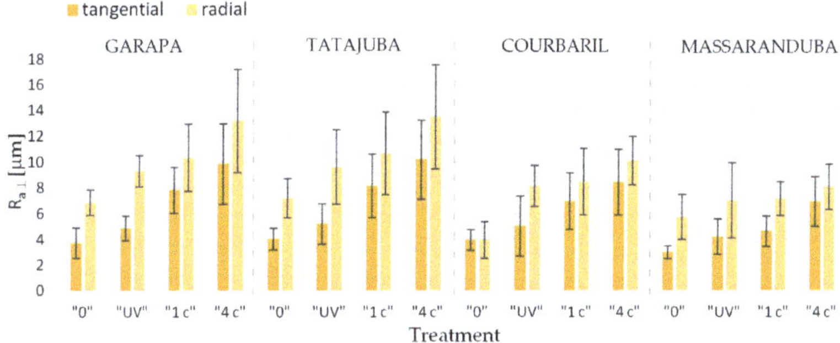

Figure 3. Roughness ($R_{a\perp}$) perpendicular to the fibres of tested wood species: "O"—fresh wood, "UV"—after 24 h of UV irradiation, "1 c"—after one cycle of artificial weathering, "4 c"—after four cycles of artificial weathering.

According to the ANOVA results (at the 0.05 confidence level), the surface roughness varied significantly depending on the examined factor (species, section, used methods of weathering) (Table 4). However, the most important influencing factors were wood section and kind of treatments used (45% and 46%, respectively).

Table 4. Statistical evaluation of the factors influencing wood surface roughness.

Feature	Factor	Sum of Squares	Mean Sum of Squares	Fisher's F-Test	Significance Level	Factor of Influence
		SS	MS	F	p	%
$R_{a\parallel}$	Intercept	513.3754	513.3754	1365.962	0.000000 *	-
	Species	12.3166	4.1055	10.924	0.000102 *	9
	Section	19.6308	19.6308	52.233	0.000000 *	45
	Treatment	59.8475	19.9492	53.080	0.000000 *	46
	Error	9.0200	0.3758	-	-	1
$R_{a\perp}$	Intercept	1700.612	1700.612	709.3115	0.000000 *	-
	Species	27.501	9.167	3.8235	0.022689 *	12
	Section	34.875	34.875	14.5462	0.000842 *	46
	Treatment	90.381	30.127	12.5657	0.000039 *	41
	Error	57.541	2.398	-	-	4

*—significant at the 0.05 level.

Generally, the surface roughness values increased due to UV irradiation when compared with the surface roughness values of the control wood samples. The differences were statistically significant at the 0.05 confidence level in most of the cases. UV treatment initiates surface oxidation (increase in the acid/base or polar component), which leads to the introduction of functional (carboxyl) groups [15]. R_a values increased after UV irradiation at a similar level as after one cycle of artificial weathering and significantly higher after a further full weathering treatment. Exposure to UV light and water caused fewer cracks. Soaking wood in water and then drying caused the raising of wood fibres. Initial roughness was not decisive in the scope of changes.

3.3. Wettability Changes

The effects of UV irradiation on the changes in the chemistry of wood surfaces were verified using contact angles with distilled water. Results are given in Table 5. The wettability decreased (contact angle increased) for all investigated surfaces with treatment by UV radiation and the full artificial weathering process. The contact angle (θ) measurements with distilled water showed the same trend for the radial and tangential surfaces of all tested wood species. In contrast to changes in surface roughness, changes in wettability caused by exposure to UV light were not statistically significant. In general, for most of all tested wood species, UV irradiation for 24 h caused an increase in surface contact angle of 8%–15% (decrease in wettability). The inclusion of water in the weathering process resulted in a much wider range of changes. Currently, one full weathering cycle caused a change in the wettability of the surface of up to 51% (for massaranduba wood). Further ageing resulted in minor changes. Carrying out four weathering cycles resulted in slightly larger wettability changes. This can be explained by the fact that leaching of extractives with water and its accumulation on the wood surface effect decreased the degree of surface hydrophilicity [40,41]. Thus, the decrease in wetting of a wood surface is related to the chemical changes after outdoor exposure. Leaching of the extractives from the surface of weathered wood reduces water repellence, while the degradation of lignin results in a more hydrophilic surface [40]. Garapa wood has shown a different nature of change. UV exposure caused a slight decrease in the contact angle (increase in wettability). This reverse direction of change can be explained by the fact that garapa wood has the lowest density in the studied group. Thus, UV penetration of structure was more possible. The degradation of hydrophobic lignin and allowing cellulose to become more abundant on the wood surface increased the degree of surface hydrophilicity [41].

Table 5. Contact angles (θ) tested wood species before and after used weathering treatments (means and standard deviations in parentheses).

Species	Section	Before Exposure (°)	After 25 h UV Irradiation (°)	After 1 Cycle of Artificial Weathering (°)	After 4 Cycles of Artificial Weathering (°)
Garapa	tangential	66.4 (5.8)	59.5 (5.2)	70.9 (4.5)	78.0 (7.5)
	radial	49.9 (5.3)	55.3 (3.1)	65.0 (6.3)	77.0 (5.6)
Tatajuba	tangential	48.4 (6.2)	52.9 (4.7)	59.8 (5.1)	63.3 (3.7)
	radial	47.8 (6.5)	53.5 (6.0)	62.2 (4.7)	65.8 (4.6)
Courabril	tangential	49.8 (5.0)	55.8 (2.5)	68.3 (4.5)	69.3 (4.4)
	radial	45.7 (5.2)	54.1 (6.4)	66.4 (4.1)	68.2 (5.8)
Massaranduba	tangential	44.4 (4.4)	49.6 (4.5)	66.8 (3.7)	74.1 (6.9)
	radial	43.2 (4.4)	45.3 (3.6)	65.1 (8.1)	70.9 (7.5)

UV light can destruct pits, which enable coatings and adhesives to penetrate deeper into the wood surface, and therefore enhance mechanical anchoring. Each of the chemical components of wood (i.e., lignin, cellulose, hemi-cellulose or extractives) is sensitive to UV radiation with a consequential deterioration effect. Of these chemical constituents, lignin, because of its strong ultraviolet absorbing characteristic due to the phenolic nature of its molecular architecture, appears to be oxidized and degraded very rapidly by UV light [42]. This information is important in the context that, in the wood working industry, UV irradiation is usually used for the curing of coating systems and a UV source mostly exists in the production line, so it seems feasible to integrate UV-modification into the finishing process. Contrary to other activation methods, UV treatment is suitable for an online manufacturing process [16].

According to ANOVA results (at the 0.05 confidence level), the surface wettability varied significantly depending on the used methods of weathering (Table 6). It was the most important influencing factor (76%). Wood section and wood species did not have a significant influence.

Table 6. Statistical evaluation of the factors influencing wood surface wettability.

Factor	Sum of Squares	Mean Sum of Squares	Fisher's F-Test	Significance Level	Factor of Influence
	SS	MS	F	*p*	%
Intercept	117,570.1	117,570.1	6298.194	0.000000 *	-
Species	518.4	172.8	9.256	0.000300 *	18
Section	49.1	49.1	2.632	0.117806 NS	5
Treatment	2112.0	704.0	37.714	0.000000 *	76
Error	448.0	18.7	-	-	1

*—significant at the 0.05 level, NS—not significant.

3.4. Structure Changes

Microstructure changes in the wood surface properties were characterized using SEM. Most of the previous conducted studies regarded softwoods. Figure 4 presents images of wood surface before weathering, after UV irradiation and after full ageing treatment. The untreated samples indicted cell walls and some easy to recognize damages caused by splitting after mechanical preparation.

As a result of weathering, some characteristic anatomical changes occurred. Both UV irradiation and full artificial weathering treatment caused the apparent erosion of wood structure elements. This also agreed with previous studies of wood surface deterioration after exposure to sunlight or UV irradiation [43]. A number of distinctive cracks formed in the specimens after exposure to UV light. Distortion and creasing of the cells wall occurred in the longitudinal direction and resulted in the delamination of the cell walls—degradation of S1 layer [44]. The deterioration of connections between cell elements were associated with the degradation of the middle lamella. At the same time, it confirmed the degradation of lignin. The SEM micrographs revealed much more degradation in wood specimens subjected to the full weathering process. Heavy damage to the pits within vessels was observed. Most of the cracks on the pits formed transverse to the cell axis, which resulted from microfibril orientation in the S3 layer of cell walls, produced from condensing compounds of degraded lignin and hemicelluloses. In all analysed specimens, thinning of parenchyma cell walls and their shrinkage or total degradation were observed. As seen in Figure 4, deterioration was present in parenchyma cells of wood rays. Similar observations were made by Mamoňová and Reinprecht [44], who tested tropical wood species weathered in natural conditions. They stated that the highest incidence of micro-cracks after weathering is correlated with wood density.

The changes caused by UV irradiation were much smaller than those caused by full artificial weathering. Thus, this confirmed the influence of desorption tensions in structure changes that caused disruptions to wooden tissue subjected to cyclic humidity changes.

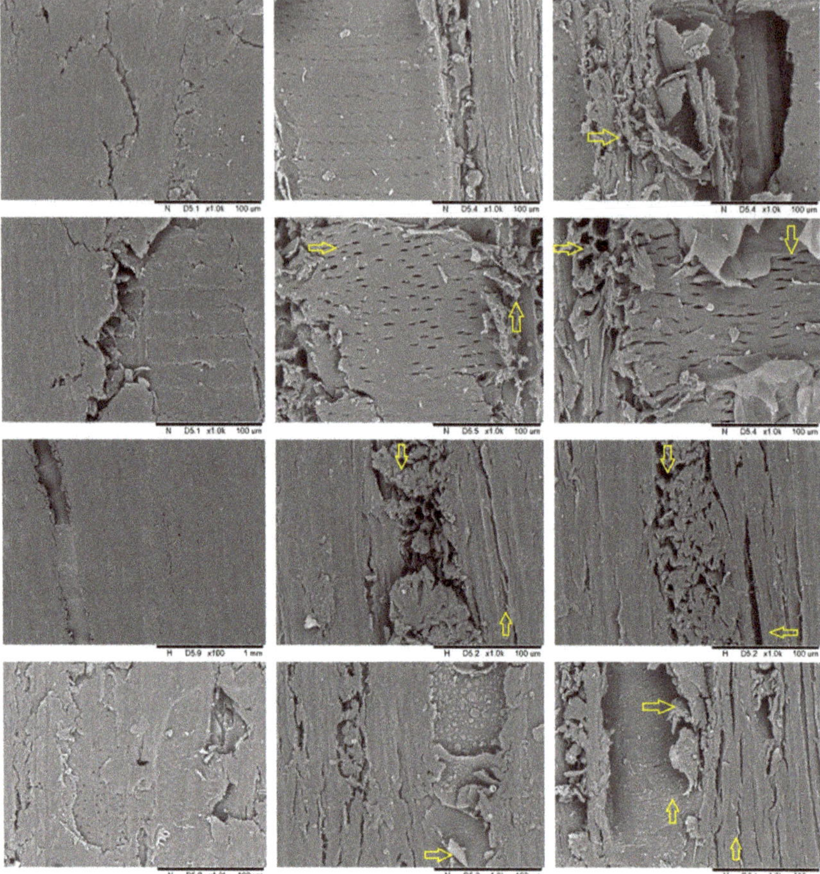

Figure 4. Comparison of microstructure exposed and non-exposed wood surface (from top: garapa, tatajuba, courbaril and massaranduba; from left: fresh wood, after 24 h of UV irradiation, after 4 cycles of artificial weathering); the described changes are marked with yellow arrows.

3.5. FTIR Analysis

The FTIR spectra of woods before and after UV irradiation and full weathering (four cycles) are shown in Figure 5. The main changes were found at 2900, 1730, 1650, 1590, 1512, 1420, 1317, 1260, 1163, and 1030 cm^{-1}. The garapa wood showed the greatest scope of changes (the highest differences in absorbance) and the smallest differences in absorbance showed massaranduba. The wood species in the studied group are characterized by the lowest and highest density. In the case of garapa and tatajuba wood, in general, the relative intensity of the bands increased more after UV treatment than after full artificial weathering. This can be explained by the fact that during the full weathering process, water-extractable compounds leached out and their accumulation on wood surfaces caused a negligible effect of weathering factors on wood. Nevertheless, the changes in the chemical structure of extractives during weathering cannot be determined by infrared spectroscopy, due to the significantly lower amount of lignin and polysaccharides [45–47]. According to the current knowledge [45–47], in the spectral range 1740–1720 cm^{-1}, various overlapping stretching vibrations of bond C=O in carbonyl and carboxyl compounds can occur. The results showed that carbonyl groups, determined at 1730 and 1650 cm^{-1}, increased more as an effect of full weathering process than after UV irradiation.

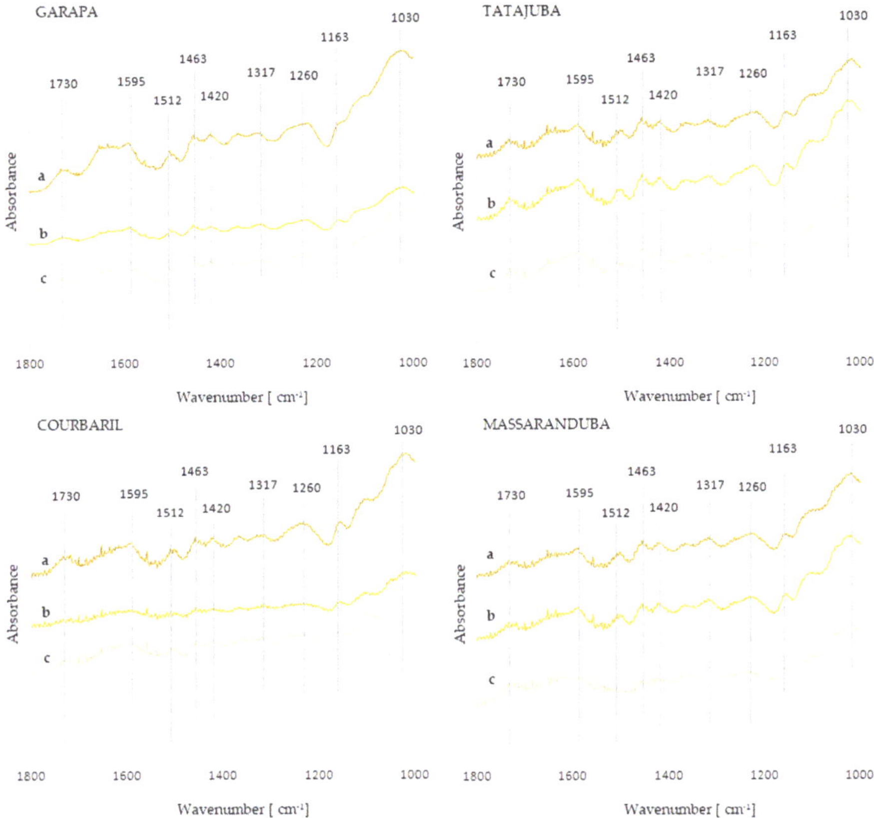

Figure 5. FTIR spectra of tested wood species before, after UV irradiation and after four full artificial weathering cycles: **a**—fresh wood, **b**—after UV irradiation, **c**—after four artificial weathering cycles.

The absorption near 1735 cm^{-1} can be attributed to vibrations in bond C=O in xylan (hemicellulose), as well as in carboxylic acids (lignin oxidation products). The occurrence of oxidation products was confirmed by the changes in intensity band, reaching 1650 cm^{-1}. These results are in accordance with the studies of other researchers, involving one or more different species and extended for longer periods of ageing [33,38,45,46]. This absorption was due to C=O carbonyl stretching in aromatic compounds, which can suggest the carbonyl compounds' formation, and the highest changes were observed in the case of garapa wood. The relative intensity of band at 1512 cm^{-1} decreased after weathering process. That peak is due to the skeletal stretching vibration of C=C in the aromatic ring of lignin [46]. The absorption at 2900 cm^{-1} was characteristic of alkane CH vibrations of methylene in cellulose, 1417 cm^{-1}—CH_2 bending crystallized, and amorphous cellulose and 1317 cm^{-1}—CH_2 wagging in crystallized cellulose [44,45]. Only in the case of massaranduba wood irradiated with UV did these peaks remain almost unchanged, which could be due to the high wood density and negligible effect of weathering factors on wood or their blocking with extractives. The peak at 1030 cm^{-1} for the C–O stretching was linked to cellulose and wood extractives [45]. The notable decrease in the relative intensity of bands at 1260 and 1230 cm^{-1} was observed after weathering. The absorbance in the range 1290–1200 cm^{-1} is characteristic for stretching C–O vibrations in lignin (guaiacyl), hemicellulose (xylans) and the conjugation of the C–O of extractives [43]. After a full weathering process, in the case of garapa and tatajuba wood, a peak at 1260 cm^{-1} was found, which might be associated with the conjugation of the C–O of extractives. Tropical woods have a relatively high number of extractives in comparison to European wood species [37]. Wood, as a complex structure, is a material for which it is difficult to formulate general conclusions. Followed by Reinprecht et al. [46], differences observed between tested wood species and changes observed on its surfaces after UV irradiation and full weathering could be explained by chemical changes taking place individually for each of them during photo-oxidation processes in the lignin-polysaccharide system, as well as extractives contained in wood. Differences between results obtained after UV treatment and full artificial weathering process might be associated with the migration of extractives such as UV-protectable extractives [45] (water-soluble extractives accumulated on the wood surfaces after full weathering process) and with the fact that extractives photooxidation products can increase the delay photodegradation of lignin [48].

4. Conclusions

Based on the results of the research into the influence of selected ageing factors such as UV irradiation and complex artificial weathering methods on the colour stability, wettability and roughness changes in garapa, tatajuba, courbaril and massaranduba, it can be concluded that:

- Regardless of ageing method, wood surfaces become darker and turn into shades of brown. The scope of changes was similar for both garapa, tatajuba of light colour and massaranduba of dark colour. Massive colour changes showed in courbaril wood, known for its high extractives content;
- The roughness of the tested wood species demonstrated variation, depending on the wood section and the measurement direction. In general, radial sections showed higher roughness due to interlocked fibres in tested wood species (fibres cut in various ways on the radial section of the wood). Roughness measured perpendicular to the wood fibres was twice as high as that measured parallel to the grain;
- The surface roughness values increased due to UV irradiation. The same time of irradiation connected with soaking wood in water and the drying caused roughness changes at a similar level, but was larger after a higher number of artificial weathering cycles;
- In contrast to changes in surface roughness, changes in wettability caused by exposure to UV light were not significant. The inclusion of water and drying in the weathering process resulted in a much wider range of changes. The wettability decreased (contact angle increased). One full weathering cycle caused a change in the surface from 24% for tatajuba wood to 51% for massaranduba wood;
- Anatomical analyses of tested tropical woods revealed structure changes after using ageing treatments. Distortion was found between cell elements associated with the degradation of the middle lamella, micro-cracks in cell walls, thinning and degradation of parenchyma cells in wood rays, and cracks along the pits in vessels formed on the S3 layer of cell walls. As a result of desorption tension, the changes caused by UV irradiation were much smaller than those caused by full artificial weathering;
- FTIR analysis indicated the occurrence of oxidative changes in lignin and hemicelluloses after the weathering process. The depolymerisation of cellulose was identified as well. The differences in FTIR results after UV treatment and full artificial weathering process must have been caused by the accumulation of water-soluble extractives on the wood surfaces.

Thus, the effect of changes caused by aging factors is the result of wood structure and its density, but also chemical composition. The results show that the observed changes may affect the long-term durability of finishes applied over wood subjected to weathering factors for a short period before finishing.

Author Contributions: Conceptualization, A.J.; methodology, A.J. and P.B.; formal analysis, A.J. and P.B.; investigation, A.J., P.B., K.R. and M.N.; resources, A.J.; data curation, A.J., P.B., K.R. and M.N.; writing—original draft preparation, A.J.; writing—review and editing, A.J. and P.B.; visualization, A.J.; supervision, A.J., M.N. and P.B. All authors have read and agreed to the published version of the manuscript.

Funding: This research received no external funding.

Acknowledgments: The APC was funded by the authors thank Warsaw University of Life Sciences—SGGW, Institute of Wood Sciences and Furniture for their financial support.

Conflicts of Interest: The authors declare no conflict of interest.

References

1. Feist, W.C. Outdoor Wood Weathering and Protection. In *Archaeological Wood: Properties, Chemistry, and Preservation, Proceedings of 196th Meeting of the American Chemical Society, Los Angeles, WA, USA, 25–28 September 1988*; Rowell, R.M., Barbour, R.J., Eds.; Advances in Chemistry Series 225; American Chemical Society: Washington, DC, USA, 1990; pp. 263–298.
2. Feist, W.C.; Hon, D.N.S. Chemistry of weathering and protection. In *The Chemistry of Solid Wood*; Rowell, R.M., Barbour, R.J., Eds.; Advances in Chemistry Series 207; American Chemical Society: Washington, DC, USA, 1984; pp. 401–451.
3. Hon, D.N.S.; Clemson, S.C.; Feist, W.C. Weathering characteristics of hardwood surfaces. *Wood Sci. Technol.* **1986**, *20*, 169–183.
4. Tolvaj, L.; Faix, O. Artificial ageing of wood monitored by DRIFT spectroscopy and CIE $L^*a^*b^*$ color measurements. 1. Effect of UV light. *Holzforschung* **1995**, *49*, 397–404. [CrossRef]
5. Fajula, X.C.; Carrillo, F.; Nogués, F.; Garriga, P. Structural analysis of photodegraded wood by means of FTIR spectroscopy. *Polym. Degrad. Stab.* **2003**, *80*, 543–549. [CrossRef]
6. Williams, R.S. Finishing of Wood. In *Wood Handbook—Wood as Engineering Material*; General Technical Report FPL-GTR-113; Department of Agriculture, Forest Service, Forest Product Laboratory: Madison, WI, USA, 1999.
7. Williams, R.S. Weathering of wood. In *Handbook of Wood Chemistry and Wood Composites*; CRC Press: Boca Raton, FL, USA, 2005; pp. 139–185.
8. Temiz, A.; Terziev, N.; Eikenes, M.; Hafrén, J. Effect of accelerated weathering on surface chemistry of modified wood. *Appl. Surf. Sci.* **2007**, *253*, 5355–5362. [CrossRef]
9. Evans, P.D.; Urban, Ć.K.; Chowdhury, M.J.A. Surface checking of wood is increased by photodegradation caused by ultrafiolet and visible light. *Wood Sci. Technol.* **2008**, *42*, 251–265. [CrossRef]
10. Follrich, J.; Teischinger, A.; Müller, U. Artificial ageing of softwood joints and its effect on internal bond strenght with special consideration of flat-to-end grain joints. *Eur. J. Wood Wood Prod.* **2011**, *69*, 597–604. [CrossRef]
11. *European Standard EN 350 Durability of Wood and Wood-Based Products—Testing and Classification of the Durability to Biological Agents of Wood and Wood-Based Materials*; European Committee for Standarization: Brussels, Belgium, 2016.
12. Feist, W.C. Weathering and Protection of Wood. In Proceedings of the 79th Annual Meeting of the American Wood Preserver's Association, Kansas City, MO, USA, 17–20 April 1983; pp. 195–205.
13. Brischke, C.; Bayerbach, R.; Rapp, A.O. Decay-influencing factors: A basis for service life prediction of wood and wood-based products. *Wood Mater. Sci. Eng.* **2006**, *1*, 91–107. [CrossRef]
14. Oberhofnerová, E.; Pánek, M.; Garcia-Cimarras, A. The effect of natural weathering on untreated wood surface. *Maderas Ciencia y Tecnología* **2017**, *19*, 173–184. [CrossRef]
15. Swann, A.L.S.; Longo, M.; Knox, R.G.; Lee, E.; Moorcroft, P.R. Future deforestation in the Amazon and consequences for South American climate. *Agric. For. Meteorol.* **2015**, *214*, 12–24. [CrossRef]
16. Gindl, M.; Sinn, G.; Stanzl-Tschegg, S.E. The effects of ultraviolet light exposure on the wetting properties of wood. *J. Adhes. Sci. Technol.* **2006**, *20*, 817–828. [CrossRef]
17. Akgul, M.; Korkut, S.; Camlibel, O.; Candan, Z.; Akbulut, T. Wettability and surface roughness characteristics of medium density fiberboard panels from rhododendron (Rhododendron ponticum) biomass. *Maderas Ciencia y Tecnología* **2012**, *14*, 185–193. [CrossRef]
18. Rathke, J.; Sinn, G. Evaluating the wettability of MUF resins and pMDI on two different OSB raw materials. *Eur. J. Wood Wood Prod.* **2013**, *71*, 335–342. [CrossRef]
19. Qin, Z.; Gao, Q.; Zhang, S.; Li, J. Surface free energy and dynamic wettability of differently machined poplar woods. *BioResources* **2014**, *9*, 3088–3103. [CrossRef]
20. Nylund, J.; Sundberg, K.; Shen, Q.; Rosenholm, J.B. Determination of surface energy and wettability of wood resins. *Coll. Surf. A Physicochem. Eng. Asp.* **1998**, *133*, 261–268. [CrossRef]

21. Nussbaum, R.M. Natural surface inactivation of scots pine and norway spruce evaluated by contact angle measurements. *Holz Als Roh-Und Werkst.* **1999**, *57*, 419–424. [CrossRef]
22. Wålinder, M. Wetting Phenomena on Wood: Factors Influencing Measurements of Wood Wettability. Ph.D. Thesis, KTH—Royal Institute of Technology, Stockholm, Sweden, 2000.
23. Baar, J.; Gryc, V. The analysis of tropical wood discoloration caused by simulated sunlight. *Eur. J. Wood Wood Prod.* **2012**, *70*, 263–269. [CrossRef]
24. Costa, J.A.; Gonçalez, J.C.; Camargos, J.A.A.; Gomes, I.A.S. Photodegradation of two tropical wood species: Jatoba' (*Hymenaea courbaril*) and tauari (*Couratari oblongifolia*) submitted to ultraviolet radiation. *Cerne* **2011**, *17*, 133–139. [CrossRef]
25. *ASTM D2395 Standard Test Methods for Specific Gravity of Wood and Wood-Based Materials*; ASTM International: West Conshohocken, PA, USA, 2002.
26. Gardner, D.J. Application of the Lifshitz-van der Waals acid-base approach to determine wood surface tension components. *Wood Fiber Sci.* **1996**, *28*, 422–428.
27. Liptáková, E.; Kudela, J.; Bastl, Z.; Spirovová, I. Influence of mechanical surface treatment of wood on the wetting process. *Holzforschung* **1995**, *49*, 369–375. [CrossRef]
28. ISO 7724-3 Paints and Varnishes. In *Colorimetry. Part 3: Calculation of Colour Differences*; International Organization for Standardization: Geneva, Switzerland, 1984.
29. Zisman, W.A. Influence of Constitution on Adhesion. *Ind. Eng. Chem.* **1963**, *55*, 18–38. [CrossRef]
30. ISO 4287 Geometrical Product Specifications (GPS). In *Surface Texture: Profile Method. Terms, Definitions and Surface Texture Parameters*; International Organization for Standardization: Geneva, Switzerland, 1997.
31. Chalkias, C.; Faka, A.; Kalogeropoulos, K. Assessment of the direct sun-light on rural road network through solar radiation analysis using GIS. *Open J. Appl. Sci.* **2013**, *3*, 224–231. [CrossRef]
32. Jankowska, A.; Kozakiewicz, P. Evaluation of wood resistance to artificial weathering factors using compressive properties. *Drv. Ind.* **2016**, *67*, 3–8. [CrossRef]
33. Jankowska, A.; Reder, M.; Golofit, T. Comparative study of wood color stability using accelerated weathering process and infrared spectroscopy. *Wood Res.* **2017**, *62*, 549–556.
34. Hon, D.N.S.; Minemura, N. Color and discoloration. In *Wood and Cellulosic Chemistry*; Hon, D.N.S., Shiraishi, N., Eds.; Marcel Dekker: New York, NY, USA, 2001; pp. 385–442.
35. Oltean, L.; Teischinger, A.; Hansmann, C. Wood surface discolouration due to simulated indoor sunlight exposure. *Holz Als Roh-Und Werkst.* **2007**, *66*, 51–56. [CrossRef]
36. Pastore, T.C.M.; Santos, K.O.; Rubim, J.A. Spestrocolorimetric study on effect on ultraviolet irradiation of four tropical hardwoods. *Biosource Technol.* **2004**, *93*, 37–42. [CrossRef] [PubMed]
37. Jankowska, A.; Boruszewski, P.; Drożdżek, M.; Rębkowski, B.; Kaczmarczyk, A.; Skowronska, A. The role of extractives and wood anatomy in the wettability and free surface energy of hardwoods. *BioResources* **2018**, *13*, 3082–3097. [CrossRef]
38. Pandey, K.K. A note on the influence of extractives on the photo-discoloration and photo-degradation of wood. *Polym. Degrad. Stab.* **2005**, *87*, 375–379. [CrossRef]
39. Richter, H.G.; Dallwitz, M.J. Commercial Timbers: Descriptions, Illustrations, Identification, and Information Retrieval. In English, French, German, Portuguese, and Spanish. Version: 9th April 2019. Available online: http://delta-intkey.com (accessed on 10 July 2020).
40. Aydin, I.; Colakoglu, G. Effects of surface inactivation, high temperature drying and preservative treatment on surface roughness and colour of alder and beech wood. *Appl. Surf. Sci.* **2005**, *252*, 430–440. [CrossRef]
41. Rowell, R.M. Handbook of wood chemistry and wood composites. In *Wood and Cellulosic Chemistry*; Hon, D.N.-S., Shiraishi, N., Eds.; CRC Press: Boca Raton, FL, USA, 2005; p. 487.
42. Gassan, J. Effects of corona discharge and UV treatment on the properties of jute-fibre epoxy composites. *Compos. Sci. Technol.* **2000**, *60*, 2857–2863. [CrossRef]
43. Srinivas, K.; Pandey, K.K. Photodegradation of thermally modified wood. *J. Photochem. Photobiol. B Biol.* **2012**, *117*, 140–145. [CrossRef]
44. Mamoňová, M.; Reinprecht, L. The impact of natural and artificial weathering on the anatomy of selected tropical hardwoods. *IAWA J.* **2020**, *41*, 333–355. [CrossRef]
45. Liu, R.; Zhu, H.; Li, K.; Yang, Z. Comparison on the aging of woods exposed to natural sunlight and artificial xenon light. *Polymers* **2019**, *11*, 709. [CrossRef] [PubMed]

46. Reinprecht, L.; Mamoňová, M.; Pánek, M.; Kačík, F. The impact of natural and artificial weathering on the visual, colour and structural changes of seven tropical woods. *Holz Als Roh-Und Werkst.* **2017**, *76*, 175–190. [CrossRef]
47. Tolvaj, L.; Molnár, Z.; Németh, R. Photodegradation of wood at elevated temperature: Infrared spectroscopic study. *J. Photochem. Photobiol. B Biol.* **2013**, *121*, 32–36. [CrossRef]
48. Chang, T.-C.; Chang, H.-T.; Wu, C.-L.; Chang, S.-T. Influences of extractives on the photodegradation of wood. *Polym. Degrad. Stab.* **2010**, *95*, 516–521. [CrossRef]

Communication

Glossiness Evaluation of Coated Wood Surfaces as Function of Varnish Type and Exposure to Different Conditions

Emilia-Adela Salca [1,*], Tomasz Krystofiak [2], Barbara Lis [2] and Salim Hiziroglu [3]

1 Faculty of Furniture Design and Wood Engineering, Transilvania University of Brasov, Universitatii 1, 500068 Brasov, Romania
2 Faculty of Forestry and Wood Technology, Poznan University of Life Sciences, Wojska Polskiego 28, 60-637 Poznan, Poland; tomkrys@up.poznan.pl (T.K.); blis@up.poznan.pl (B.L.)
3 Department of Natural Resource Ecology and Management, Oklahoma State University, 303-G Agricultural Hall, Stillwater, OK 74078-6013, USA; salim.hiziroglu@okstate.edu
* Correspondence: emilia.salca@unitbv.ro; Tel.: +40-268-415-315

Abstract: The objective of this study was to evaluate the glossiness of black alder wood (*Alnus glutinosa* L.) samples coated with two varnish types as a function of exposure to dry heat and artificial aging. The chemical resistance of the coated samples to cold liquids was also evaluated. Based on the findings in this work, it appears that the varnish types and their structural differences influenced the overall glossiness of the coated samples. The UV varnish exhibited higher gloss values than those coated with the water-borne product within the range of silky gloss and silky matte grades. The heat exposure influenced the surface glossiness of the UV-coated samples more than the samples coated with water-borne varnish. The overall gloss values of the samples decreased with the exposure time to artificial aging, resulting in no layer cracks. The cold household liquids left less visible traces on the surfaces and alcohol was found to be the strongest agent. This study could have practical applications in the furniture industry to produce value-added furniture units according to their specific conditions of indoor use.

Keywords: artificial aging; coating; gloss; properties; varnish; black alder wood

Citation: Salca, E.-A.; Krystofiak, T.; Lis, B.; Hiziroglu, S. Glossiness Evaluation of Coated Wood Surfaces as Function of Varnish Type and Exposure to Different Conditions. *Coatings* **2021**, *11*, 558. https://doi.org/10.3390/coatings11050558

Academic Editor: Marko Petric

Received: 19 April 2021
Accepted: 6 May 2021
Published: 9 May 2021

1. Introduction

The glossiness of any surface is used to evaluate the quality of a finished product as a result of reflection due to the incident light from different directions [1,2]. High gloss surfaces are in demand in the furniture industry, but matte gloss still has its importance in a specific solid wood furniture market. A reflection structure image of high-gloss composite products was developed as an alternative method to describe the visual human perception of gloss [3]. However, recent studies have shown off the comfortable feelings given by both the coated and uncoated surfaces of wood products [4,5]. A model based on the value generation of wooden furniture has been validated by these qualitative, innovative, and ecological products. These products are used as reliable and environmentally friendly furniture [5,6].

The glossiness and color, apart from the visible wood texture, represent important aesthetic properties that influence the choice of any furniture customer. Several factors including wood species, surface roughness, chemical composition of the varnish, coating system, number of layers, angle of incident light, and the direction of gloss measurement influence the gloss quality of the finished wood product [7–10]. The anisotropic texture of the wood makes the reflection from a surface a complicated process. The variation of different structure patterns can be obtained from the most common lengthwise measurements on radial and tangential directions [11]. The diameters of vessels and tracheids in the earlywood and latewood are not the same because of the annual growth rate.

The roughness of the two areas of an annual ring differs under the same machining conditions and some irregular reflections of the properties appear [11]. The wood grain may raise, twist, and lift during machining, with effects during the finishing step [12,13]. The surface quality influences the glossiness of the wood [8,14]. A certain correlation may exist between the surface gloss and its roughness. Such a correlation is valid when the dominant effect influences the reflection originating from the surface structure of the unit.

Different resin clear coatings could alter the reflection properties depending on the coating thickness [11]. Transparent coatings preserve the natural color of wood and improve its glossiness. However, they reduce stability over time when compared to that of pigmented coatings, which are less affected by the sunlight [15–17]. However, they still maintain their popularity. Different levels of gloss can be achieved as a function of varnish type and its application system [6,9,10]. Alternative ecological products including water-based and UV varnish are widely used due to their low emissions, rapid application, and good gloss retention. In the case of water-borne coatings, the curing time and penetration are longer because of their lower water absorption [18].

Volatile organic compounds (VOC) regulations are applied to interior coatings as well, in order to reach abrasion and chemical resistance [19]. VOC emissions from indoor materials and finishing products are still current subjects of concern among studies on indoor air quality [20,21]. The acceptable total VOC level in the air for human health ranges from 0.3 to 0.5 mg/m^3 concentration [20].

The cost of the UV equipment could be considered high for small companies [10]. In terms of gloss, a UV varnish applied by a roller system produces surfaces with a higher gloss than when applied by spraying. The gloss of water-based, nitrocellulose, and polyurethane varnish has already been studied by several authors [7,22]. Successive layers of coating and polishing could also contribute to a higher gloss. They are applied on furniture parts depending on the surface visibility and exposure during use, and therefore different values of gloss are recorded [10,23]. For beech wood, the gloss level increased by 5−20 gloss units with the increase in varnish layers of the water-based product compared to polyurethane [9].

The direction of viewing or measurement influences the glossiness of a wood surface [11]. The reflection properties of a surface depend on the angle of incident light. There are three standardized measuring angles: 20°, 60°, and 85°. Generally, the 60° geometry is recommended for wooden surfaces, but it provides limited information.

Comparative measurements on the same surface using different measuring angles and gloss correlations between the angles may help to better evaluate the wood surface [8]. The scattering of the incident light strongly depends on the direction of gloss measurement relative to the wood grain; parallel gloss is higher than perpendicular gloss [10,11].

Wood exposed to the outdoors suffers due to the photo-degradation of the lignin and, to some extent, the hemicelluloses [24]. Changes in color and gloss, and also cracks that occur on the surface during weathering, limit the wood's utilization [25]. These changes appeared after a few hours of accelerated exposure or after a few days of natural exposure [26,27]. The lignin content in softwoods ranges from 25−35%, and consists mostly of guaiacyl units, while hardwoods have a lignin content of 15−28%, constituted mainly by guaiacyl and syringyl units [28]. It was shown that hardwoods underwent faster degradation than softwoods. The syringyl structure in hardwoods degraded faster than the guaiacyl structures in softwoods [29]. The natural, accelerated, and simulated weathering tests have proved very useful in the wood protection industry [30].

When used indoors, wooden products are subjected to less intensive UV radiation than when used outdoors [23]. Slow degradation of the coating layer by changes in the surface gloss and color, apart from some fine cracks, are noticed [31]. In a previous study, it was proven that the most relevant color changes of coated beech were generated during the first 100 h of artificial aging [23]. Retarding effects on the surface photo-degradation could be obtained with protective agents against UV radiation. Panek et al. [17] showed that the gloss of exposed surfaces decreased with the exposure time to radiation.

The coating resistance to different chemicals has been previously studied for wooden flooring. They are the most exposed elements of an interior, along with some horizontal visible furniture parts [32–36]. To evaluate the chemical resistance of coatings applied to lignocellulosic materials, a rating scale, included in the standards and test procedures, is commonly used [34]. Oils enhance the natural wood appearance, but they produce limited quality in terms of resistance to various chemicals [35]. No major differences in the surface resistance to cold liquids such as coffee, ethanol, red wine, water, or paraffin oil have been observed for oak parquet covered with different coatings [34]. It was found that the resistance to cold liquids depends on the properties of the topcoat used [34].

Currently, there is limited information about the glossiness of varnished alder wood. Therefore, the objective of this study was to evaluate the glossiness of varnished alder wood surfaces when using two varnish types and after their exposure to various test conditions including the dry heat test and artificial aging. The resistance to different cold liquids of the coated surfaces was also tested. The results of this study can provide a better understanding of the properties of the coatings of such wood species, and show its potential for furniture manufacturing.

2. Materials and Methods

In this study, planed specimens of black alder (*Alnus glutinosa* L.) wood, supplied by a local sawmill in Buzau County, Romania, were used. Black alder wood is a native species in Romania, less-utilized but known mostly for its good workability, properties, and pleasant appearance. The average basic density of the samples was 520 ± 10 kg/m^3, and their moisture content was 8 ± 1%. A total of 29 samples were used for the experiment and they were grouped as presented in Table 1. The samples were conditioned for 7 days in a room with a temperature of 20 ± 2 °C and relative humidity of 50% ± 5% before any tests were carried out.

Table 1. Experimental design.

Wood Species	**Black Alder**
Dimension of samples (mm)	L = 300; R = 6; T = 95
Number of samples and their distribution	Total: 29 5 control and 12 samples per each varnish type, and, out of them, 3 per each test and varnish type
Processing	Planed samples were sanded with 100 and 150 grit size
Coating System	Spraying
Varnish Products	A2: 100% UV varnish–2 layers
	B2: water-borne varnish–2 layers
A light 220 grit sanding between layers	
Tests	Dry heat test
	Artificial aging
	Chemical resistance
Decorative property	Glossiness

2.1. Surface Preparation of the Samples

The specimens (Figure 1) were subjected to parallel sanding by employing a portable sander (FESTOOL ETS 125, FESTOOL GmbH, Wendlingen, Germany) shown in Figure 2 with the technical data displayed in Table 2. Two types of sandpaper with aluminum oxide grains (FESTOOL Rubin 2) of 100 and 150 grit size were used.

Figure 1. Wood samples.

Figure 2. Festool ETS 125 sander.

Table 2. Technical characteristics of the FESTOOL portable sander ETS 125.

Characteristics						
Power Consumption, W	Eccentric Motion Speed, min^{-1}	Sanding Stroke, mm	Sanding Pad Diameter, mm	Dust Extraction Connection Diameter, mm	Weight, kg	Drive Type
250	6000–12,000	2.00	125.00	27	1.20	Hand

2.2. Coating of the Samples

The sanded samples were coated by spraying with two types of varnish, namely a UV acrylic varnish (A2) and a water-borne varnish (B2) at a room temperature of 20 °C and 40% RH in two layers. An industrial low-pressure spray gun at a pressure of 0.25 bar and a spread rate of 120 ± 5 g/m^2 was employed as shown in Figure 3. A light sanding using a foam pad of 220 grit size made of electrocorundum grains (Klingspor Abrasives, Bielsko-Biala, Poland) was applied between the two varnish layers. The varnish parameters are presented in Table 3. UVC-250 × 2-type UV curing equipment (MIKON UV Ltd., Warsaw, Poland) was used to cure the samples coated with the UV varnish. The samples coated with the water-borne varnish were cured at a room temperature of 20 °C and 40% RH. Dry film thicknesses of 90 ± 5 μm and 30 ± 5 μm were determined for the UV and water-borne varnishes, respectively [37].

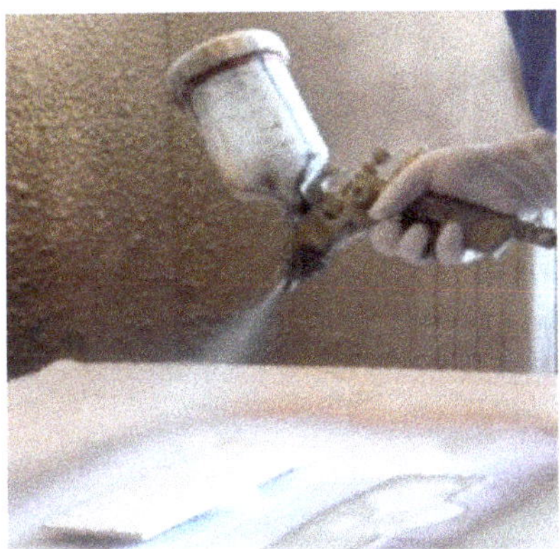

Figure 3. Application with the spray gun.

Table 3. Parameters of the varnish products.

Varnish Product	VOC-EU (Volatile Organic Compounds), g/L	Density (g/cm^3), 20 °C	Conventional Viscosity (s), 20 °C [38]	Organic Solvents (%)	Solid Content (%)
UV acrylic (A2)	55.2	1.229	42	6.5	93.5
Water-borne (B2)	55.2	1.024	65	5.4	27.9

2.3. Gloss Measurement of the Samples

The gloss of the control and coated samples was determined employing a PICO GLOSS 503 gloss meter (ERICHSEN GmbH, Hemer, Germany) as illustrated in Figure 4. The gloss measurements were conducted at a degree level of 20°, 60°, and 85° geometry, both in parallel and perpendicular to the wood grain. Five measurements per sample were taken for each standardized measuring angle and direction according to the ISO 2813 standard [39]. The method was applied for the samples before and after the dry heat test and artificial aging.

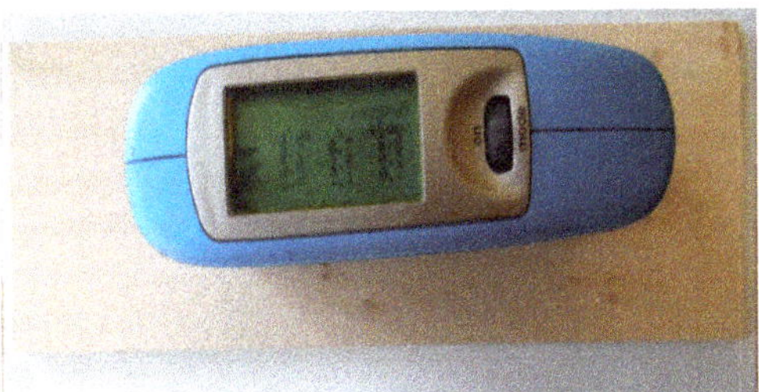

Figure 4. PICO GLOSS 503 gloss meter.

2.4. Dry Heat Test of the Coated Samples

The dry heat test was carried out by employing a device, heated to the temperature of 70 °C, applied to the coated samples for 20 min according to the EN 12722 standard [40]. As the standard device did not fit the sample size, a new one with a smaller diameter of about 70 mm was used (Figure 5).

Figure 5. Hot device (70 °C).

2.5. Artificial Aging of the Coated Samples

The coated samples positioned at an angle of 45° were exposed to intensive ultraviolet light and infrared radiation (UV + IR). The artificial aging test was carried out with a special quartz lamp (VT-800, FAMED Lodz S.A., Lodz, Poland) having radiation energy of 740 W (Figure 6). The radiation was applied from a distance of 40 cm to the samples for 30 min, 1, 4 and 8 h aging time. The temperature of 65 °C at the surface of the coated samples was determined with the help of a temperature detector (DT 8662 Dual Laser Infrared Thermometer, CEM, Shenzhen, China).

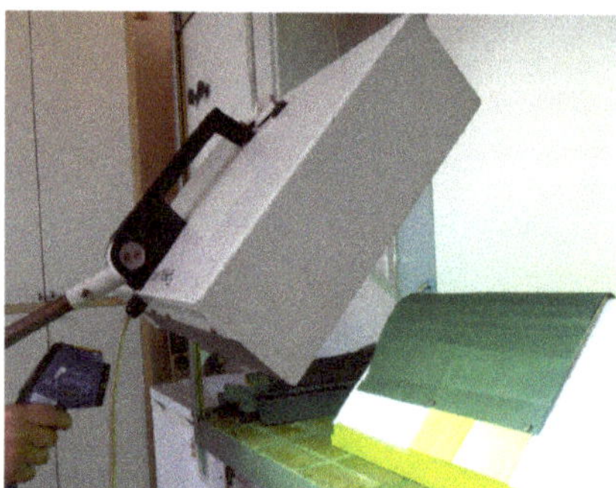

Figure 6. VT-800 quartz lamp (UV + IR).

2.6. Chemical Resistance of the Coated Samples

The chemical resistance of the coated surfaces was determined by using four types of liquids, namely water and fat (liquid paraffin) applied for 24 h and alcohol (48%) and coffee for 6 h, according to the EN 12720 standard [41]. Soft filter paper disks of 25 mm diameter were soaked for 30 s in the above mentioned liquids and then placed on the coated samples and each covered with a glass rim (Figure 7). After the exposure time, the glass rims and paper disks were removed and the samples were carefully cleaned with a soft paper towel. The surfaces were evaluated visually under the laboratory light environment according to an assessment scale from 1 to 5 for the varnish structure (1—severe damage, 2—traces with no change, 3—slight traces, 4—slight change, 5—no visible change) [41].

Figure 7. Test of cold liquids.

2.7. Processing of the Data

All data in this study have been processed using the Minitab 17.0 software. The mean values of the gloss have been used to represent the variation for each test. The regression fit equations, along with the gloss correlations and the response optimization, have also been provided.

3. Results and Discussion

3.1. Gloss Evaluation of the Coated Samples

The glossiness for the 20°, 60°, and 85° geometry was determined for the control and coated samples by respecting two measurement directions, parallel and perpendicular to the grain direction. The gloss variations of the control and coated samples are displayed in Figure 8. The varnish type and their structural differences influenced the glossiness of the coated samples [42]. The gloss values of the samples coated with the UV varnish for both directions of measurement were found to be higher than the gloss values obtained when using the water-borne product to varnish the samples. As expected, the UV-cured varnish produced an enhanced coating layer [10]. The coating structure of the UV varnish was more cured due to the influence of the UV energy when compared to the water-based varnish, and this can therefore explain the gloss differences. The direction of the gloss measurement at a 20° angle did not influence the gloss values of the same sample type and varnish, while in the case of the gloss at 60° and 85°, the values from along the grain were found higher than those from across the grain. The parallel gloss value at 60° geometry increased after coating from 2.95 to 34.87 gloss units (GU) in the case of the UV varnish, while the surfaces coated with water-borne varnish reached 27.21 GU. For the 85° geometry the gloss did not show much difference between the two varnishes when considering the same gloss direction. Sonmez et al. [12] reported that the water-borne varnish reduced the glossiness of the coated wood surface. Similar results have been found in a previous study for beech samples [9].

To obtain good interpretations and to give insights into the diversity of the results, it is best to use the correlations of gloss. Such correlations present interest in terms of their practical applications in furniture manufacturing [11]. The matrix plots of such correlations are presented in Figures 9–11. Table 4 also displays the general regression equations for the correlations of gloss. It appears that strong correlations were obtained for the gloss at 20° and 60°, and 60° and 85° (R-sq = 0.83 and R-sq = 0.88, respectively). A moderate correlation between the gloss at 20° and the gloss at 85° (R-sq = 0.6) was noticed. These results are also supported by the Pearson coefficients displayed in Table 4 (p-value = 0.000). The good value of correlation presented in Figure 9 could be explained by the incident angles used, which were large enough to be released from the surface microstructure effect. There was very little difference between the gloss readings of the samples for each varnish type and gloss direction, as depicted in Figure 9. It was determined that gloss readings at 20° are practically the same in both directions. The gloss along and across the grain for the UV varnished samples varied over a wide range, corresponding to the silky gloss grade (25–40 GU), while the water-borne varnish produced a perpendicular gloss of silky matte grade (15–25 GU), but silky gloss along the grain (Figures 9 and 11). The gloss readings perpendicular to the grain changed in a narrower range and they were much lower than the ones parallel to the grain for each varnish type (Figure 11). The literature provides information on the gloss correlations mainly for old oil and wax-treated furniture or flooring with a clear high-gloss resin. An oak-veneered old cabinet having a thin clear coating presented a gloss in the range of a silky matte grade (15–25 GU), while an old Biedermeier cabinet polished with shellac in several layers showed a high gloss grade (70–100 GU) [11].

The response optimization for glossiness as a function of varnish type and gloss direction is displayed in Table 5. As already found before, the UV varnish produced the best gloss value when measured parallel to the wood grain.

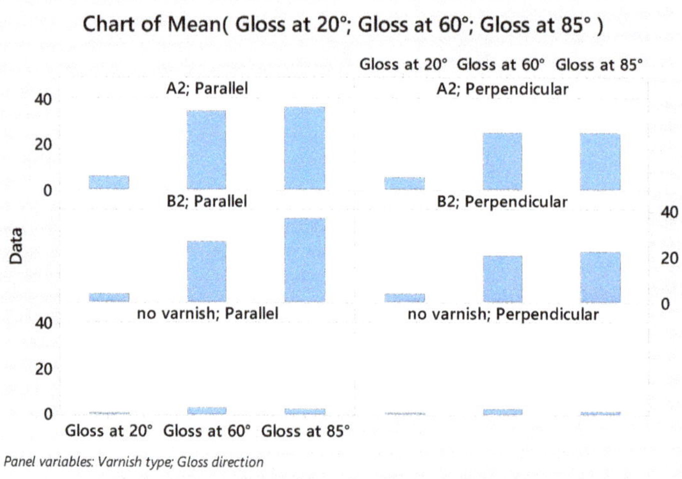

Figure 8. Gloss variation of the coated samples as a function of incident angle, measuring direction, and varnish type.

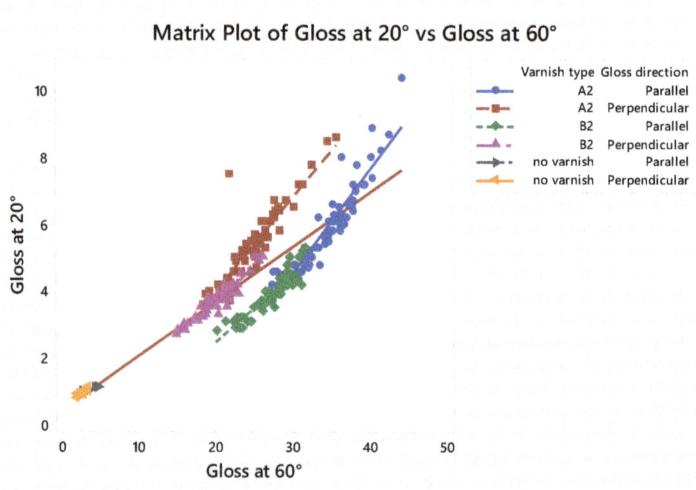

Figure 9. Correlation of gloss at 20° and 60° geometry of the coated samples.

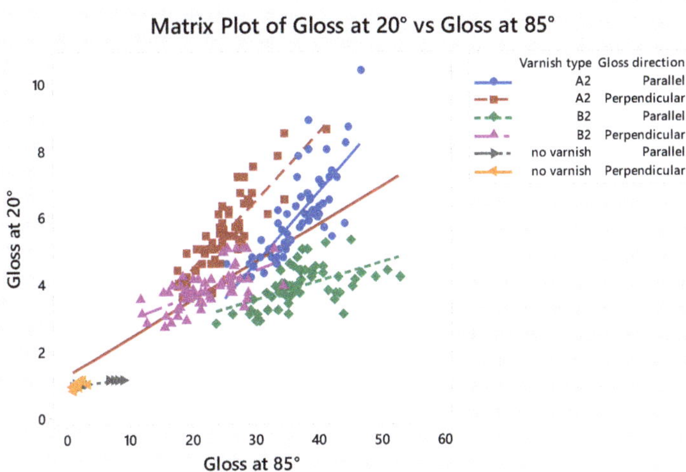

Figure 10. Correlation of gloss at 20° and 85° geometry of the coated samples.

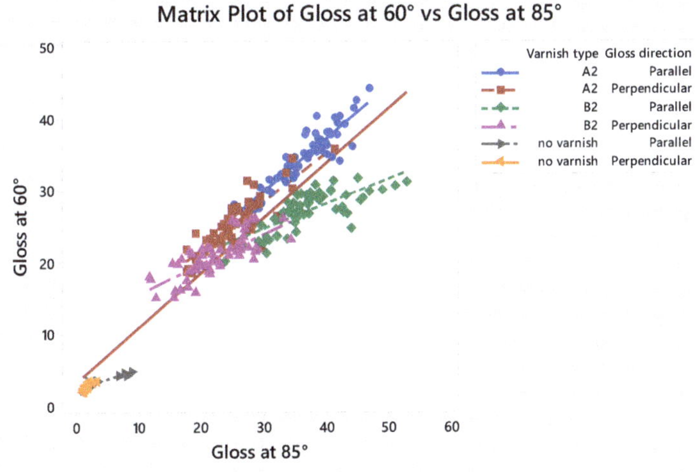

Figure 11. Correlation of gloss at 60° and 85° geometry of the coated samples.

Table 4. Regression fit equations for the correlations of gloss.

No.	Correlation of Gloss	*R*-sq, %	Equation	Pearson Correlation Coefficient
1	Gloss at 20° and gloss at 60°	83.7	Gloss at 20° = 0.4365 + 0.1627 gloss at 60°	0.915
2	Gloss at 20° and gloss at 85°	60.7	Gloss at 20° = 1.287 + 0.1128 gloss at 85°	0.779
3	Gloss at 60° and gloss at 85°	88.4	Gloss at 60° = 3.401 + 0.7658 gloss at 85°	0.940

Table 5. Response Optimization for Gloss at 85°, Gloss at 60°, and Gloss at 20°.

Solution	Varnish Type	Gloss Direction	Gloss at 20° Fit	Gloss at 60° Fit	Gloss at 85° Fit	Composite Desirability
1	A2	parallel	5.9	33.4	36.5	0.648015

3.2. Gloss Evaluation of the Coated Samples after the Dry Heat Test

The test for resistance to dry heat was carried out to evaluate the effect produced by the contact of the coated surface with a hot object heated to a temperature of 70 °C.

The results after the dry heat test showed that the high temperature applied to the coated wood surface influenced the surface glossiness (Figure 12). Overall, very little or no increase in glossiness values between the samples coated with both varnish types and the tested samples at 20° gloss geometry was found. The parallel gloss at 60° and 85° geometry for the UV-coated samples was highly influenced by the dry heat test when compared to water-borne varnish samples; a gloss increase of 15.83% and 43.4% for UV, and 4.79% and 5.21% for water-borne, respectively, were found. In regards to the perpendicular gloss at an 85° angle, the dry heat test produced a gloss increase in the same range for the two varnish types. In another study, the thermal test produced a high gloss in the case of polyurethane resin [43], while a low gloss was produced by powder coating [44].

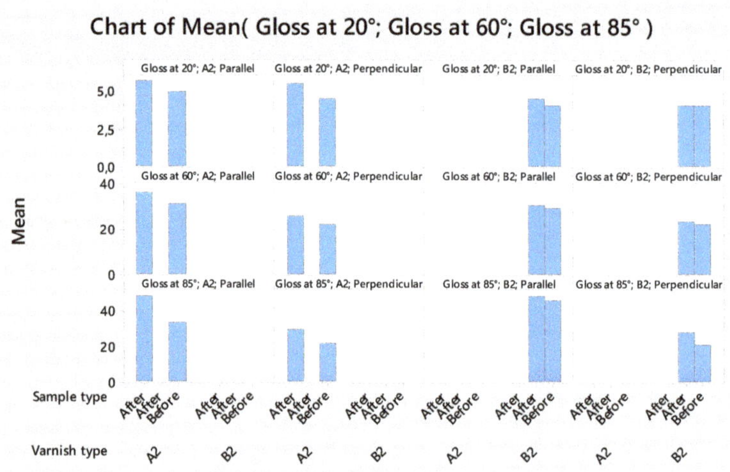

Figure 12. Gloss variation of the coated samples before and after the dry heat tests as a function of incident angle, measuring direction and varnish type.

3.3. Gloss Evaluation of the Coated Samples after the Artificial Aging

The gloss variation of the coated samples before and after the artificial aging for the two measurement directions is presented in Figure 13. The gloss measurement direction and the exposure time to radiation had almost no influence on the gloss values at 20° geometry. However, there were small differences in glossiness between the two varnish products. The gloss values recorded at 60° and 85° geometry for the coated samples showed a subsequent decrease and increase in a parallel direction with the increase of the exposure time to radiation. The gloss in the perpendicular direction was almost constant for 1 h radiation, and it then decreased for the next 8 h of exposure. Overall, the gloss of the coating layer decreased with the exposure time to radiation, predicting the degradation of the surface layer [17]. In a previous study, Irmouli et al. [31] estimated the surface degradation by quantifying the cracks at the surface layer. In the present study, no cracks were found on the coating. The findings of this test are similar to the results determined in two past studies [14,42]. The temperature plays an important role in the degradation of the varnish molecules on the surface. It is also stated that the changes on the surface are due not only to changes in the coating layer but also in the wood [14].

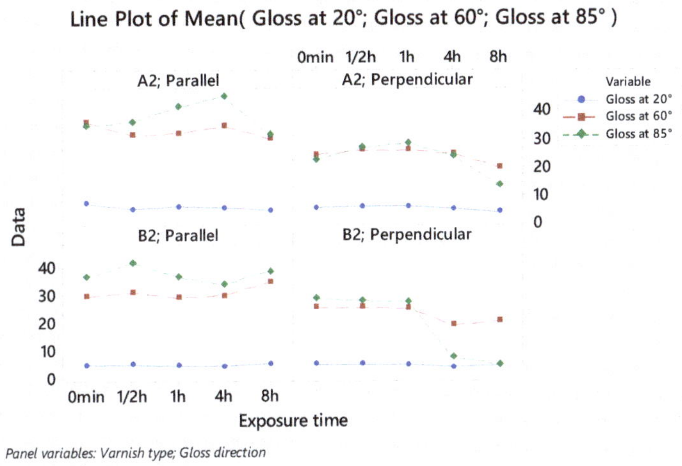

Figure 13. Gloss variation of the coated samples before and after the artificial aging.

Kudela and Kubovski [23] found the best color stability after aging in the case of the pre-treated beech samples before coating with a varnish-containing UV filter. Another study showed that the UV-accelerated weathering of the coated beech and spruce wood samples produced significant degradation of the oil-based coating compared to the acrylic coating [17]. Reduced gloss values are usually connected with the surface micro-roughness changes and diverse formulations of the varnishes [45].

3.4. Evaluation of the Coated Samples Resistance to Cold Liquids

The chemical resistance of the coated surfaces was determined by using four types of liquid: paraffin, water, alcohol, and coffee. The cold liquids used in the household left both visible and less visible traces on the tested surfaces. Alcohol was noticed to be the strongest agent because it produced surface deterioration very fast, while coffee, paraffin, and water did not produce much change, as displayed in Table 6. The results of the resistance to chemical tests are similar to other findings in the literature for wood surfaces coated with UV and water-borne varnishes [36]. Even though oils enhanced the wood's natural appearance, they had limited resistance to chemicals [35]. In their study, Pavlic et al. [34] showed that the resistance to cold liquids including coffee, ethanol, red wine, water, and paraffin oil, among others, depended on the properties of the topcoat. In terms of surface chemical resistance, no major differences were found [34]. In previous work, Nejad et al. [32] showed that household chemicals including vegetable oil, ketchup, and mustard increased the gloss of coated oil-heat-treated samples made of maple, beech, and hemlock.

Table 6. Assessment of surface resistance to cold liquids.

Cold Liquids	Time, h	Varnish Type	Scale 1–5	Description
Paraffin	24 h	A2 B2	4	Slight change on the varnish layer, only visible under reflected light
Water		A2 B2	4	
Alcohol (48%)	6 h	A2 B2	1	Severe damage on the varnish layer
Coffee		A2B2	4	Slight change on the varnish layer, only visible under reflected light

4. Conclusions

The present work evaluated the glossiness of alder wood surfaces coated with two varnish types and the effect of their exposure to the specific conditions of dry heat and artificial aging on the gloss. The chemical resistance of the coated surfaces was also assessed. The findings of this work are useful in furniture manufacturing for selecting the best varnish type. The specific conclusions of this study are presented as follows:

1. The varnish types and their structural differences influenced the glossiness of the coated samples tested in this study. The samples coated with the UV varnish exhibited higher gloss values than the samples coated with the water-borne product in both gloss directions. The gloss readings across the grain were found lower than those recorded along the grain for each varnish type. The two varnish types produced glossiness in the range of silky gloss and silky matte grades. The incident angles used were large enough to be relieved of the surface microstructure effect, and therefore very good correlations were found for the gloss at 20° and 60°, and 60° and 85°.
2. The high temperature applied to the coated wood surface influenced the surface glossiness. The parallel gloss at the 60° and 85° geometry for the UV-coated samples was highly influenced by the dry heat test when compared to the samples coated with the water-borne varnish.
3. The overall gloss values of the samples decreased with the exposure time to artificial aging, predicting the degradation of the surface layer. No cracks were noticed on the coating layer. The glossiness in the parallel direction at the same geometry for both varnish types showed a subsequent decrease and increase with the increase of the exposure time to radiation.
4. The cold liquids used in the household left both visible and less visible traces on the tested surfaces. Alcohol was found to be the strongest agent because it produced surface deterioration very fast.
5. The findings of this study could have practical applications in the furniture industry for producing value-added furniture units according to their specific conditions of indoor use.

Author Contributions: E.-A.S. and T.K. conceived and designed the experiments; E.-A.S., T.K., and B.L. performed the experiments; E.-A.S. and S.H. analyzed the data; E.-A.S. wrote the paper. All authors have read and agreed to the published version of the manuscript.

Funding: This research received no external funding.

Institutional Review Board Statement: Not applicable.

Informed Consent Statement: Not applicable.

Data Availability Statement: The data presented in this study are available on request from the corresponding author. The data are not publicly available due to privacy issues.

Acknowledgments: The authors would like to give thanks for the support received from the Division of Gluing and Finishing of Wood from the Faculty of Wood Technology in Poznan, and Remmers Company in Poland.

Conflicts of Interest: The authors declare no conflict of interest.

References

1. Ged, G.; Obein, G.; Silvestri, Z.; Rohellec, J.; Vienot, F. Recognizing real materials from their glossy appearance. *J. Vis.* **2010**, *10*, 18. [CrossRef] [PubMed]
2. Vardi, J.; Golan, A.; Levy, D.; Gilead, I. Tracing sickle-blade levels of wear and discard patterns: A new sickle gloss quantification method. *J. Archaeol. Sci.* **2010**, *37*, 1716–1724. [CrossRef]
3. Ettwein, F.; Rohrer-Vanzo, V.; Langhthaler, G.; Werner, A.; Stern, T.; Moser, O.; Leitner, R.; Regenfelder, K. Consumer's perception of high gloss furniture: Instrumental gloss measurement versus visual gloss evaluation. *Eur. J. Wood Prod.* **2017**, *75*, 1009–1016. [CrossRef]
4. Ikey, H.; Song, C.; Miyazaki, Y. Physiological effects of touching coated wood. *Int. J. Environ. Res. Public Health* **2017**, *14*, 773. [CrossRef] [PubMed]
5. Demirkir, C.; Aydin, I.; Çolak, S.; Çolakoğlu, G. Effects of plasma treatment and sanding process on surface roughness of wood veneers. *Turk. J. Agric. For.* **2014**, *38*, 663–667. [CrossRef]
6. Salca, E.A.; Krystofiak, T.; Lis, B. Evaluation of selected properties of alder wood as functions of sanding and coating. *Coatings* **2017**, *7*, 176. [CrossRef]
7. Cakicier, N.; Korkut, S.; Korkut, D. Varnish layer hardness, scratch resistance, and glossiness of various wood species as affected by heat treatment. *BioResources* **2011**, *6*, 1648–1658. [CrossRef]
8. Bekhta, P.; Proszyk, S.; Lis, B.; Krystofiak, T. Gloss of thermally densified alder (*Alnus glutinosa* Goertn.), beech (*Fagus sylvatica* L.), birch (*Betula verrucosa* Ehrh.), and pine (*Pinus sylvestris* L.) wood veneers. *Eur. J. Wood Wood Prod.* **2014**, *72*, 799–808. [CrossRef]
9. Slabejova, G.; Smidriakova, M.; Fekiac, J. Gloss of transparent coating on beech wood surface. *Acta Fac. Xylol.* **2016**, *58*, 37–44. [CrossRef]
10. Salca, E.A.; Krystofiak, T.; Lis, B.; Mazela, B.; Proszyk, S. Some coating properties of black alder wood as function of varnish type and applications method. *BioResources* **2016**, *11*, 7580–7594. [CrossRef]
11. Csanady, E.; Magos, E.; Tolvaj, L. Gloss of Colour Surfaces. In *Quality of Machined Wood Surfaces*; Springer International Publishing: Cham, Switzerland, 2015; pp. 109–181, ISBN 978-3-319-22419-0. [CrossRef]
12. Sönmez, A.; Budakçı, M.; Pelit, H. The effect of the moisture content of wood on the layer performance of water-borne varnishes. *BioResources* **2011**, *6*, 3166–3177.
13. Richter, K.; Feist, W.C.; Knaebe, M.T. The effect of surface roughness on the performance of finishes. Part 1. Roughness characterisation and stain performance. *For. Prod. J.* **1995**, *45*, 91–97.
14. Demirci, Z.; Sonmez, A.; Budakci, M. Effect of thermal ageing on the gloss and adhesion strength of the wood varnish layers. *BioResources* **2013**, *8*, 1852–1867. [CrossRef]
15. Cogulet, A.; Blanchet, P.; Landry, V. The multifactorial aspect of wood weathering: A review based on a holistic approach of wood degradation protected by clear coating. *BioResources* **2018**, *13*, 2116–2138. [CrossRef]
16. Evans, P.D.; Vollmer, S.; Kim, J.D.W.; Chan, G.; Gibson, S.K. Improving the performance of clear coatings on wood through the aggregation of marginal gains. *Coatings* **2016**, *6*, 66. [CrossRef]
17. Panek, M.; Simunkova, K.; Novak, D.; Dvorak, O.; Schonfelder, O.; Sedivka, P.; Kobeticova, K. Caffeine and TiO_2 nanoparticles treatment of spruce and beech wood for increasing transparent coating resistance against UV-radiation and mould attacks. *Coatings* **2020**, *10*, 1141. [CrossRef]
18. Kesik, H.I.; Akyildiz, M.A. Effect of the heat treatment on the adhesion strength of water based wood varnishes. *Wood Res. Slovak.* **2015**, *60*, 987–994.
19. Philipp, C. The future of wood coatings. *Eur. Coat. J.* **2010**, *1*, 18–21.
20. de Gennaro, G.; Loiotile, A.D.; Fracchiolla, R.; Palmisani, J.; Saracino, M.R.; Tutino, M. Temporal variation of VOC emission from solvent and water based wood stains. *Atmos. Environ.* **2015**, *115*, 53–61. [CrossRef]
21. Palmisani, J.; Di Gilio, A.; Cisternino, E.; Tutino, M.; de Gennaro, G. Volatile Organic Compound (VOC) emissions from a personal care polymer-based item: Simulation of the inhalation exposure scenario indoors under actual conditions of use. *Sustainability* **2020**, *12*, 2577. [CrossRef]
22. Bila, N.F.; Trianoski, R.; Egas, A.F.; Iwakiri, S.; Rocha, M.P. Evaluation of the quality of surface finish of messassa wood for three types of varnishes. *J. Biotechnol. Biodivers.* **2020**, *8*, 122–130. [CrossRef]
23. Kudela, J.; Kubovsky, I. Accelerated-aging-induced photo-degradation of beech wood surface treated with selected coating materials. *Acta Fac. Xylo* **2016**, *58*, 27–36. [CrossRef]
24. Kubovski, I.; Kacik, F. Changes of the wood surface colour induced by CO_2 laser and its durability after the xenon lamp exposure. *Wood Res.* **2013**, *58*, 581–590.
25. Liu, R.; Zhu, H.; Li, K.; Yang, Z. Comparison on the aging of woods exposed to natural sunlight and artificial Xenon Light. *Polymers* **2019**, *11*, 709. [CrossRef]

26. Muller, U.; Ratzsch, M.; Schwanninger, M.; Steiner, M.; Zobi, H. Yellowing and IR changes of spruce wood as a result of UV-irradiation. *J. Photochem. Photobiol. B Biol.* **2003**, *69*, 97–105. [CrossRef]
27. Evans, P.D.; Thay, P.D.; Schmalzl, K.J. Degradation of wood surfaces during natural weathering. Effects on lignin and cellulose and on the adhesion of acrylic latex primers. *Wood Sci. Technol.* **1996**, *30*, 411–422. [CrossRef]
28. Lourenco, A.; Pereira, H. Compositional Variability of Lignin in Biomass. Lignin-Trends and Applications. Matheus Poletto, IntechOpen. Available online: https://www.intechopen.com/books/lignin-trends-and-applications/compositional-variability-of-lignin-in-biomass (accessed on 24 April 2021). [CrossRef]
29. Pandey, K.K. Study of the effect of photo-irradiation on the surface chemistry of wood. *Polym. Degrad. Stab.* **2005**, *90*, 9–20. [CrossRef]
30. Kropat, M.; Hubbe, M.A.; Laleicke, F. Natural, accelerated, and simulated weathering of wood: A review. *Bioresources* **2020**, *15*, 9998–10062.
31. Irmouli, Y.; George, B.; Merlin, A. Artificial ageing of wood finishes monitored by IR analysis and color measurements. *J. Appl. Polym. Sci.* **2012**, *124*, 1938–1946. [CrossRef]
32. Nejad, M.; Shafaghi, R.; Ali, H.; Cooper, P. Coating performance on oil-heat treated wood for flooring. *BioResources* **2013**, *8*, 1881–1892. [CrossRef]
33. Williams, R.S. Finishing of Wood. In *Wood Handbook—Wood as an Engineering Material*; General Technical Report FPL–GTR–113; U.S. Department of Agriculture, Forest Service, Forest Products Laboratory: Madison, WI, USA, 1999; 463p.
34. Pavlic, M.; Petric, M.; Zigon, J. Interactions of coating and wood flooring surface system properties. *Coatings* **2021**, *11*, 91. [CrossRef]
35. Bulian, F.; Graystone, J.A. *Wood Coatings—Theory and Practice*; Elsevier: Amsterdam, The Netherlands, 2009.
36. Vidholdova, Z.; Slabejova, G.; Smidriakova, M. Quality of oil- and wax-based surface finishes on thermally modified oak wood. *Coatings* **2021**, *11*, 143. [CrossRef]
37. Paprzycki, O.; Proszyk, S.; Przybylak, A. *Materials for the Exercises from the Technologies of the Finishing of Wood and Wood Based Materials Surfaces*; Agriculture Academy: Poznan, Poland, 1985; p. 141. (In Polish)
38. DIN 53211. *Testing of Paints, Varnishes and Similar Products. Determination of the Flow Time by the DIN 4 Cup*; German Institute for Standardization: Berlin, Germany, 1974.
39. ISO 2813. *Paints and Varnishes—Determination of Gloss Value at 20 Degrees, 60 Degrees and 85 Degrees*; International Organization for Standardization: Geneva, Switzerland, 2014.
40. EN 12722. *Furniture-Assessment of Surface Resistance to Dry Heat*; European Standardization Organizations: Brussels, Belgium, 2009.
41. EN 12720. *Furniture-Assessment of Surface Resistance to Cold Liquids*; European Standardization Organizations: Brussels, Belgium, 2009.
42. Pelit, H.; Budakci, M.; Sonmez, A.; Burudurlu, E. Surface roughness and brightness of scots pine (*Pinus sylvestris*) applied with water-based varnish after densification and heat treatment. *J. Wood Sci.* **2015**, *61*, 586–594. [CrossRef]
43. Saeed, A.; Shabir, G. Synthesis of thermally stable high gloss water dispersible polyurethane/polyacrylate resins. *Prog. Org. Coat.* **2013**, *76*, 1135–1143. [CrossRef]
44. Lee, S.S.; Koo, J.H.; Lee, S.S.; Chai, S.G.; Lim, J.C. Gloss reduction in low temperature curable hybrid powder coatings. *Prog. Org. Coat.* **2003**, *46*, 266–272. [CrossRef]
45. Herrera, R.; Sandak, J.; Robles, E.; Krystofiak, T.; Labidi, J. Weathering resistance of thermally modified wood finished with coatings of diverse formulations. *Prog. Org. Coat.* **2018**, *119*, 145–154. [CrossRef]

Article

Hybrid Approach for Wood Modification: Characterization and Evaluation of Weathering Resistance of Coatings on Acetylated Wood

Anna Sandak [1,2], Edit Földvári-Nagy [1], Faksawat Poohphajai [1,3], Rene Herrera Diaz [1,4], Oihana Gordobil [1], Nežka Sajinčič [1], Veerapandian Ponnuchamy [1,5] and Jakub Sandak [1,5,*]

1 InnoRenew CoE, Livade 6, 6310 Izola, Slovenia; anna.sandak@innorenew.eu (A.S.); edit.foldvari-nagy@innorenew.eu (E.F.-N.); faksawat.poohphajai@innorenew.eu (F.P.); rene.herdiaz@innorenew.eu (R.H.D.); oihana.gordobil@innorenew.eu (O.G.); nezka.sajincic@innorenew.eu (N.S.); veerapandian.ponnuchamy@innorenew.eu (V.P.)
2 Faculty of Mathematics, Natural Sciences and Information Technologies, University of Primorska, Glagoljaška 8, 6000 Koper, Slovenia
3 Department of Bioproducts and Biosystems, School of Chemical Engineering, Aalto University, P.O. Box 16300, 00076 Aalto, Finland
4 Department of Chemical and Environmental Engineering, University of the Basque Country, Plaza Europa, 1, 20018 Donostia-San Sebastian, Spain
5 Andrej Marušič Institute, University of Primorska, Titov trg 4, 6000 Koper, Slovenia
* Correspondence: jakub.sandak@innorenew.eu; Tel.: +386-40282959

Citation: Sandak, A.; Földvári-Nagy, E.; Poohphajai, F.; Diaz, R.H.; Gordobil, O.; Sajinčič, N.; Ponnuchamy, V.; Sandak, J. Hybrid Approach for Wood Modification: Characterization and Evaluation of Weathering Resistance of Coatings on Acetylated Wood. *Coatings* **2021**, *11*, 658. https://doi.org/10.3390/coatings11060658

Academic Editor: Salim Hiziroglu

Received: 15 May 2021
Accepted: 28 May 2021
Published: 30 May 2021

Abstract: Wood, as a biological material, is sensitive to environmental conditions and microorganisms; therefore, wood products require protective measures to extend their service life in outdoor applications. Several modification processes are available for the improvement of wood properties, including commercially available solutions. Among the chemical treatments, acetylation by acetic anhydride is one of the most effective methods to induce chemical changes in the constitutive polymers at the cellular wall level. Acetylation reduces wood shrinkage-swelling, increases its durability against biotic agents, improves UV resistance and reduces surface erosion. However, even if the expected service life for external cladding of acetylated wood is estimated to be 60 years, the aesthetics change rapidly during the first years of exposure. Hybrid, or fusion, modification includes processes where the positive effect of a single treatment can be multiplied by merging with additional follow-up modifications. This report presents results of the performance tests of wood samples that, besides the modification by means of acetylation, were additionally protected with seven commercially available coatings. Natural weathering was conducted in Northern Italy for 15 months. Samples were characterized with numerous instruments by measuring samples collected from the stand every three months. Superior performance was observed on samples that merged both treatments. It is due to the combined effect of the wood acetylation and surface coating. Limited shrinkage/swelling of the bulk substrate due to chemical treatment substantially reduced stresses of the coating film. Hybrid process, compared to sole acetylation of wood, assured superior visual performance of the wood surface by preserving its original appearance.

Keywords: wood modification; natural weathering; acetylation; coatings; service life performance; aesthetic

1. Introduction

Biodegradability, identified as an important advantage of biomaterials from a sustainability perspective, is its biggest weakness when considering materials durability. It is especially relevant to all wooden elements or timber structures that are exposed outdoor to environmental factors like humidity/temperature variation, wetting by rain or UV radiation [1]. The selection of naturally resistant species or improvement of the less durable by means of modification processes is a common solution to minimize degradation of biomaterials in service [2]. However, all actions related to the chemical treatment of biomaterials should be carefully implemented by balancing the benefits with costs and other possible negative consequences. An optional solution is to program a sequence of maintenance actions where façade elements may undertake cleaning, painting or replacement operations. In the majority of cases, wood weathering is an uneven process resulting in diverse deterioration kinetics in different zones on the building [3]. The surface protected by the roof or other architectural detail changes slower than the fully exposed. The same trend can be associated with the exposure side, with cardinal directions resulting in diverse doses of deteriorating factors [4–6]. Wood, as well as other hygroscopic materials, is subject to dimensional distortions caused by changes in the moisture content, air humidity and temperature. Even if a properly designed structure may not be affected by the moisture-induced shrinkage, the dimensional changes may affect the surface coatings creating excessive stresses followed by surface cracks [7,8]. It alters the protection function of the coating and may promote even more excessive decay through the moisture bridges. It is essential to implement best practices of the "protection by design" paradigm to project and construct façade details that allow or compensate for dimensional distortions and minimize related treats [9].

The most critical limitation of using non-protected wood on the building façade is its aesthetical instability, evidenced as appearance changes over time [10]. All materials used on the façade change their outlook during service, but alterations of the majority of bio-based materials are usually more pronounced and occur more rapidly compared to ones that are not bio-based. Under certain circumstances, this may be considered as an advantage, especially when it is a part of the design strategy or intended rapid homogenization with the local context [11]. However, in the majority of cases, the visual changes on the wooden façade claddings are perceived negatively [12]. Unprotected wood exposed to UV rays ultimately changes its colour turning toward grey tonality. In some cases, it may crack or splits at the ends, especially when the façade is not properly designed or boards are wrongly installed. There are several factors affecting weathering intensity of biomaterials surface, such as location, exposure direction, specific microclimate, architectural details, orientation/position of the assembly, use condition or material intrinsic properties [10]. The biotic attack changes surface colour resulting in green or grey-blue spots on the surface of the material [13,14]. Glossy and shiny appearance may turn over time into the matte surface. After a sufficiently long period of moulds' growth, the component may require replacement due to its aesthetical appearance, even though it can still fulfil other functions such as load-bearing capacity, protecting the building envelope from weathering or providing shelter [15].

Modification processes lead to the enhancement of selected wood properties by means of chemical, biological or physical agents. Several alternative treatments, including both active and passive modifications, are recently available [16,17]. The active modification changes the chemical nature of material through chemical, thermo-hygro-mechanical or enzymatic treatments. Conversely, the passive modification does not alter material chemical composition but rather deposits selected functional molecules by means of bulk impregnation or surface treatments [16]. Consequently, various properties of wood are changed to a different extent depending on the modification process and its intensity.

Chemical modification is the result of a chemical agent reaction with wood chemical components resulting in the formation of covalent bonds. Acetylation is the most well-established treatment where acetic anhydride reacts with hydroxyl groups of the cell wall polymers by forming ester bonds. Acetylation slightly improves UV resistance and reduces surface erosion by 50%, which is important while using wood as façade material [18]. The mechanical strength properties of acetylated wood are not considerably different than not-treated wood; however, its durability is substantially improved [19]. Besides bulk modification processes, several techniques affect properties of the surface only without interfering with the interior of the piece [20,21]. The changes of the surface functionalities affected by the exterior treatments include UV stabilization (e.g., surface esterification), an increase of hydrophobicity (e.g., reaction with silicone polymers) or improvement of the adhesion (e.g., enzymatic treatment, plasma discharge) [22–24]. In addition, alternative processes can be applied with the purpose to improve biomaterial's surface resistance against biotic and abiotic factors. This includes surface coating [8,20] as well as other treatments such as surface densification [25] or surface carbonization (e.g., Shou Sugi Ban) [26]. Surface finishing by diverse coatings, waxes, oils or stains is the most common procedure of the surface treatment, which is highly influencing its service life performance [27]. The resistance of the surface against deterioration in service highly depends on the finishing product quality (chemical formulation), surface preparation (oxidation stage, roughness, wettability, surface free energy) and the application procedure (industrial coating, immersing, brush or spray) [28]. A large variety of commercially available products for surface finishing allow obtaining different appeal, including colour variation, transparency or gloss. The proper use of the surface finishing technologies may contribute significantly to the aesthetical attractiveness of the structure as well as the appearance changes along the service life of the façade [21]. The cost of the finish, including proper surface pre- and post-treatment, may be considerable. However, in a majority of cases, it is economically viable to increase the initial cost of the façade by increasing the thickness of the coating layer, as it may significantly increase the time of maintenance-free use. It has been reported that the change of the coating layer from 30 μm to 50 μm increases the time of the surface resistance for cracking, and the service life period, by a factor of 1.2 [29].

The bulk and surface modification processes may affect one or more functionalities of the biomaterial used for building façade. Even if it improves certain material assets, the positive effect can be multiplied by merging two or more modification processes. Such an approach turns out to be a "hybrid process" and becomes an optimal solution frequently implemented by bio-materials producers [10]. An example of a successful hybrid modification is the surface coating of acetylated or thermally treated wood. The combined effect of the reduced shrinkage/swelling of the bulk substrate and water protecting coating significantly reduces stresses of the coating film preventing it from cracking. Consequently, the surface façade remains intact for a longer period, preserving its original attractive appearance. Benefits obtained by merging different materials and treatments are highly useful to address design limitations and biomaterials' deficiencies. If properly implemented, hybrid modifications can contribute to reducing the environmental burden and economic cost of the façade. It has to be mentioned, however, that some of the modification processes cannot be merged or may affect undesired changes of other material properties. An example may be an increase of the biomaterial brittleness after some modifications that affect its machinability or paint-ability [17]. For that reason, special attention should be directed toward selecting appropriate treatment combinations and extensive quality control of the hybrid modification processes. The goal of this research was to evaluate the combined effect of wood acetylation followed by seven industrial coatings on wood performance when exposed to natural weathering.

2. Materials and Methods

2.1. Experimental Samples

Acetylated wooden boards manufactured from radiata pine (*Pinus radiata* D. Don), were used for the preparation of experimental samples. The acetylation and coating application was performed in industrial environments by the collaborating companies. One hundred forty–four small blocks (150 × 75 × 20 mm^3, length × width × thickness, respectively) were cut out and randomly separated into two sets. The first set included eighteen samples that were not surface finished and used for the following tests without additional treatments. The second set contained hundred twenty–six samples that were subjected to the supplementary surface coating procedure. It was implemented as a part of the industrial process by applying selected coatings usually used for building façades. Seven alternative coating products manufactured by three producers were tested and confronted; 144 (= 8 × 6 × 3) samples were analysed in total, including 8 sample types (#A, #B, #C, #D, #E, #F, #G, #H), 6 natural weathering scenarios (0, 3, 6, 9, 12 and 15 months) and 3 replicas for each testing scenario. The summary of technical details regarding experimental samples is presented in Table 1.

Table 1. Summary of acetylated Radiata pine samples tested within natural weathering experiment.

Sample Code	Surface Coating	Appearance
A	no treatment	natural wood
B	hybrid hydro-oil (producer #1)	transparent brown, mat
C	preserving stain emulsified with hydro-oil (producer #1)	transparent brown, glossy
D	water-based (producer #1)	white, opaque
E	waterborne high-gloss (producer #2)	transparent yellow, glossy
F	water-based VV MAT (producer #2)	semi-opaque yellow, glossy
G	water-based (producer #3)	white, opaque
H	water-based impregnating agent (producer #3)	transparent brown, glossy

2.2. *Weathering Tests*

Natural weathering tests were performed in San Michele, Italy (46°11′15″ N, 11°08′00″ E). The objective of this test was to collect a set of reference data regarding material performance at the different exposure time. Samples were exposed on vertical stands representing a building façade. The stand was oriented to face the southern direction. The weathering experiment was carried out for a total period of 15 months, starting in March 2017. Three replicas were collected from the stand each third month to stop the deterioration progress. Consequently, a collection of samples exposed to 0, 3, 6, 9, 12 and 15 months was gathered. Each exposure period corresponds to a single experimental scenario, and three replica samples assured insight on the reliability and repeatability of results. All samples were stored in a climatic chamber (20 °C, 65% RH) to stabilize conditions before follow-up measurements.

2.3. *Characterization Methods*

2.3.1. Digitalization and Colour Measurement

Samples after conditioning were scanned with an office scanner HP Scanjet 2710 (300 dpi, 24 bit) and saved as TIF files. Colour changes were assessed by means of a spectrometer following the *CIE Lab* system where colour is expressed with three parameters: L^* (lightness), a^* (red-green tone) and b^* (yellow-blue tone). *CIE* $L^*a^*b^*$ colours were measured using a MicroFlash 200D spectrophotometer (DataColor Int, Lawrenceville, Illinois, USA). The selected illuminant was D65 and the viewer angle was 10°. All specimens were measured on five randomly selected spots over the weathered surface. The mean values were considered as a representative colour, even if the maximum and minimum readings were preserved to assess the natural variation of the colour distribution.

2.3.2. Gloss

The mode of light reflection from the surfaces was measured using a REFO 60 (Dr. Lange, Düsseldorf, Germany) gloss meter with incidence and reflectance angles of 60°. Ten measurements along and across the fibre direction were taken on each specimen to address the optical heterogeneity of the light reflectance from the wood surface.

2.3.3. Microscopic Observation and 3D Roughness Measurement

Keyence VHX-6000 digital microscope (Keyence, Osaka, Japan) was used for microscopic observation, high magnification image acquisition and 3D surface topography scanning. Colour images were collected with an optical configuration corresponding to ×30 and ×200 magnifications. The light direction and intensity were adjusted to assure a wide dynamic range of the image and avoid saturated pixels generation. Part of the high magnification images was acquired in the real-time 3D depth reconstruction mode. It allowed post-processing of data to determine surface profile as well as surface roughness indicators. For that reason, an area of 2×2 mm^2 was assessed while series of 3D images stacked together. The proprietary software of the Keyence microscope was used for roughness data post-processing. The protocol included removal of the error of form (plane extraction) and filtering of the surface topography data with a Gaussian band pass filter (2 μm < λ < 0.8 mm). The surface roughness quantifiers determined on such prepared data included arithmetical mean height (*Sa*), skewness (*Ssk*) and kurtosis (*Skt*).

2.3.4. Contact Angle and Surface Free Energy

Dynamic contact angle measurements were performed using optical tensiometer Attention Theta Flex Auto 4 (Biolin Scientific, Gothenburg, Sweden). Five replica measurements were performed implementing the sessile drop method on each specimen with a sequence of distilled water and formamide droplets. The dosing volume of each drop was 4 μL controlled by both precise dispenser and drop image analysis. The measurement of the drop shape started at the moment of the initial drop contact with the assessed sample surface and lasted for 20 s. The series of images collected were post-processed with the proprietary software of the tensiometer. The Laplace equation was used for the estimation of the contact angle for both investigated liquids. The series of contact angles observed at 3 s after dispensing were averaged to reduce the scatter of results. However, the range (minimum to maximum) of the observed contact angles was also recorded for further analysis. The surface free energy was computed following OWRK/Fowkes method [30]. The total surface free energy (γ^{tot}), as well as its polar (γ^{p}) and disperse (γ^{d}) components, were determined for all investigated samples.

3. Results

The appearance alteration of investigated samples after natural weathering conducted at diverse exposure times is shown in Figure 1. The colour changes are mainly observed for uncoated samples (#A), appearing lighter after three months of exposure. The colour is maintained for the three following months. However, a presence of mould is noticed at month 9 (December). The mould area gradually increased with the subsequent weathering duration. The overall tonality of uncoated samples at the end of the natural weathering test (month 15) became grey. It should be mentioned that in parallel to the above-mentioned colour alterations, samples #A exhibit minor disintegration of surface, evidenced by raised fibres and small cracks. All coated samples performed well, considering the overall appearance alteration as evaluated by the visual assessment. No sign of deterioration was observed in samples #D and #G that were coated with a white and non-transparent film. Similarly, dark brown samples #B and #H did not noticeably change. Conversely, samples #E and #F were coated with a product of original yellow tonality. The colour of coated sample #E become lighter during the initial weathering period to maintain the same appearance afterwards. The colour of coating #C appeared to be constant along the whole testing period.

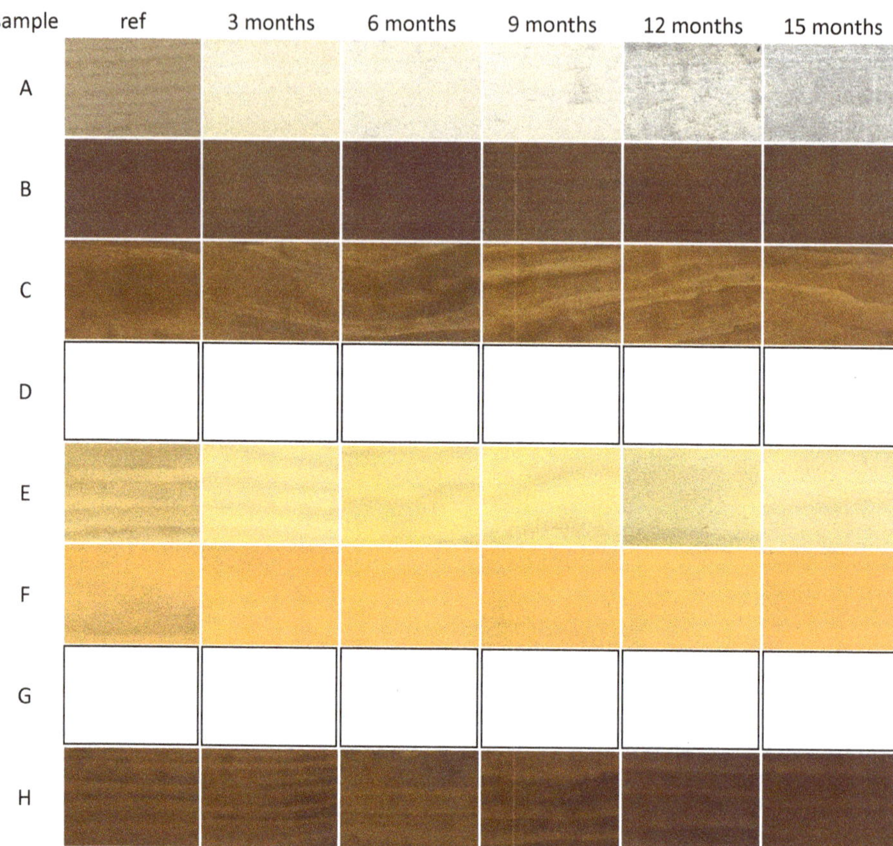

Figure 1. Macro colour images of acetylated wood samples exposed to natural weathering.

Figure 2 present colourimetry changes of acetylated wood samples exposed to natural weathering. In analogy to the visual observations, *CIE* L^*, a^* and b^* are nearly constant for the majority of coated samples along the entire period of the test. It was noticed particularly for samples #D, #G and #H, which are considered here as the most colour-stable. These contain white opaque coatings (provided by producers 1 and 3) as well as a transparent glossy brown coating (producer 3). Slightly higher fluctuations were observed for samples #B and #F. Similar trends were noticed for coated samples #C and #E. *CIE* L^* and *CIE* b^* values slightly increased during the weathering period in the case of sample #C, while *CIE* a^* was relatively constant. A similar tendency was observed in sample #E, where *CIE* L^* and *CIE* a^* values gradually decreased. *CIE* b^* slowly decreased after the initial increase observed at month 3, to reach almost initial values afterwards. The highest changes of colour indicators were noticed for uncoated samples (#A). The apparent lightness (*CIE* L^*) noticed for the acetylated radiata pine was relatively stable, with only a slight and steady rise at the initial phase of the weathering test. *CIE* a^* gradually decreased for all uncoated samples, reaching more stable values from month 6 of the exposure. Values of *CIE* b^* progressively dropped after a slight gradual increase at the beginning of the weathering test.

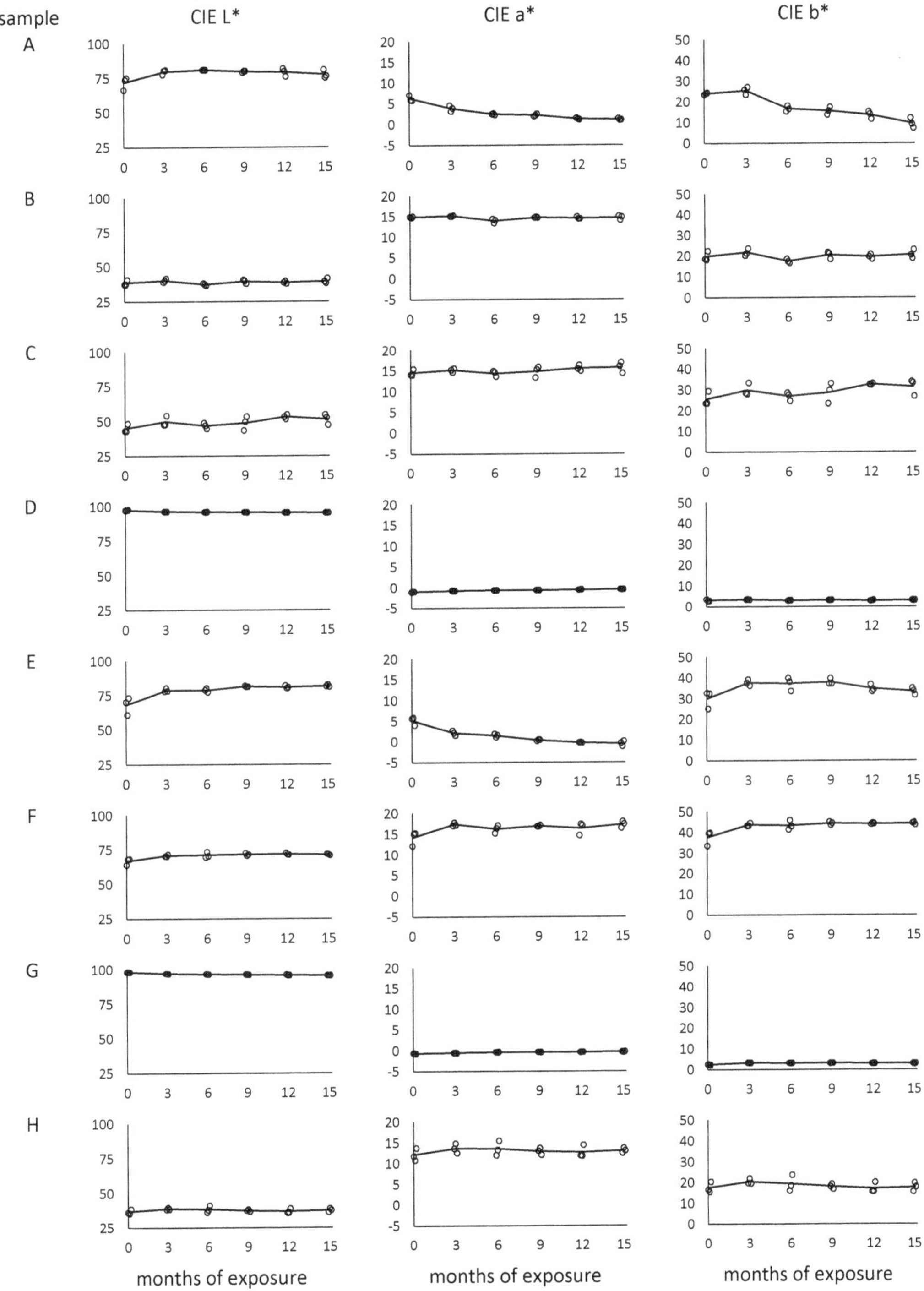

Figure 2. Colour changes of acetylated wood samples exposed to natural weathering. A–H: sample type.

Gloss measured along and perpendicular to the fibres (Figure 3) was, in a majority of the investigated cases, very similar. An exception was sample #F where gloss values measured along the fibres were approximately 10 units higher than perpendicular. For coated sample #B, as well as for uncoated samples #A values were rather constant along the whole weathering period. A slight decrease was observed for coated samples #C, #D and #H, contrary to #E, where gloss value increased for about 10 units during the 15 months. Samples #F and #G showed some fluctuations during the weathering period, even if the gloss at the end of the test was close to the initial value. The gloss mainly depended on the performance of the coating and no interaction with the acetylated substrate was evidenced.

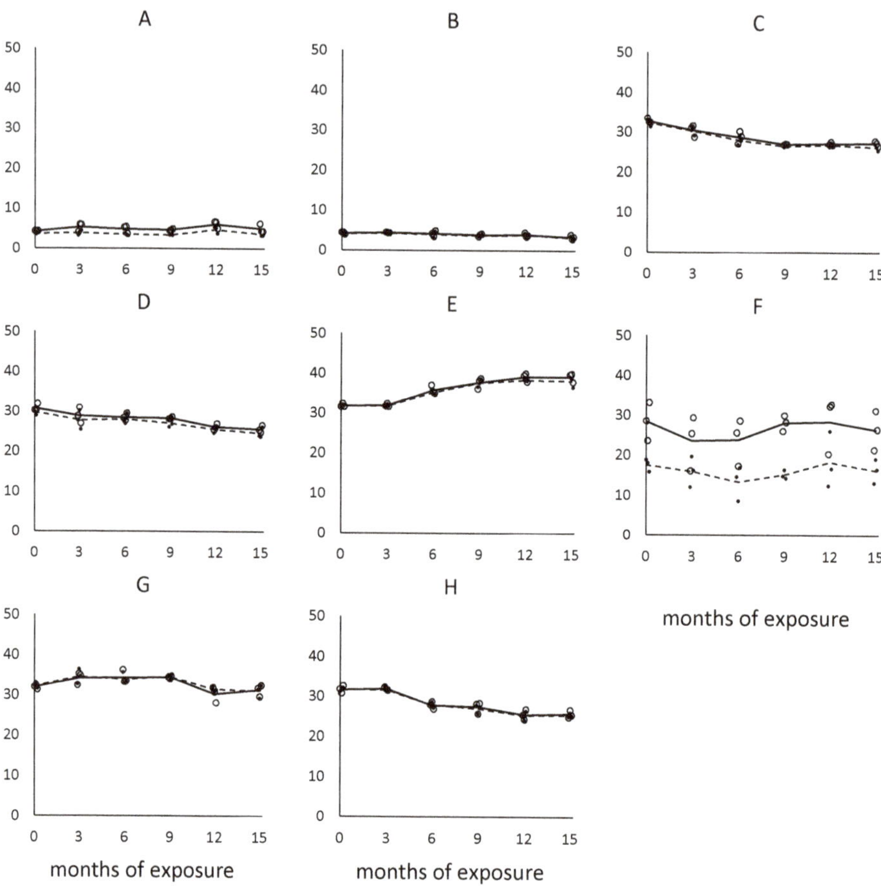

Figure 3. Gloss changes of acetylated wood samples exposed to natural weathering (note: solid line–gloss along fibres, dash line–gloss perpendicular to fibres). A–H: sample type.

The surface roughness of the coated wood samples (#B–#H) was not measurably altered during the natural weathering test period. However, an apparent increase in the surface roughness was noticed in uncoated specimens. The result of surface topography analysis performed for samples #A is presented in Figure 4. The 3D topography map aligned with the colour image allows identification of the large tracheids as a major surface irregularity source. However, clear progress of the surface erosion is evident in the analysis of the surface profile outline. It is confirmed as steady progress of the surface roughness parameters. Three-dimensional (3D) areal surface texture assessment is a favourable approach to characterise surfaces of heterogenic and anisotropic materials, such as wood. Arithmetical mean surface height (*Sa*) is the basic irregularity quantifier corresponding to the arithmetical mean height of the roughness profile (*Ra*) traditionally determined from the two-dimensional surface roughness profiles. Values of *Sa* remained relatively constant for acetylated radiata pine. Skewness (*Ssk*) represents the degree of the roughness distribution bias or asperity. Negative skewness observed for all measured samples indicates a deviation to the higher side of the topography map that is typical for porous materials, such as wood. The steady decrease of skewness indicates a shift of the top material ratio toward a central distribution that can be associated with the loss of fibres and general progress of the uncoated wood surface erosion. Kurtosis (*Skt*) is a measure of the sharpness of the topography histogram profile. Values of *Skt* > 3 correspond to leptokurtic distribution with tails fatter than in normal/Gaussian curves. The spiked nature of the surface irregularity histogram is a consequence of the flat samples' surfaces generated by the planning operation. The trend of *Skt* alterations indicates that the curve tended toward platykurtic distribution and a more balanced contribution of the surface peaks and valleys.

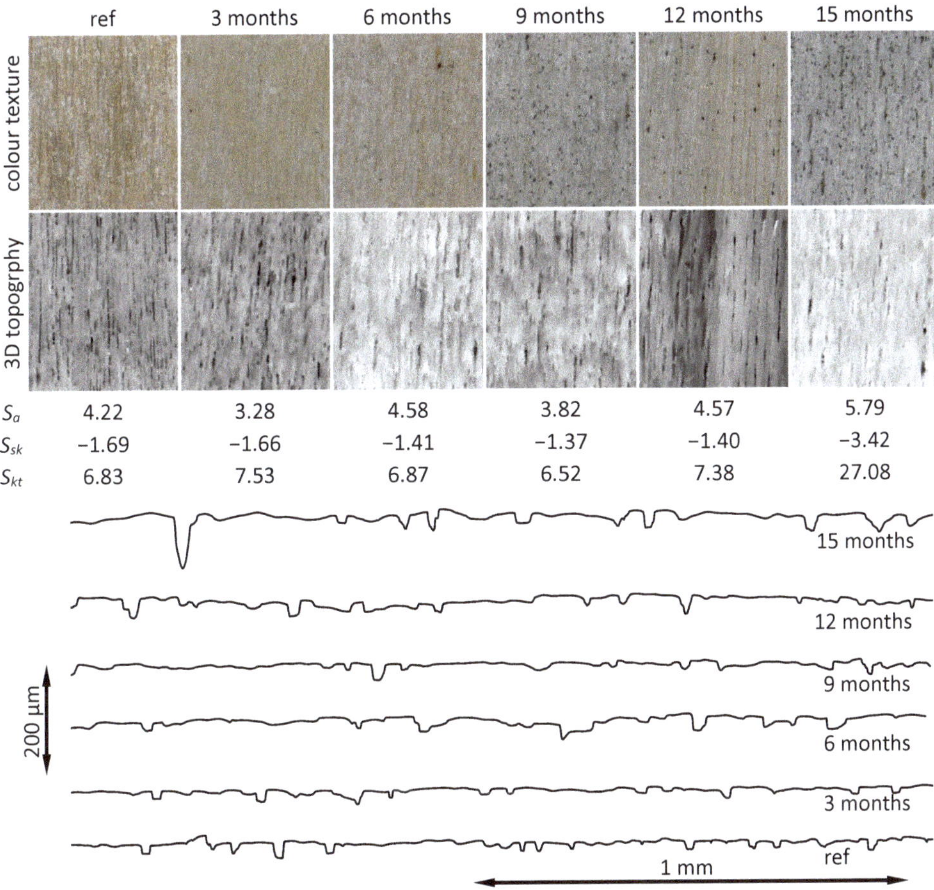

Figure 4. 3D surface topography map, typical surface profiles and high magnification images of the acetylated radiata pine wood (sample #A) exposed to natural weathering for a period of 15 months.

An interesting observation is related to the dark spots present on the high magnification colour images. These correspond to the surface contamination and presence of mould spores. Such spores gradually appeared with the progress of the weathering test that was also perceived on the macro-scale images presented in Figure 1. An evident increase of the infestation kinetics was noticed for all samples exposed for one year or more.

Contact angle (θ) is a quantitative measure of the wetting ability of a solid surface by a liquid. It is defined geometrically as the angle formed by a liquid at the three-phase boundary where the liquid, gas and solid intersect. Results of the dynamic contact angle measurement when wetting the surface of experimental samples with water and formamide are presented in Figure 5. Low contact angle values indicate that the liquid easily spreads over the assessed surface. Conversely, high contact angle implies poor spreading and physical affinity. The high value of θ observed for most of the coated wood samples indicates low wettability by both tested liquids. In all cases, the contact angle measured with water was higher than that with formamide. No clear pattern of the θ changes can be observed on the coated wood samples, even if an initial increase can be noticed in samples #C, #D, #G and #H. A change of the water contact angle at the initial period of the weathering test was observed on the uncoated wood samples (#A). In fact, the surface was so easily wetted by both water and formamide that sessile drops disappeared after a few seconds of the test.

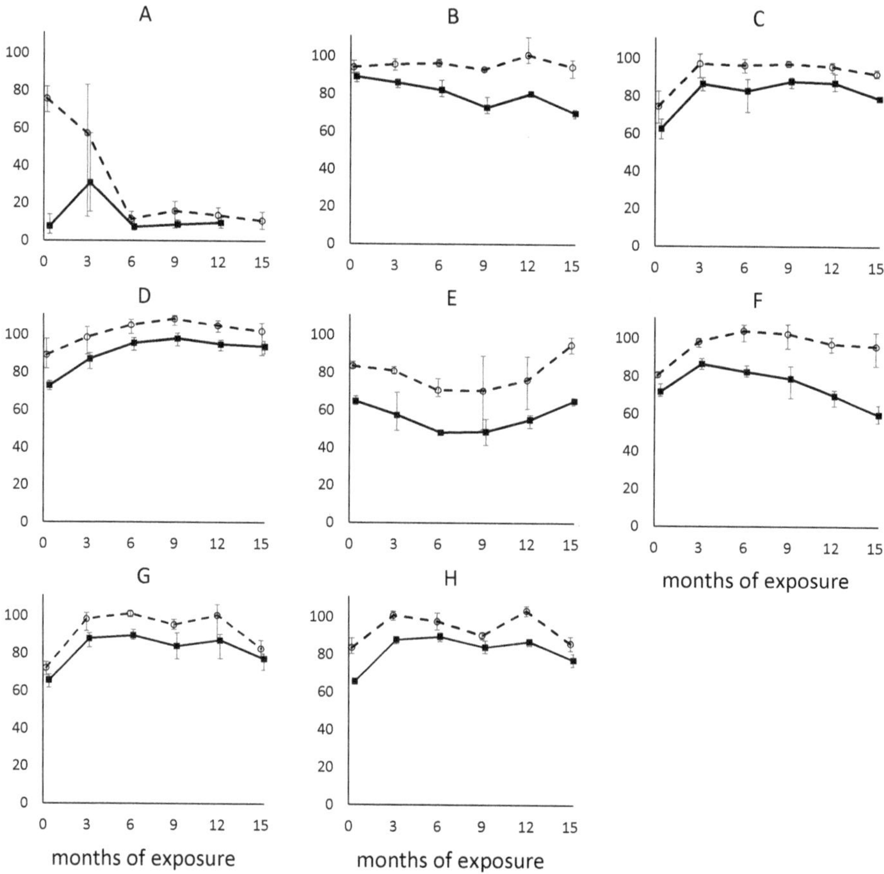

Figure 5. Contact angle measured after 3 s of wetting with water (white circle) and formamide (black square). A–H: sample type.

A direct implication of the contact angle measurement is the possibility of the surface free energy (SFE) estimate. SFE was determined following OWRK/Fowkes energy theory that is a recommended method for the coated surfaces. Results of the SFE determined for each measured sample are presented in Figure 6, including total surface free energy (γ^{tot}) that contains polar (γ^{p}) and disperse (γ^{d}) components. Values of the SFE were highest for the uncoated wood samples and oscillated around 70 mJ·m^{-2}. Even with that, the disperse component dominated in samples #A and #F. An increase of the SFE with the progress of weathering was noticed for coated sample #B and #F. Conversely, the SFE value decreased along with the weathering progress for coated samples #C, #D and #G, while it fluctuated in the case of #E and #H. An increased dispersive part of the modified substrate (acetylated radiata pine wood) and a lower surface free energy after coating application, positively contributed to a stable performance of the coated samples after weathering.

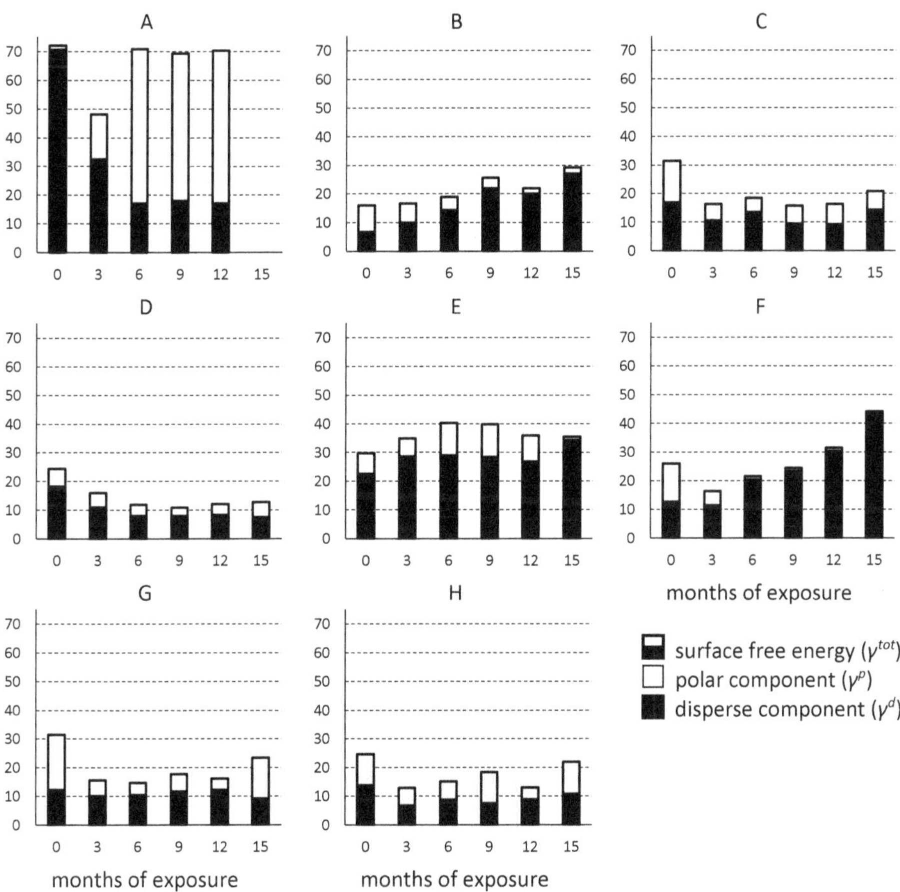

Figure 6. Surface free energy (γ^{tot}) including polar (γ^{p}) and disperse (γ^{d}) components. A–H: sample type.

4. Discussion

All evaluated hybrid modified systems revealed a satisfactory performance during natural weathering tests conducted within this study. Four of seven coatings were transparent, one semi-transparent and two opaques. Transparent coatings are frequently considered attractive solutions for customers since they maintain the natural texture of wood. Though, from the practical implementation perspective, such coating solutions are not recommended unless regular maintenance is accepted and regularly carried out [31]. Previous studies demonstrated that clear (transparent) coatings failed to give satisfactory performance after just eight months of exposure [21,32]. A detailed overview of the clear coatings' performance applied on the wood surface was provided by Estrada [33]. It was stated that unacceptable appearance is commenced in a relatively short period of 12 to 24 months, depending on the climatic conditions. The main disappointing issue disqualifying clear coatings is associated with the unsatisfactory appeal and necessity for excessive maintenance related to the removal of the old coating layers before re-coating. Clear coatings contain binders that are usually susceptible to photodegradation and/or hydrolysis. That leads to the failure not only to the coating layer but also to the degradation of the underlying wood. It results in discolouration (yellowing) of the surface outlook, loss of glossiness and alteration of the mechanical and physical properties of the coating film [21]. An effective method for improvement of the coating performance is dimension stabilization of the substrate material. The swelling and shrinkage of wood due to changes in its moisture content leads to the generation of checks and cracks within the coating film. Several methods successfully implemented to stabilize the wood and improve the performance of clear coatings were identified and tested experimentally [16,21]. Bulking the wood cell wall with polyethylene glycol or pre-treatment with water repellents as well as various wood modification and impregnation treatments are the best performing solutions. The chemical modification of wood by means of acetylation researched in this study is an example of such an innovative technological approach. It leads to the drastic reduction of native wood hygroscopicity and, consequently, minimizing its shrinkage/swelling. Wood species used as a substrate, type of applied coating, architectural detailing and exposure configuration are known to determine the mould growth on the weathered surfaces [34,35]. No apparent mould presence was noticed in this experiment for all coating systems investigated. Though, evident mould growth was detected after nine months of exposure on the surfaces of uncoated wooden specimens. A similar natural weathering testing campaign with acetylated radiata pine cladding samples was performed in Rotorua (New Zealand) [36]. The first sign of mould presence, as well as slight surface discoloration (lightening), were reported there after only eight weeks of exposure. Conversely, lightening of the acetylated wood surface is declared in the experiment presented here as occurring after the initial three months of exposure. It is related to the methodological approach adopted. Investigated samples were assessed here with scientific instruments after the weathering test (post–factum) by using specimens collected at the three-month interval. Nevertheless, high-resolution images of the same weathered materials exposed on the stand were additionally taken with the frequency of one month [37]. The following image analysis allowed the determination of the red/green/blue (RGB), as well as hue/saturation/luminosity (HSL) colour descriptor variations. It was found that RGB and HSL paraments measured on uncoated acetylated wood samples revealed noticeable alterations after just one month of exposure. Similar research on the natural weathering performance of five wood species coated with 12 diverse formulations was conducted in Oregon (US) for a period of 18 months [38]. It was reported that most of the tested coatings lost their protective effects within the first year of exposure due to combinations of diverse biological and abiotic factors. No single coating system was identified as a superb protective solution for the untreated wood substrate.

Changes in the surface roughness along the weathering progress are associated with the vulnerability to accumulate dirt, moisture and pollution as well as enhancing the spore attachment and germination [14]. Likewise, higher abrasion resistance and improved ability to shed environmental residues deposited on the coating are limiting the ability of fungi to grow on and into the coating [39]. The roughness of hybrid-modified wood surfaces investigated in this research was constant along with the exposure duration. It resulted in negligible appearance changes and minimal dirt/mould presence observed on the coated surface. However, an increase in the surface roughness was observed in uncoated wood specimens. Two from three parameters describing the surface topography

(*Ssk* and *Sku*) changed during weathering progress. It was associated with the removal of the single fibres, leaching of photodegradation components and general erosion of the wood surface. Similar observations were derived in other reports where the roughness of uncoated samples gradually increased with the progress of the weathering process [1].

The change of the surface roughness directly affects the physical characteristics of the light reflected from the surface. It is revealed as gloss changes associated with the abrasion of wood surfaces and the accompanying erosion processes [40,41]. Loss of gloss is an indicator of degradation at its early stage. It is caused by either non-chemical changes (e.g., cracking, checking) or by chemical changes occurring in the topmost portion of the coating layer [42,43]. Therefore, observation of the gloss changes is highly useful for assessing and evaluating coating systems. It was evidenced by Pánek et al. that gloss change is more sensitive to coating degradation than total colour difference measured on the same surface set [44]. Oberhofnerová et al. observed decreased gloss values during natural and artificial weathering of eight transparent and pigmented coating systems [45]. The gloss variation recorded in the hybrid-modified wood samples evaluated in this experiment were minor. It indicates the high stability of the coating systems applied and their satisfactory performance.

The contact angle (θ) measurement implies the wettability by water (or other liquids) of the exposed material surface. The extent of θ changes is an important indicator of weathering progress [46]. The kinetic and trends of contact angle changes observed in this research for formamide were similar to those of water. Though, the nominal values were higher when wetting the wood surface with water. It is explained by the strong hydrogen bonding of formamide, which reduces the interfacial free energy at the liquid-solid interface through the acid-base interactions [47]. Surface free energy can be calculated by using the θ data when combing at least two wetting liquids of diverse physical characteristics. This parameter can be utilized to better understand the interactions between solid and liquid. In general, the weathering process of unprotected wood increases wettability. Leaching of the extractives from the surface of weathered wood reduces water repellence, while the degradation of lignin results in a more hydrophilic surface [48]. Different performance trends regarding wettability were reported in relation to the coating transparency and composition [45]. Even if applying the hydrophobic layers significantly increase the hydrophobicity of natural wood surfaces, it does not necessarily indicate long-term functionality during the prospective service life [49].

Bulk and surface modification processes highly affect the suitability and functionality of wood-derived components used in construction of the building façade. Conventional wood modification systems use chemical treatments, impregnation or thermal modification, resulting in various intrinsic properties achieved. Innovative modification approaches focus on adding some extra functionalities, such as UV stabilisation, fire retardancy or enhanced suitability for paints and coatings [50]. Although each modification process improves certain material properties on its own, the positive effect can be multiplied by merging two or more modification solutions. An example of successfully combined modifications is the surface coating of acetylated or thermally treated wood. The synergic effect of the reduced shrinkage/swelling of the bulk substrate and water protecting coating substantially reduces stresses of the coating film, thus preventing it from cracking [51,52] and provides UV-VIS light protection [53]. Such a "hybrid process" of wood modification seems to be an optimal solution frequently implemented by biomaterials producers. Superior service life performance with extended maintenance-free periods is reflected in environmental impact analysis. A recent review of Hill et al. clearly demonstrates a positive impact on the global warming potential (GWM) of wood modification combined with coatings [54]. It is anticipated that a lifetime of unmodified wooden façade is 20 years, assuming coated surface maintenance applied every five years. Conversely, a lifetime of 60 years is assumed for the modified wood solution with the renovation of the coating applied every 10 years. The potential environmental benefit to be realized when utilizing modified wood is evident, especially when considering an extended time between re-coating [54]. It is expected that the maintenance frequency can be even lower than once per 10 years when implementing the hybrid modification approach [55]. Such an alternative solution was investigated here, where appropriately selected coating systems were applied on the acetylated wooden elements. Consequently, the façade surface remains intact for a longer period by preserving its original and attractive appearance.

5. Conclusions

Hybrid modification includes processes where the positive effect of a single treatment can be multiplied by merging with additional modifications. All coated samples performed satisfactorily, considering both, the overall appearance evaluated by the visual assessment as well as deterioration determined with sensors. No sign of degradation was observed in samples #D and #G that were coated with a white and non-transparent film. Dark brown samples #B, #C and #H did not change noticeably. Samples #E and #F were coated with a product of original yellow tonality. The colour of sample #E become lighter during the initial weathering period to maintain the same appearance afterwards. Results of visual evaluations correspond to tendencies observed in *CIE* L^*, a^* and b^* alterations. Gloss values measured along and perpendicular to fibres were relatively constant. Only a slight decrease was observed for coated samples #C, #D and #H. On the contrary, the gloss value of #E increased for about 10 units during the 15 months exposure. The high values of θ observed for coated samples indicate low wettability, in contrary to uncoated wood (#A). Surface free energy was highest for the uncoated wood samples and corresponded to ~70 mJ·m^{-2}. It varied between 15 and 40 mJ·m^{-2} for coated samples. An increased disperse component γ^d combined with a lower surface free energy after coating application resulted in a stable performance of the coated samples during weathering.

It was demonstrated that the hybrid solution of merging wood acetylation process followed by the surface coating, as investigated here, amplified the advantages of both treatments when applied separately. Acetylated wood samples without protective coating altered their appearance noticeably after a relatively short period of exposure to the natural weathering. It was evidenced as a considerable change of the colour, decrease of gloss and increase of roughness. Even if possessing different colours and textures, all coated samples preserved their functionality as a cladding material protecting the building envelope. Limited shrinkage/swelling of the bulk substrate due to chemical treatment substantially reduced stresses within the coating film, therefore no cracks developed in the coating layer. Hybrid processes, compared to solely acetylated wood, assured superior visual performance of the wood surface by preserving its original appearance during whole test duration.

Author Contributions: Conceptualization, A.S., E.F.-N., F.P., R.H.D., O.G., N.S., V.P. and J.S.; methodology, A.S., E.F.-N., F.P., R.H.D., O.G., N.S., V.P. and J.S.; software, J.S.; validation, A.S., E.F.-N., F.P., R.H.D., O.G., N.S., V.P. and J.S..; formal analysis, A.S., J.S.; investigation, A.S., E.F.-N., F.P., R.H.D., O.G., N.S., V.P. and J.S.; resources, A.S. and J.S.; data curation, A.S. and J.S.; writing-original draft preparation, A.S.; writing—review and editing, A.S., E.F.-N., F.P., R.H.D., O.G., N.S., V.P. and J.S.; visualization, J.S.; supervision, A.S. and J.S.; project administration, A.S.; funding acquisition, A.S. and J.S. All authors have read and agreed to the published version of the manuscript.

Funding: The authors gratefully acknowledge the European Commission for funding the InnoRenew project (grant agreement #739574 under the Horizon2020 Widespread-2-Teaming program), the Republic of Slovenia (investment funding from the Republic of Slovenia and the European Regional Development Fund) and infrastructural ARRS program IO-0035. Part of this work was conducted during project BIO4ever (RBSI14Y7Y4), funded within call SIR by MIUR-Italy; the project Multi-spec (BI-IT/18-20-007), funded by ARRS-Slovenia; Archi-BIO (BI/US-20-054) funded by ARRS-Slovenia, BI-AT/20-21-014 funded by ARRS-Slovenia, J7-9404 (C) funded by ARRS-Slovenia, and CLICK DESIGN, "Delivering fingertip knowledge to enable service life performance specification of wood", (No. 773324) supported under the umbrella of ERA-NET Cofund ForestValue by the Ministry of Education, Science and Sport of the Republic of Slovenia. ForestValue has received funding from the European Union's Horizon 2020 research and innovation programme.

Institutional Review Board Statement: Not applicable.

Informed Consent Statement: Not applicable.

Data Availability Statement: The data presented in this study are available on request from the corresponding author.

Acknowledgments: The experimental samples were provided by Accsyss The Netherlands and coated by Teknos/Drywood The Netherlands, ICA Italy, and Renner Italy. Authors would also like to thank Ferry Bongers for valuable comments and discussion.

Conflicts of Interest: The authors declare no conflict of interest.

References

1. Williams, R.R. Weathering of wood. In *Handbook of Wood Chemistry and Wood Composites*, 1st ed.; Rowell, R.M., Ed.; CRC Press: Boca Raton, FL, USA, 2005; pp. 139–185.
2. Hill, C. *An Introduction to Sustainable Resource Use*; Taylor and Francis: London, UK, 2011.
3. Rüther, P.; Time, B. External wood claddings—Performance criteria, driving rain and large-scale water penetration methods. *Wood Mater. Sci. Eng.* **2015**, *10*, 287–299. [CrossRef]
4. Sandak, J.; Sandak, A.; Burud, I. Modelling of weathering. In *Performance of Bio-based Building Materials*, 1st ed.; Jones, D., Brischke, C., Eds.; Woodhead Publishing: Sawston, UK, 2017; pp. 502–510.
5. Thiis, T.K.; Burud, I.; Kranitis, D.; Gobakken, L.R. Simulation of surface climate and mould growth on wooden facades. In Proceedings of the International Research Group on Wood Protection IRG/WP 16-20585, Lisbon, Portugal, 15–19 May 2016.
6. Sandak, J.; Sandak, A.; Riggio, M. Characterization and monitoring of surface weathering on exposed timber structures with multi-sensor approach. *Int. J. Arch. Heritage* **2015**, *9*, 674–688. [CrossRef]
7. Podgorski, L.; Arnold, M.; Hora, G. A reliable artificial weathering test for wood coatings. *Coat. World* **2003**, *2*, 39–48.
8. Grüll, G.; Tscherne, F.; Spitaler, I.; Forsthuber, B. Comparison of wood coating durability in natural weathering and artificial weathering using fluorescent UV-lamps and water. *Eur. J. Wood Prod.* **2014**, *72*, 367–376. [CrossRef]
9. Brischke, C.; Thelandersson, S. Modelling the outdoor performance of wood products—A review on existing approaches. *Constr. Build. Mater.* **2014**, *66*, 384–397. [CrossRef]
10. Sandak, A.; Sandak, J.; Brzezicki, M.; Kutnar, A. *Bio-based Building Skin*; Springer: Singapore, 2019.
11. Kotradyová, V.; Teischinger, A. Exploring the Contact Comfort within Complex Well-Being in Microenvironment. *Adv. Mater. Res.* **2014**, *899*, 356–362. [CrossRef]
12. Sandak, A.; Brzezicki, M.; Sandak, J. Trends and perspectives in the use of timber and derived products in building façades. In *New Materials in Civil Engineering*; Samui, P., Ed.; Butterworth-Heinemann: Oxford, UK, 2020; pp. 333–374.
13. Gobakken, L.R.; Høibø, O.A.; Solheim, H. Mould growth on paints with different surface structures when applied on wooden claddings exposed outdoors. *Int. Biodeterior. Biodegrad.* **2010**, *64*, 339–345. [CrossRef]
14. Gobakken, L.R.; Vestøl, G.I. Surface mould and blue stain fungi on coated Norway spruce cladding. *Int. Biodeterior. Biodegrad.* **2012**, *75*, 181–186. [CrossRef]
15. Flores-Colen, I.; de Brito, J. A systematic approach for maintenance budgeting of buildings façades based on predictive and preventive strategies. *Constr. Build. Mat.* **2010**, *24*, 1718–1729. [CrossRef]
16. Hill, C. *Wood Modification Chemical, Thermal and Other Processes*; John Willey & Sons Ltd.: Chichester, UK, 2006.
17. Gérardin, P. New alternatives for wood preservation based on thermal and chemical modification of wood—A review. *Ann. For. Sci.* **2016**, *73*, 559–570. [CrossRef]
18. Rowell, R.M. Chemical modification of wood. In *Handboook of Wood Chemistry and Wood Composite*; Rowell, R.M., Ed.; CRC Press (Taylor and Francis Group): Boca Raton, FL, USA, 2005; pp. 381–420.
19. Rowell, R.M. Acetylation of wood—Journey form analytical technique to commercial reality. *For. Prod. J.* **2006**, *56*, 4–12.
20. Herrera, R.; Sandak, J.; Robles, E.; Krystofiak, T.; Labidi, J. Weathering resistance of thermally modified wood finished with coatings of diverse formulations. *Prog. Org. Coat.* **2018**, *119*, 145–154. [CrossRef]
21. Evans, P.D.; Haase, J.G.; Seman, A.S.B.M.; Kiguchi, M. The Search for Durable Exterior Clear Coatings for Wood. *Coatings* **2015**, *5*, 830–864. [CrossRef]
22. Díaz, R.H.; Gordobil, O.; De Hoyos-Martinez, P.L.; Sandak, A.; Labidi, J. Hydrophobization and Photo-Stabilization of Radiata Pinewood: The Effect of the Esterification on Thermal and Mechanical Properties. *Forests* **2020**, *11*, 1243. [CrossRef]
23. Mai, C.; Militz, H. Modification of wood with silicon compounds. Inorganic silicon compounds and sol-gel systems: A review. *Wood Sci. Technol.* **2004**, *37*, 339–348. [CrossRef]
24. Peng, X.; Zhang, Z. Improvement of paint adhesion of environmentally friendly paint film on wood surface by plasma treat-ment. *Prog. Org. Coat.* **2019**, *134*, 255–263. [CrossRef]
25. Sadatnezhad, S.H.; Khazaeian, A.; Sandberg, D.; Tabarsa, T. Continuous Surface Densification of Wood: A New Concept for Large-scale Industrial Processing. *BioResources* **2017**, *12*, 3122–3132. [CrossRef]
26. Wang, X.; Wu, Y.; Chen, H.; Zhou, X.; Zhang, Z.; Xu, W. Effect of surface carbonization on mechanical properties of LVL. *BioResources* **2019**, *14*, 453–463.

27. Williams, R.S.; Knaebe, M.T.; Feist, W.C. *Finishes for Exterior Wood, Forest Products Society*; Forest Products Research: Madison, WI, USA, 1996.
28. Feist, W.C. *Finishing Exterior Wood*; Federation Series on Coatings Technology; Federation of Societies for Coatings Technology: Blue Bell, PA, USA, 1996.
29. Servo-wood Improved Service Life Prediction and Test Capability for Wood Coatings. Available online: www.servowood.eu (accessed on 5 May 2021).
30. Fowkes, F.M. Attractive Forces at Interfaces. *Ind. Eng. Chem.* **1964**, *56*, 40–52. [CrossRef]
31. *Finishes for Exterior Timber*; Timber Research and Development Association: High Wycombe, UK, 1984; p. 8.
32. Sarvis, J.C. *Exterior Clear Finishes*; Western Pine Association Research Laboratory: Portland, OR, USA, 1953.
33. Estrada, N. Exterior durability of catalyzed clear coatings on redwood. *J. Paint Technol.* **1967**, *39*, 655–662.
34. Gobakken, L.R.; Jenssen, K.M. *Growth and Succession of Mould on Commercial Paint Systems in Two Field Sites*; IRG/WP 07–30421; The International Research Group on Wood Protection: Stockholm, Sweden, 2007.
35. Viitanen, H. Modelling the time factor in the development of mould fungi—Effect of critical humidity and temperature conditions in pine and spruce sapwood. *Holzforschung* **1997**, *51*, 6–14. [CrossRef]
36. De Meijer, M.; Bongers, F. Coating performance on acetylated wood—A review paper. In Proceedings of the PRA's 8th International Wood Coatings Congress, Amsterdam, The Netherlands, 30–31 October 2012.
37. Sandak, J.; Sandak, A.; Cocchi, M. Multi-sensor data fusion and parallel factor analysis reveals kinetics of the wood weathering. *Talanta* **2021**, *225*, 122024. [CrossRef]
38. Davis, K.; Leavengood, S.; Morrell, J.J. Performance of Exterior Wood Coatings in Temperate Climates. *Coatings* **2021**, *11*, 325. [CrossRef]
39. Goodell, B.; Winandy, J.E.; Morrell, J.J. Fungal Degradation of Wood: Emerging Data, New Insights and Charging Perceptions. *Coatings* **2020**, *10*, 1210. [CrossRef]
40. Petrillo, M.; Sandak, J.; Grossi, P.; Sandak, A. Chemical and appearance changes of wood due to artificial weathering—Dose–response model. *J. Near Infrared Spectrosc.* **2019**, *27*, 26–37. [CrossRef]
41. Sandak, A.; Sandak, J.; Noël, M.; Dimitriou, A. A Method for Accelerated Natural Weathering of Wood Subsurface and Its Multilevel Characterization. *Coatings* **2021**, *11*, 126. [CrossRef]
42. Wood, K.A.; Cypcar, C.; Hedhli, L. Predicting the exterior durability of new fluoropolymer coatings. *J. Fluor. Chem.* **2000**, *104*, 63–71. [CrossRef]
43. Bobadilha, G.D.S.; Stokes, C.E.; Ohno, K.M.; Kirker, G.; Lopes, D.J.V.; Nejad, M. Physical, Optical, and Visual Performance of Coated Cross-Laminated Timber during Natural and Artificial Weathering. *Coatings* **2021**, *11*, 252. [CrossRef]
44. Pánek, M.; Reinprecht, L. Critical view on the possibility of color changes prediction in the surfaces of painted wood exposed outdoors using accelerated weathering in Xenotest. *J. Coat. Technol. Res.* **2019**, *16*, 339–352. [CrossRef]
45. Oberhofnerová, E.; Šimůnková, K.; Dvořák, O.; Štěrbová, I.; Hiziroglu, S.; Šedivka, P.; Pánek, M. Comparison of Exterior Coatings Applied to Oak Wood as a Function of Natural and Artificial Weathering Exposure. *Coatings* **2019**, *9*, 864. [CrossRef]
46. Žlahtič, M.; Humar, M. Influence of Artificial and Natural Weathering on the Hydrophobicity and Surface Properties of Wood. *Bioresources* **2016**, *11*, 4964–4989. [CrossRef]
47. Stehr, M.; Gardner, D.J.; Wålinder, M.E.P. Dynamic Wettability of Different Machined Wood Surfaces. *J. Adhes.* **2001**, *76*, 185–200. [CrossRef]
48. Jankowska, A.; Rybak, K.; Nowacka, M.; Boruszewski, P. Insight of Weathering Processes Based on Monitoring Surface Characteristic of Tropical Wood Species. *Coatings* **2020**, *10*, 877. [CrossRef]
49. Pánek, M.; Oberhofnerová, E.; Zeidler, A.; Šedivka, P. Efficacy of Hydrophobic Coatings in Protecting Oak Wood Surfaces during Accelerated Weathering. *Coatings* **2017**, *7*, 172. [CrossRef]
50. Spear, M.J.; Curling, S.F.; Dimitriou, A.; Ormondroyd, G.A. Review of Functional Treatments for Modified Wood. *Coatings* **2021**, *11*, 327. [CrossRef]
51. Rowell, R.; Bongers, F. Role of Moisture in the Failure of Coatings on Wood. *Coatings* **2017**, *7*, 219. [CrossRef]
52. Rowell, R.; Bongers, F. Coating Acetylated Wood. *Coatings* **2015**, *5*, 792–801. [CrossRef]
53. Schaller, C.; Rogez, D. Light stabilization of modified wood species. In Proceedings of the Third European Conference on Wood Modification, Cardiff, UK, 15–16 October 2007.
54. Hill, C.; Hughes, M.; Gudsell, D. Environmental Impact of Wood Modification. *Coatings* **2021**, *11*, 366. [CrossRef]
55. Bongers, F.; Creemers, J.; Kattenbroek, B.; Homan, W. Performance of coatings on acetylated Scots pine after more than nine years outdoor exposure. In Proceedings of the Second European Conference on Wood Modification, Göttingen, Germany, 6–7 October 2005; pp. 125–129.

Article

Bioinspired Living Coating System in Service: Evaluation of the Wood Protected with Biofinish during One-Year Natural Weathering

Faksawat Poohphajai [1,2], Jakub Sandak [1,3,*], Michael Sailer [4], Lauri Rautkari [2], Tiina Belt [5] and Anna Sandak [1,6]

1 InnoRenew CoE, Livade 6, 6310 Izola, Slovenia; faksawat.poohphajai@innorenew.eu (F.P.); anna.sandak@innorenew.eu (A.S.)
2 Department of Bioproducts and Biosystems, School of Chemical Engineering, Aalto University, P.O. Box 16300, 00076 Aalto, Finland; lauri.rautkari@aalto.fi
3 Andrej Marušič Institute, University of Primorska, Titov trg 4, 6000 Koper, Slovenia
4 Xylotrade B.V, Brinkpoortstraat 28, 7411HS Deventer, The Netherlands; michael@xyhlo.com
5 Production Systems, Natural Resources Institute Finland (Luke), Tietotie 2, 02150 Espoo, Finland; tiina.belt@luke.fi
6 Faculty of Mathematics, Natural Sciences and Information Technologies, University of Primorska, Glagoljaška 8, 6000 Koper, Slovenia
* Correspondence: jakub.sandak@innorenew.eu or jakub.sandak@upr.si; Tel.: +386-40282959

Abstract: The service life performance of timber products exposed to natural weathering is a critical factor limiting the broad use of wood as an external building element. The goal of this study was to investigate the in-service characterization of an innovative biofinish coating system. It is a novel surface finishing solution based on the bioinspired concept of living fungal cells designed for effective wood protection. The performance of Scots pine (*Pinus sylvestris* L.) wood coated with biofinish was compared with uncoated references. Samples were exposed to natural weathering for 12 months under the climatic conditions of northern Italy. The visual appearance, colour, gloss, wettability, and 3D surface topography of the wood surface were examined. Results revealed that the total colour changes (ΔE) of biofinish-coated wood were negligible. Untreated Scots pine wood revealed the changes in colour after just three months of exposure. The gloss changes of both surface types were small. The contact angle measured on biofinish-coated wood was higher compared to that of uncoated Scots pine. Surface roughness increased in uncoated wood due to the erosion effect caused by the weathering progress. Conversely, the surface roughness of biofinish-coated samples decreased along the exposure time. This phenomenon was explained by two self-healing mechanisms: migration of non-polymerized oil to the cracked surface, where it polymerizes and creates a closed layer, and local regrowth to cover damaged spots by living fungal cells present in the coating. The obtained results revealed the superior aesthetic performance of the biofinish surface treatment against natural weathering. By considering the fully bio-based nature of the investigated coating, it was concluded that this solution can be an attractive alternative for state-of-the-art wood protection technologies.

Keywords: natural weathering; bio-based coating; service life performance; aesthetics; living fungal cells; bioinspired materials design

Citation: Poohphajai, F.; Sandak, J.; Sailer, M.; Rautkari, L.; Belt, T.; Sandak, A. Bioinspired Living Coating System in Service: Evaluation of the Wood Protected with Biofinish during One-Year Natural Weathering. *Coatings* **2021**, *11*, 701. https://doi.org/10.3390/coatings11060701

Received: 15 May 2021
Accepted: 7 June 2021
Published: 11 June 2021

1. Introduction

Weathering is the general term used to describe the slow degradation processes that occur when the material is exposed to the outdoor environment [1–3]. In that case, the deterioration is caused primarily by environmental or abiotic factors and not due to microorganisms. However, the uncoated wood weathering is usually combined with biotic attacks that together may alter the surface appearance. This leads to the creation of green or grey-blue spots on the surface of weathered material which, in association with other weathering effects, changes the colour of wood towards a grey tonality. Particularly, the weathering processes of wood are triggered by the mutual action of solar radiation and wind-driven rain in combination with several other environmental factors, such as temperature changes, relative humidity (*RH*), wind, air pollutants, molecular oxygen (O_2) concentration, or human activities, among others. Surface discolorations, loose fibres, raised grain, checks, cracks, or general roughening are the usual results of wood exposure to natural weathering. Solar radiation is absorbed during weathering by the constitutive woody polymers, including lignin, cellulose, and hemicellulose. Among these components, lignin is recognized as the most sensitive to photodegradation due to its complex structure, revealed to be a variety of functional groups and bonds [1,4]. The photolysis of lignin caused by sunlight leads to the generation of free radicals, which can interact with oxygen to produce hydroperoxides that are converted finally to new lignin chromophores [1,5]. The formation of such chromophores induces the initial yellowing of wood followed by surface colour changes to a silver-grey shade when these chromophore groups are leached out by rain or wind [6,7]. The depth of photodegradation increases with the irradiating light wavelength. Blue light can penetrate wood to a greater extent than violet light. It is capable of bleaching the wood, even if does not alter lignin. Violet light, however, deepens the photodegradation toward wood bulk beyond the zone already affected by the ultraviolet (UV) radiation. Short-wavelength UV radiation causes the photo-yellowing that is directly correlated with the lignin degradation [8]. The detailed mechanisms of natural weathering were the subject of several proceedings studies, including the thorough observation of the alterations at different scales, kinetics, or reactions as well as dose–response models [9,10]. Similarly, diverse approaches were proposed for the characterization of the weathering extent and the determination of the resulting material properties. The colour space coordinates defined by the International Commission on Illumination *CIE* $L^*a^*b^*$ and derived ΔE are frequently used to quantify colour alterations due to lignin degradation and chromophores generation [11–14]. Spectroscopic techniques such as Fourier Transform Infrared Spectroscopy (FTIR) [15,16], Fourier Transform Infrared Spectroscopy–Attenuated Total Reflectance (FTIR-ATR) [17], and/or Fourier Transform Near Infrared Spectroscopy (FT-NIR) [18–20] are often used to investigate changes in functional groups of wood chemical components. Other aspects frequently assessed while evaluating the weathering process of materials include the determination of the surface roughness [21,22], wettability [23,24], or glossiness [18,25,26].

Several surface treatment solutions have been developed to protect wood against weathering-induced deterioration. The application of a coating is one of the most common methods to prevent wood alteration due to its economic- and convenience-based advantages [27]. A coating is defined here as a physical barrier layer that protects wood bulk from environmental factors and microorganisms. The majority of coating solutions can be classified into two categories, including film-forming and various impregnation approaches [2]. Paints and solid-body stains are classic examples of film-forming techniques. Impregnation induces the saturation of the finished subsurface with chemical mixtures containing hydrophobic substances that can penetrate the wood and cures. The most frequently used impregnation products include waxes as well as oil- or resin-based solutions. Although the application of coatings has been widely used for wood protection for a long time, the degradation of the coating layer due to weathering can emerge in a variety of deterioration processes. Some of the most relevant include the photofading of dyed and pigmented polymers [27,28], a loss of gloss and an increase in surface roughness [29], the yellowing of clear coating [30,31], crack formation and peeling [2], or aesthetical problems caused by microorganisms [32–34], among many others. Moreover, an important drawback of modern coating formulations is their composition. Many surface finishing products contain toxic chemicals that might cause negative impacts on the environment and/or can be harmful to living organisms, including humans [35]. Due to the increasing awareness of environmental protection issues, the use of toxic substances allowed until now is increasingly restricted by international legal regulations [36].

Systems and solutions found in nature are a valuable source of inspiration for several advanced applications. Scientists and researchers from different fields use concepts of biomimicking and bioinspiration to solve the most complex challenges and technical tasks. The possibility to benefit from solutions developed by nature recently became interesting for use in cutting-edge materials science and sustainable architecture [37]. Biomimicry aims to understand the fundamental principles of biological processes and adapt these concepts for bio-inspired product applications [38]. New materials and methods for structural design, thermal insulation, and waterproofing can be developed by mimicking the structural, behavioural, functional, and morphological aspects of natural organisms [39]. The largest area of biomimicry research is related to materials, accounting for around 50% of published research [38]. Materials' surface modification includes a wide range of changes related to adjustments of the expected surface energy, appearance, adhesion, and/or inducing unique properties, such as self-regeneration or self-healing. Bioinspired modification can be related here to the generation of a particular surface topography, patterning, activation, or functionalization. Most surfaces are exposed during service life to a diversity of damaging threats, which reduce the surface functionality and require certain actions, such as cleaning, reparation, or replacement. The self-repairing phenomena observed in several biological systems is challenging for replication in traditional building materials since they are simply not alive as biological organisms. However, the development of bio-concrete (where limestone-producing bacteria are activated when a crack occurs) [40] as well as bio-coatings (where living moulds are re-grooving on a wooden surface after damage) provides a pioneering proof of concept for bioinspired solutions demonstrating the novel capacities of building materials [41].

Biofinish technology represents such a bioinspired and environmentally friendly wood treatment solution. The idea of using the functional biofilm produced by a black, yeast-like fungus, *Aureobasidium pullulans*, for the protection of wood, was inspired by the growth of microorganisms on leaves. In industrial applications, the substrate wood was first impregnated with vegetable oil, which serves as a future nutrition source for living *A. pullulans*. Afterwards, the wood's surface was coated with a proprietary biofinish emulsion containing a complete set of ingredients necessary for mould to establish a long-lasting, self-sufficient living biofilm. The proposed solution increases the hydrophobicity of wood substrate due to oil treatment which results in the enhanced dimensional stability of the timber element [41]. The presence of *A. pullulans* living cells plays an important role in protecting wood against other decay from fungi infestation. The colour-giving melanin pigment produced by the fungus provides an aesthetically appealing dark surface on wood, and at the same time, protects the substrate from UV radiation. One of the most attractive aspects of using *A. pullulans* as a living coating system is an induced self-healing ability of such a finished surface. Biofinish is solely composed of natural substances and it is considered a fully sustainable wood treatment solution with minimal environmental impact and low maintenance requirements [42]. Timber products are currently treated at an industrial scale and the do-it-yourself (DIY) coating formulation is available on the market [43]. A combined effect of wood species, oil types, and material climate conditions on the formation of a uniform biofilm were studied previously [42,44,45]. Nevertheless, several aspects regarding protection mechanisms and the durability of biofinish are still poorly understood.

The goal of this study is, therefore, to investigate the in-service performance of biofinish coating during a natural weathering test. A special focus was directed toward the assessment of the alteration of appearance and aesthetical properties, to be evaluated with an objective multi-sensor approach and confronted with a reference timber product frequently used in a similar configuration.

2. Materials and Methods

2.1. Experimental Samples and Natural Weathering

Thirty wooden blocks made of Scots pine (*Pinus sylvestris* L.) wood with dimensions of 150 mm × 75 mm × 20 mm (length × width × thickness, respectively) were used as experimental samples. Half of the sample set (15 blocks) was coated with biofilm (Biofinish, Xyhlo B.V. Deventer, The Netherlands) following the industrial procedure of the coating application (0.125 $L \cdot m^{-2}$). Two layers of coating were applied penetrating wood subsurface to the depth of approximately 350 μm. The remaining set of fifteen Scots pine blocks was left uncoated and served as a reference. Thirty samples were analysed in total, including two finishing types, five natural weathering scenarios, and three replicas for each scenario.

Natural weathering tests were performed in San Michele, Italy (46°11′15″ N, 11°08′00″ E), for a total period of 12 months, starting in March 2017. The objective of this test was to collect a set of reference data regarding material performance in a function of the exposure time. Samples were exposed on a vertical stand corresponding to a typical building façade configuration. The stand was oriented to face the southern direction. The representative meteorological data for the studied location are summarized in Table 1. Three replicas were collected from the stand each third month to terminate the deterioration progress. Thus, a collection of samples exposed for 0, 3, 6, 9, and 12 months was gathered as a result of the experimental campaign. Each exposure period corresponds to a single experimental scenario. Three replica samples allowed insight into the reliability and repeatability of results. All samples were stored after collection from the stand in a climatic chamber (20 °C, 65% *RH*) to stabilize conditions before follow-up measurements.

Table 1. Weather data (average values) acquired during weathering progress.

Exposure Month	1	2	3	4	5	6	7	8	9	10	11	12
Mean Temp (°C)	12	14	19	24	25	25	17	14	7	1	4	4
Days of Frost	0	0	0	0	0	0	0	8	5	24	16	10
Total Rain (mm)	32	108	50	36	27	17	6	1	28	3	1	7
RH (%)	49	45	51	51	50	52	68	62	69	74	82	55
Mean Wind Speed (km/h)	7	9	7	9	8	6	5	5	4	3	3	5

2.2. Colour Measurement

Experimental samples were scanned with the office scanner HP Scanjet G2710 (Palo Alto, CA, USA). The colour changes were determined using portable MicroFlash 200D spectrophotometer (DataColor Int, Lawrenceville, NJ, USA). The selected illuminant was D65, and the viewer angle was 10°. The colour was measured at ten given spots that were randomly selected over the surface of investigated wood specimens. Colour changes were evaluated following the *CIE* $L^*a^*b^*$ colour space system where colour is expressed by three parameters: L^* (lightness), a^* (red-green tone), and b^* (yellow-blue tone). The total colour change ΔE was calculated according to Equation (1):

$$\Delta E = \sqrt{(\Delta L^*)^2 + (\Delta a^*)^2 + (\Delta b^*)^2}, \tag{1}$$

where ΔL, Δa, Δb correspond to differences between colour coordinate values measured at the given time and referenced to the corresponding value of initial colour.

2.3. Gloss Measurement

The mode of light reflection from the evaluated surfaces was measured using a REFO 60 (Dr. Lange, Düsseldorf, Germany) portable gloss meter with incidence and reflectance angles of 60°. Ten measurements were performed on each wood sample in parallel and perpendicular to the fibre directions. An average value was considered as a measure of the gloss, with both minimum and maximum values, determining the observation range and repeatability of the assessment.

2.4. Wettability

Dynamic contact angle measurement when wetting the surface with distilled water was performed with optical tensiometer Attension Theta Flex Auto 4 (Biolin Scientific, Gothenburg, Sweden). Five measurements were performed on each specimen, implementing the sessile drop measurement approach. The nominal volume of each drop was 4 μL. The volume was precisely controlled by both droplet dispenser and a digital image analysis tool integrated with instrument software. The measurement of the drop contour was initiated at the moment of the drop contact with the assessed sample surface. The droplet image acquisition lasted for 20 s. Images were post-processed with the proprietary software (OneAttension v.4.0.5) of the tensiometer. The contact angle was determined by implementing the Laplace equation. The contact angle observed at the third second of the test was assumed as a representative quantifier. Five replica measurement results were averaged to reduce the scatter of results. The range of observed values (minimum–maximum) was used to define the variability and reliability of the contact angle assessment.

2.5. Microscopic Observation and 3D Roughness Measurement

Keyence VHX-6000 digital microscope (Keyence, Osaka, Japan) was used for microscopic observations and 3D surface topography assessment. Colour images were acquired with magnifications of ×30 and ×1000. The light configuration and its intensity were adjusted to assure an optimal image quality and elimination of the saturated pixels. High magnification images were acquired in 3D depth reconstruction mode. It allowed post-processing of data to extract representative surface profiles as well as computation of the surface roughness indicators. An area of 20 mm × 20 mm was assessed by implementing a stitching algorithm for acquired 3D images. The proprietary software (VHX-6000 v.3.2.0.121) of the Keyence microscope was used for the surface topography post-processing. The error of form was initially removed by extracting an average plane. The resulting 3D topography map was filtered with a Gaussian band pass filter (2 μm < λ < 2.5 mm). Arithmetical mean height (S_a), skewness (S_{sk}), and kurtosis (S_{kt}) were computed as the surface irregularity quantifiers.

3. Results and Discussions

3.1. Surface Visual Appeal

Appearance images of the investigated samples are presented in Figures 1 and 2 for reference (uncoated) and biofinish-coated samples, respectively. Natural Scots pine wood becomes darker after the initial three months of exposure, followed by a slight lightening and converting to the grey tonality. The grey colour becomes dominant after weathering for one year. Small surface cracks become visible to the naked eye from month 9. At the same time, big cracks were present together with raised fibres and eroded surface marks. The appearance of samples coated with biofinish does not change, including consistent impressions of the colour, roughness and overall surface integrity.

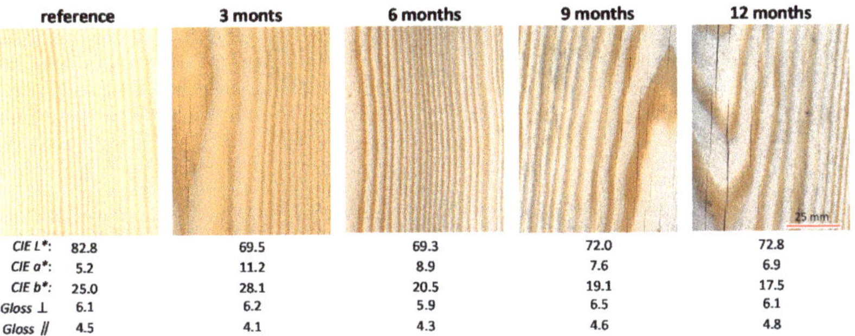

Figure 1. Surface texture, *CIE L*a*b** colour coordinates and glossiness parameters altered during weathering progress of uncoated Scots pine samples.

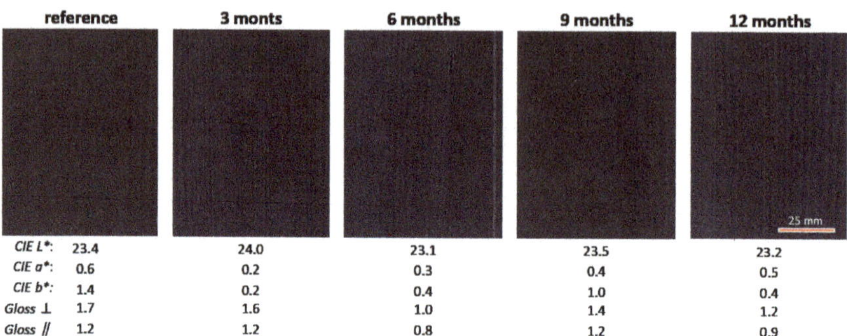

Figure 2. Surface texture, *CIE L*a*b** colour coordinates and glossiness parameters altered during weathering progress of biofinish-coated Scots pine samples.

3.2. Surface Colour

The progress of *CIE L*a*b** changes over the duration of the natural weathering test is shown in Figure 3. The average *CIE L** of the uncoated Scots pine samples decreased during the first three months of exposure, to become constant afterwards (Figure 3a). The average *CIE a** of the uncoated wood increased at the initial stage of exposure, but after that, it steadily decreased until the end of the natural weathering process (Figure 3b). A similar trend was noticed in the case of the *CIE b** coordinate (Figure 3c). Both, *CIE a** and *b** asymptotically followed the horizontal axis, confirming the overall tendency for greying the weathered surface outlook. A very different trend of the *CIE L*a*b** colour changes was noticed for coated samples. All investigated colour indicators were constant along the whole test duration. Correspondingly, the scatter of the *CIE L*a*b** values within all replicas was substantially smaller than the same scatter observed in the uncoated wood samples. It was confirmed by the total colour changes (ΔE) analysis present in Figure 3d. The ΔE of uncoated Scots pine rapidly increased during the initial three months of exposure. The same quantifier steadily decreased afterwards until the end of the exposure period of twelve months. Biofinish-coated specimens revealed much lower values of ΔE compared to those noticed in uncoated Scots pine. ΔE was relatively constant with minor variations. The nature of ΔE changes in the coated wood samples suggests the source of the analysed discoloration to be related to the intrinsic variance of the colour within tested samples rather than the weathering-induced alterations.

The decrease in *CIE L** of uncoated Scots pine at the initial stage of exposure indicates that the darkening of the wood surface is due to the accumulation of degradation products of lignin [2] followed by the infestation of the wood surface by fungi [2,27]. The initial increase in *CIE b** is also associated with the degradation of lignin [46]. Consequently, the follow-up decreases in yellowness (*CIE b**) may be attributed to the leaching of decomposed lignin and its extractives by water [47]. The changes in *CIE a** values are primarily determined by the changes of the chromophore groups present in some wood extractive components [48]. The formation of additional chromophoric systems was found to be triggered by the solar radiation interaction with the wood chemical components [2,49]. The overall mechanism of wood discolouration induced by the weathering processes follows three overlapping phases. An initial darkening, yellowing or browning of wood is observed as a result of the quinone formation and accumulation of degradation products of lignin. The greyish tendency of weathered wood after a long-term exposure period is associated with the leaching out of extractives and cleavage products of lignin photodegradation. Finally, the dark-grey colour of the long-term weathered wood surface is a result of surface infestation by fungal hyphae, spores and derived pigments [49,50].

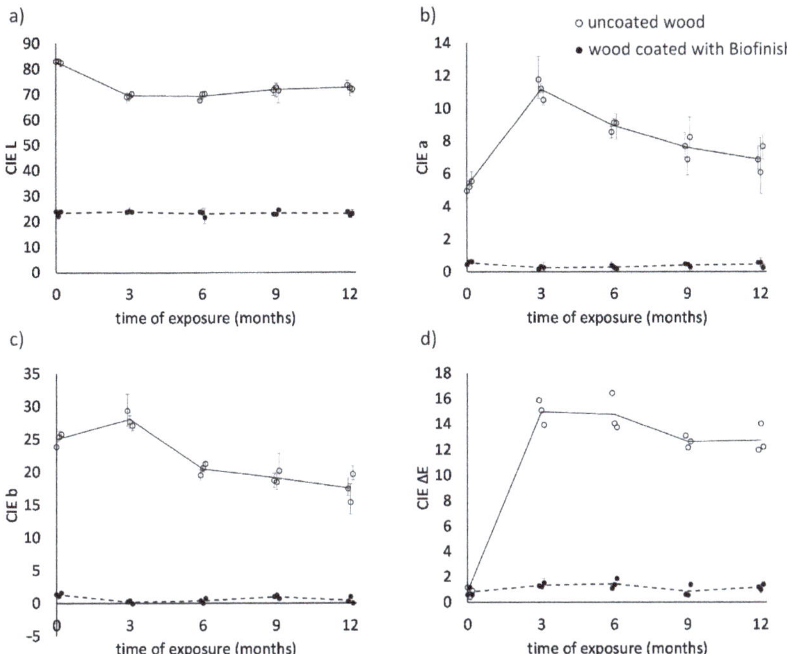

Figure 3. *CIE L*a*b** colour coordinates of uncoated Scots pine and wood coated with biofinish as a function of exposure time. Note: Error bars correspond to the range (minimum–maximum) of results observed: (**a**) *CIE L**, (**b**) *CIE a**, (**c**) *CIE b**, (**d**) *CIE ΔE*.

Presence of moulds depends on the intrinsic properties of the specific wood substrate. Heartwood is considered to be less susceptible to decay fungi than sapwood. In the case of Scots pine heartwood, it is due to the presence of extractive components possessing fungicide capacity, which can suppress fungal growth [51]. The relative humidity and temperature of the ambient air are two factors regulating the moisture content of wood. Wood moisture is known as a critical factor determining the ability of microorganisms for initial surface colonization and further growth. The threshold limit for mould growth is usually at the level of wood moisture content of 20% and corresponding air relative humidity of 75–80%. However, the actual fungal infestation risk highly depends on the exposure time to the elevated moisture, ambient air temperature, inherent susceptibility of the exposed surface to moulds, and presence of specific mould species in the ambient air [52].

Negligible variations of the *CIE L*a*b** colour coordinates of wood coated with biofinish can be explained by the presence of colour-giving melanin pigment produced by *A. pullulans*. Melanin, in addition to colour determination, functions as a barrier layer protecting the wood surface against alien microorganisms. Melanin production by the fungi does not affect their growth or development but enhances survival under environmental stresses. These include protection against UV irradiation, the accumulation of pollutants (such as heavy metals), tolerance to extreme temperatures, desiccation or high salts concentrations, as well as resistance to the cell wall-degrading enzymes and other pathogens [53–56]. Several studies confirm the stimuli effect of UV radiation on the enhanced melanin production by the *A. pullulans* [57–59].

3.3. Gloss Evaluation

The gloss measured parallel and perpendicular to fibres on uncoated and biofinish-coated Scots pine specimens was constant along the natural weathering process (Figure 4).

The average gloss value measured in parallel to fibres of uncoated Scots pine oscillated around 6, while those measured perpendicularly was approximately 5. Gloss values were smaller for samples coated with biofinish and corresponded to 1 in both measurement directions. The spread of data defined as the difference between the minimum and maximum recorded values was higher for gloss measured parallel to the fibres in the case of both surface finish configurations.

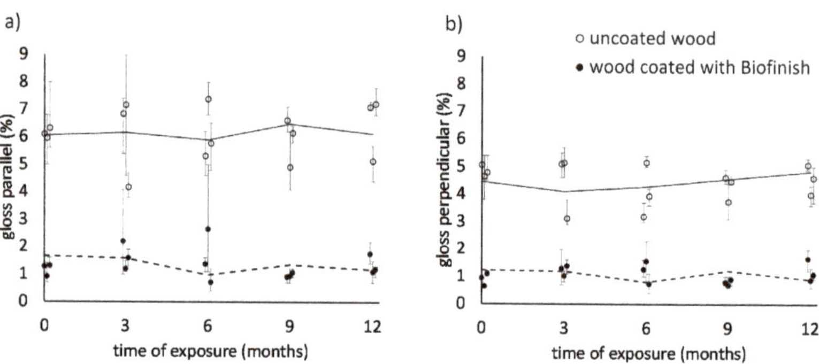

Figure 4. Gloss changes of uncoated Scots pine and wood coated with biofinish as a function of exposure time. Note: Error bars correspond to the range (minimum–maximum) of the results observed: (**a**) gloss parallel, (**b**) gloss perpendicular.

Several studies are reporting a decrease in wood glossiness associated with exposure to natural weathering [18,20,60,61]. It was not directly confirmed by the present experimental results that are related to the relatively short period of the test as well as the initial surface configuration that was intended to be moderately rough. The reduction in the wood surface glossiness is associated with the abrasion of wood surfaces and the accompanying surface erosion process. The wood surface usually becomes rougher as a result of the weathering which enhances the diffuse (Lambertian) light reflectance component. Consequently, the scatter of light on the surface become higher, which is recorded as a glossiness value reduction [49].

3.4. Wettability

The change of average wettability values expressed as a contact angle θ induced by the natural weathering process are shown in Figure 5. The contact angle of biofinish-coated wood was twice as high (~120°) as that measured on uncoated Scots pine surface (~60°). The contact angle of both types of specimens decreased during the initial three months of exposure. However, the drop was very apparent in the case of uncoated wood, resulting in $\theta < 20°$, which was stable until the end of the weathering test. It is associated with chemical changes of the weathered wood surface at the initial period of exposure, as well as the leaching of extractives possessing hydrophobic nature. Such a small θ corresponds to the instant surface wetting and spill out of the water droplet after contact with the measured surface. The contact angle measured on the surface of biofinish-coated wood was steadily decreasing along the natural weathering test duration. This leads to the conclusion that wood hydrophobicity decreased along the exposure time, even if the surface coating substantially reduced the kinetic of such changes. Similar observations were reported by other researchers for Scots pine [62] and other wood species [62–64]. The increased wettability of the weathered wood surface even further accelerates the subsurface deterioration. This eases the leaching of extractives and other degradation products of lignin. Elevated surface moisture content triggers the loosening of the cellulose fibres that trigger micro-cracks presence and an increase in the roughness [65].

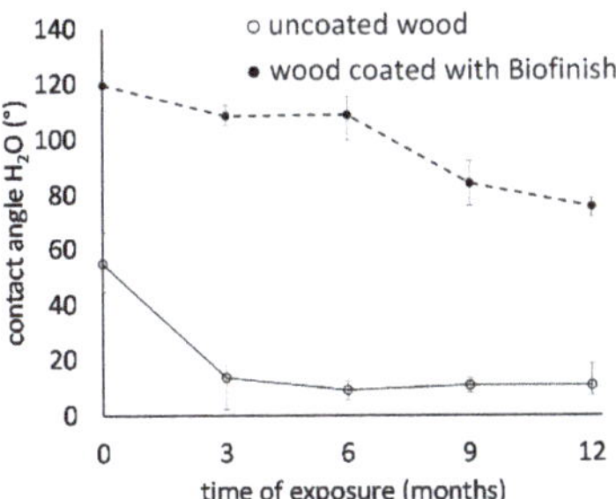

Figure 5. Changes in the contact angle of uncoated Scots pine and biofinish-coated wood as a function of exposure time. Note: Error bars correspond to the range (minimum–maximum) of results observed.

3.5. Surface Topography

Changes in surface topography for uncoated and biofinish-coated samples are presented in Figures 6 and 7, respectively. An increase in the surface roughness was mostly observed in uncoated pine wood specimens, where all surface topography descriptors (S_a, S_{sk} and S_{kt}) changed during weathering progress (Figure 6). The overall curvature changes of the surface topography profiles were related to the wood deformations induced by the moisture content variations and consequently to wood shrinkage/swelling. The biofilm coating layer affected the thermodynamic properties of the wood subsurface limiting the absorption and diffusion of moisture. An increase in the roughness was related to the natural weathering processes and resultant photolysis and/or hydrolysis of the constitutive wood polymers, particularly lignin. As a result, single fibres were removed from the wood matrix and degradation components are leached out after rain events. This leads to the wood surface erosion revealed as an increase in the surface roughness [66].

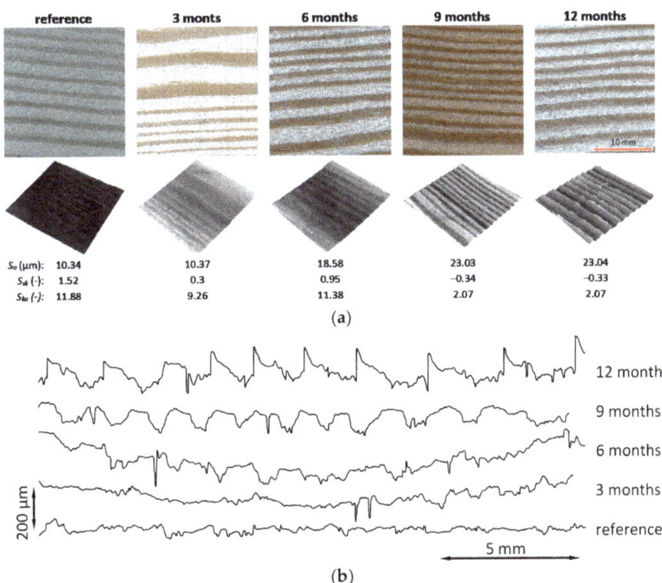

Figure 6. 3D surface topography map, typical surface profiles and high magnification images of the Scots pine wood exposed to natural weathering for 12 months, (**a**) high magnification images and 3D surface topography map, (**b**) typical surface profiles.

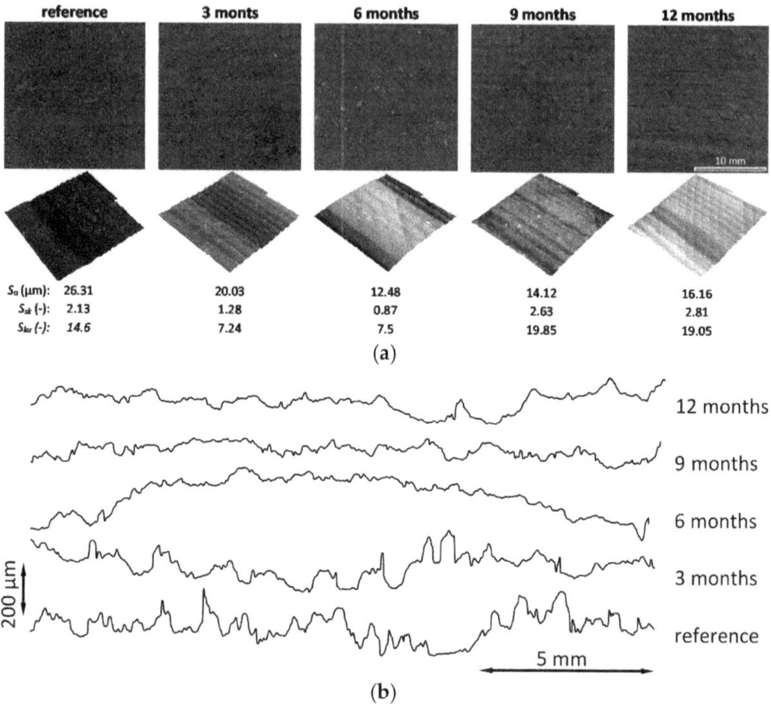

Figure 7. 3D surface topography map, typical surface profiles and high magnification images of the Scots pine wood coated with biofinish exposed to natural weathering for 12 months, (**a**) high magnification images and 3D surface topography map, (**b**) typical surface profiles.

Opposite trends of surface roughness alteration were observed for wooden samples coated with biofinish. The arithmetical mean surface height (S_a) decreased with the weathering progress, indicating that the surface became smoother than at the initial stage (Figure 7). Such flattening of the wood surface can be explained by the gradual secretion of linseed oil that was used for sample impregnation before coating application. It should be mentioned that the preferred surface for the application of the biofinish product is rough. The presence of surface irregularities eases adhesion to the coating emulsion as well as fungi spores. Furthermore, living cells of *A. pullulans* are capable of regenerating and regrowing on the wooden surface after the initial application of the coating. This leads to the apparent roughness reduction by filling surface micro-voids and other irregularities caused by the fungal maturation and development. This is confirmed by the visual analysis of surface topography profiles. While an evident roughening of the surface outline is noticeable for uncoated wood, the excessive initial waviness disappears on the biofinish-coated profiles.

Low values of skewness, $S_{sk} \approx 0$, as observed for the uncoated pine samples, indicates a normal distribution of the irregularity deviation. However, a shift in the skewness toward the higher side of the topography map is typical of porous materials, such as softwoods. The decrease in skewness observed for the uncoated wood can be associated with the loss of fibres and the general progress of the weathered wood surface erosion. Kurtosis (S_{kt}) is a measure of the topography histogram profile sharpness. Values of S_{kt} of uncoated pine wood gradually decreased from platykurtic ($S_{kt} \approx 12$) to leptokurtic ($S_{kt} \approx 2$). The kurtosis of the coated wood surface was continuously elevated, indicating its strong platykurtic nature. The coated surface possessed a balanced distribution of peaks (loose fibres) and valleys (surface cracks).

3.6. Microscopic Observations

Microscopic images of investigated samples are shown in Figure 8. The development of mould fungi with dark-coloured hyphae and spores is a very common phenomenon observed on coated and uncoated wooden façades. Moulds and blue stain fungi are considered to be undesirable elements, especially on light-coloured wooden façades [21]. Consequently, the presence of such microorganisms may substantially reduce the aesthetical service life of the whole building. Moreover, the wood decay process can be initiated by the presence of fungal spores or mycelia fragments on wood substrate. Spores, as indicated by arrows in Figure 8, will germinate under favourable conditions to produce fine hair-like structures known as fungal hyphae. Hyphal fragments landing on the wood surface can in many cases initiate fungal growth, leading to a broader colonization. *Ascomycetes* fungi, including the *A. pullulans* studied in this research, belong to soft rot species. These present a mixture of the white and brown rot characteristics digesting both carbohydrates (cellulose and hemicellulose) as well as lignin. It was found that the structural damage caused by the *A. pullulans* fungus is constrained to the external few millimetres of the exposed wooden element outline [67]. *A. pullulans* possesses nutritional versatility, which is associated with its wide enzymatic profile [68]. Therefore, the potential damage of a wooden surface can be eliminated by providing a preferable nutrition source that is different from the substrate wood. Linseed oil was used in the case study. It should also be mentioned that *A. pullulans* is known to possess antagonistic properties towards other yeasts and fungi. The biofilm developed by this fungus may inhibit the growth of non-desired microorganisms, assuring supplementary protection against biotic factors [69].

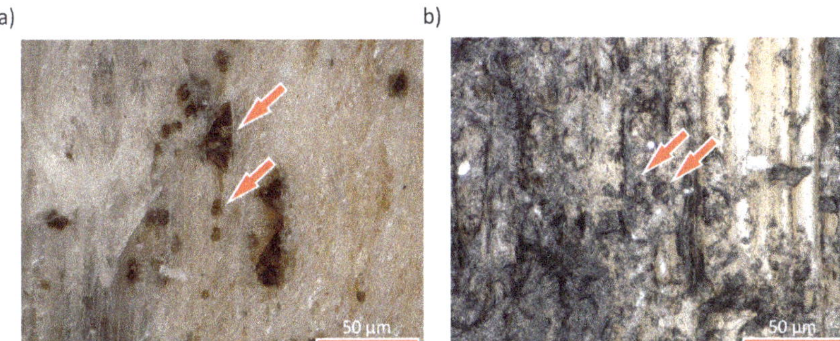

Figure 8. Chlamydospores of *A. pullulans* on the surface of uncoated Scots pine after 9 months of exposure (**a**) and a biofinish-coated reference sample not exposed to natural weathering (**b**), observed under microscope with ×1000 magnification.

4. Conclusions

A majority of the common surface treatments, coating products, or finishing techniques rely on mineral oil binders and diverse ingredients that may not be considered environmentally friendly. This study focused on the assessment of the in-service performance of the innovative coating system based on a bioinspired concept. The performance of Scots pine (*Pinus sylvestris* L.) wood coated with biofinish was confronted with uncoated reference wood following a 12-month natural weathering trial. Experimental samples were objectively characterized following a multi-sensor approach, with a special focus on aesthetical deterioration aspects. Performance of the wood treated with biofinish was superior when compared to the uncoated reference regarding all examined aspects. An interesting phenomenon related to the dual self-healing mechanism was observed as the surface texture consistency and proved by an evidenced decrease in the surface roughness along with the weathering progress. Such a trend is contrary to the majority of materials exposed to weathering that are gradually deteriorating and eroding. An entirely bio-based composition of the biofinish coating solution is a great advantage assuring unique sustainability and compatibility to the natural environments. It can be considered, therefore, as an attractive alternative for state-of-the-art wood surface protection solutions.

5. Patents

Ecologically protected material EP1704028 B1 & US7951363 B2.
PCT patent application 2020: PCT/NL2020/050585.

Author Contributions: Conceptualization, F.P., J.S., M.S., L.R., T.B., and A.S.; methodology, J.S. and A.S.; software, J.S.; validation, F.P., J.S., and A.S.; formal analysis, F.P., J.S., and A.S.; investigation, F.P., J.S., M.S., L.R., T.B., and A.S.; resources, J.S. and A.S.; data curation, J.S. and A.S.; writing—original draft preparation, F.P., J.S., and A.S.; writing—review and editing, F.P., J.S., M.S., L.R., T.B., and A.S.; visualization, J.S.; supervision, J.S., L.R., T.B., and A.S.; project administration, A.S.; funding acquisition, J.S. and A.S. All authors have read and agreed to the published version of the manuscript.

Funding: The authors gratefully acknowledge the European Commission for funding the InnoRenew project (Grant Agreement #739574 under the Horizon2020 Widespread-2-Teaming program), the Republic of Slovenia (investment funding from the Republic of Slovenia and the European Regional Development Fund) and infrastructural ARRS program IO-0035. Part of this work was conducted during project BIO4ever (RBSI14Y7Y4), funded within call SIR by MIUR-Italy; the project Multi-spec (BI-IT/18-20-007), funded by ARRS-Slovenia; Archi-BIO (BI/US-20-054) funded by ARRS-Slovenia, J7-9404 (C) funded by ARRS-Slovenia, and CLICK DESIGN, "Delivering fingertip knowledge to enable service life performance specification of wood", (No. 773324) supported under the umbrella of ERA-NET Cofund ForestValue by the Ministry of Education, Science and Sport of the Republic of Slovenia. ForestValue has received funding from the European Union's Horizon 2020 research and innovation programme.

Institutional Review Board Statement: Not applicable.

Informed Consent Statement: Not applicable.

Data Availability Statement: The data presented in this study are available upon request from the corresponding author.

Acknowledgments: The experimental samples were provided by Xyhlo B.V., the Netherlands. The support of the industrial partner is highly appreciated.

Conflicts of Interest: The authors declare no conflict of interest.

References

1. Feist, W.C.; Hon, D.N.-S. Chemistry of weathering and protection. *Chem. Solid Wood* **1984**, 401–451. [CrossRef]
2. Rowell, R.M. *Handbook of Wood Chemistry and Wood Composites*; CRC Press: New York, NY, USA, 2005; ISBN 0849315883.
3. Evans, P.; Chowdhury, M.J.; Mathews, B.; Schmalzl, S.K.; Ayer, M.; Kataoka, K.Y.; Kutz, M. *Handbook of Environmental Analysis*; William Andrew Publishing, Norwich: Norwich, UK, 2005; ISBN 0815515006.
4. Hayoz, P.; Peter, W.; Rogez, D. A new innovative stabilization method for the protection of natural wood. *Prog. Org. Coat.* **2003**, *48*, 297–309. [CrossRef]
5. Šimkovic, I. Free radicals in wood chemistry. *J. Macromol. Sci. Part C* **1986**, *26*, 67–80. [CrossRef]
6. Müller, U.; Rätzsch, M.; Schwanninger, M.; Steiner, M.; Zöbl, H. Yellowing and IR-changes of spruce wood as result of UV-irradiation. *J. Photochem. Photobiol. B Biol.* **2003**, *69*, 97–105. [CrossRef]
7. Läszlo, T.; Faix, O. Artificial ageing of wood monitored by DRIFT spectroscopy and CIE *L*a*b** color measurements. I. Effect of UV light. *Holzforschung* **1995**, *49*, 397–404. [CrossRef]
8. Kataoka, Y.; Kiguchi, M.; Williams, R.S.; Evans, P.D. Violet light causes photodegradation of wood beyond the zone affected by ultraviolet radiation. *Holzforschung* **2007**, *61*, 23–27. [CrossRef]
9. Sandak, J.; Sandak, A.; Cocchi, M. Multi-sensor data fusion and parallel factor analysis reveals kinetics of wood weathering. *Talanta* **2021**, *225*, 122024. [CrossRef] [PubMed]
10. Sandak, A.; Sandak, J.; Noël, M.; Dimitriou, A. A method for accelerated natural weathering of wood subsurface and its multilevel characterization. *Coatings* **2021**, *11*, 126. [CrossRef]
11. Pánek, M.; Reinprecht, L. Critical view on the possibility of color changes prediction in the surfaces of painted wood exposed outdoors using accelerated weathering in xenotest. *J. Coat. Technol. Res.* **2019**, *16*, 339–352. [CrossRef]
12. Bobadilha, G.D.S.; Stokes, C.E.; Ohno, K.M.; Kirker, G.; Lopes, D.J.V.; Nejad, M. Physical, optical, and visual performance of coated cross-laminated timber during natural and artificial weathering. *Coatings* **2021**, *11*, 252. [CrossRef]

13. Temiz, A.; Terziev, N.; Eikenes, M.; Hafren, J. Effect of accelerated weathering on surface chemistry of modified wood. *Appl. Surf. Sci.* **2007**, *253*, 5355–5362. [CrossRef]
14. Ayadi, N.; Lejeune, F.; Charrier, F.; Charrier, B.; Merlin, A. Color stability of heat-treated wood during artificial weathering. *Holz Als Roh-Und Werkst.* **2003**, *61*, 221–226. [CrossRef]
15. Colom, X.; Carrillo, F.; Nogués, F.; Garriga, P. Structural analysis of photodegraded wood by means of FTIR spectroscopy. *Polym. Degrad. Stab.* **2003**, *80*, 543–549. [CrossRef]
16. Pandey, K.K. Study of the effect of photo-irradiation on the surface chemistry of wood. *Polym. Degrad. Stab.* **2005**, *90*, 9–20. [CrossRef]
17. Ganne-Chédeville, C.; Jääskeläinen, A.S.; Froidevaux, J.; Hughes, M.; Navi, P. Natural and artificial ageing of spruce wood as observed by FTIR-ATR and UVRR spectroscopy. *Holzforschung* **2012**, *66*, 163–170. [CrossRef]
18. Petrillo, M.; Sandak, J.; Grossi, P.; Sandak, A. Chemical and appearance changes of wood due to artificial weathering-dose-response model. *J. Near Infrared Spectrosc.* **2019**. [CrossRef]
19. Zanetti, M.; Rials, T.G.; Rammer, D. NIR-Monitoring of in-service wood structures. In Proceedings of the Structures Congress 2005, New York, NY, USA, 20–24 April 2005; American Society of Civil Engineers: New York, NY, USA, 2005; Volume 9.
20. Sandak, J.; Sandak, A.; Riggio, M. Characterization and monitoring of surface weathering on exposed timber structures with a multi-sensor approach. *Int. J. Archit. Herit.* **2015**, *9*, 674–688. [CrossRef]
21. Gobakken, L.R.; Vestøl, G.I. Surface mould and blue stain fungi on coated Norway spruce cladding. *Int. Biodeterior. Biodegrad.* **2012**, *75*, 181–186. [CrossRef]
22. Jankowska, A.; Rybak, K.; Nowacka, M.; Boruszewski, P. Insight of weathering processes based on monitoring surface characteristic of tropical wood species. *Coatings* **2020**, *10*, 877. [CrossRef]
23. Žlahtič, M.; Humar, M. Influence of artificial and natural weathering on the hydrophobicity and surface properties of wood. *BioResources* **2016**, *11*, 4964–4989. [CrossRef]
24. Pánek, M.; Oberhofnerová, E.; Zeidler, A.; Šedivka, P. Efficacy of hydrophobic coatings in protecting oak wood surfaces during accelerated weathering. *Coatings* **2017**, *7*, 172. [CrossRef]
25. Oberhofnerová, E.; Šimunková, K.; Dvořák, O.; Štěrbová, I.; Hiziroglu, S.; Šedivka, P.; Pánek, M. Comparison of exterior coatings applied to oak wood as a function of natural and artificial weathering exposure. *Coatings* **2019**, *9*, 864. [CrossRef]
26. Salca, E.-A.; Krystofiak, T.; Lis, B.; Hiziroglu, S. Glossiness evaluation of coated wood surfaces as function of varnish type and exposure to different conditions. *Coatings* **2021**, *11*, 558. [CrossRef]
27. Cogulet, A.; Blanchet, P.; Landry, V. The multifactorial aspect of wood weathering: A review based on a holistic approach of wood degradation protected by clear coating. *BioResources* **2018**, *13*, 2116–2138. [CrossRef]
28. Pintus, V.; Wei, S.; Schreiner, M. Accelerated UV ageing studies of acrylic, alkyd, and polyvinyl acetate paints: Influence of inorganic pigments. *Microchem. J.* **2016**, *124*, 949–961. [CrossRef]
29. Christensen, P.A.; Dilks, A.; Egerton, T.A.; Temperley, J. Infrared spectroscopic evaluation of the photodegradation of paint Part I the UV degradation of acrylic films pigmented with titanium dioxide. *J. Mater. Sci.* **1999**, *34*, 5689–5700. [CrossRef]
30. Singh, R.P.; Tomer, N.S.; Bhadraiah, S.V. Photo-oxidation studies on polyurethane coating: Effect of additives on yellowing of polyurethane. *Polym. Degrad. Stab.* **2001**, *73*, 443–446. [CrossRef]
31. Rosu, D.; Rosu, L.; Cascaval, C.N. IR-change and yellowing of polyurethane as a result of UV irradiation. *Polym. Degrad. Stab.* **2009**, *94*, 591–596. [CrossRef]
32. Gobakken, L.R.; Westin, M. Surface mould growth on five modified wood substrates coated with three different coating systems when exposed outdoors. *Int. Biodeterior. Biodegrad.* **2008**, *62*, 397–402. [CrossRef]
33. Schmidt, O. *Wood and Tree Fungi*; Springer: Berlin/Heidelberg Germany; Springer: New York, NY, USA, 2006; ISBN 9783540321385.
34. Horvath, R.S.; Brent, M.M.; Cropper, D.G. Paint deterioration as a result of the growth of aureobasidium pullulans on wood. *Appl. Environ. Microbiol.* **1976**, *32*, 505–507. [CrossRef]
35. Gesthuizen, J. Bio-Based Coatings Overview: Increasing Activities | European Coatings. Available online: https://www.european-coatings.com/articles/archiv/bio_based-coatings-overview-increasing-activities (accessed on 11 May 2021).
36. Chemicals Strategy. Available online: https://ec.europa.eu/environment/strategy/chemicals-strategy_sl (accessed on 11 May 2021).
37. López, M.; Rubio, R.; Martín, S.; Croxford, B. How plants inspire façades. From plants to architecture: Biomimetic principles for the development of adaptive architectural envelopes. *Renew. Sustain. Energy Rev.* **2017**, *67*, 692–703. [CrossRef]
38. Lurie-Luke, E. Product and technology innovation: What can biomimicry inspire? *Biotechnol. Adv.* **2014**, *32*, 1494–1505. [CrossRef] [PubMed]
39. Imani, M.; Donn, M.; Balador, Z. Bio-Inspired Materials: Contribution of Biology to Energy Efficiency of Buildings. In *Handbook of Ecomaterials*; Springer International Publishing: Berlin/Heidelberg, Germany, 2018; pp. 1–24.
40. Vijay, K.; Murmu, M.; Deo, S.V. Bacteria based self healing concrete—A review. *Constr. Build. Mater.* **2017**, *152*, 1008–1014. [CrossRef]
41. Sailer, M.F.; van Nieuwenhuijzen, E.J.; Knol, W. Forming of a functional biofilm on wood surfaces. *Ecol. Eng.* **2010**, *36*, 163–167. [CrossRef]
42. Peeters, L.H.M.; Huinink, H.P.; Voogt, B.; Adan, O.C.G. Oil type and cross-linking influence growth of aureobasidium melanogenum on vegetable oils as a single carbon source. *Microbiologyopen* **2018**, *7*, 1–12. [CrossRef] [PubMed]

43. Xyhlo Biofinish | Sustainable and Eco-Friendly Wood Protection. Available online: https://www.xyhlo.com/en/ (accessed on 29 March 2021).
44. Van Nieuwenhuijzen, E.J.; Houbraken, J.A.M.P.; Meijer, M.; Adan, O.C.G.; Samson, R.A. Aureobasidium melanogenum: A native of dark biofinishes on oil treated wood. *Antonie Van Leeuwenhoek* **2016**, *109*. [CrossRef]
45. Van Nieuwenhuijzen, E.J.; Sailer, M.F.; van den Heuvel, E.R.; Rensink, S.; Adan, O.C.G.; Samson, R.A. Vegetable oils as carbon and energy source for aureobasidium melanogenum in batch cultivation. *Microbiologyopen* **2019**, *8*, 1–12. [CrossRef]
46. Evans, P.D.; Michell, A.J.; Schmalzl, K.J. Studies of the degradation and protection of wood surfaces. *Wood Sci. Technol.* **1992**, *26*, 151–163. [CrossRef]
47. Turkoglu, T.; Baysal, E.; Toker, H. The effects of natural weathering on color stability of impregnated and varnished wood materials. *Adv. Mater. Sci. Eng.* **2015**, *2015*. [CrossRef]
48. Oberhofnerová, E.; Pánek, M.; García-Cimarras, A. The effect of natural weathering on untreated wood surface. *Maderas Cienc. Y Tecnol.* **2017**, *19*, 173–184. [CrossRef]
49. Hon, D.N.S.; Shiraishi, N. *Wood and Cellulosic Chemistry, Revised, and Expanded*, 2nd ed.; Marcel Dekker: New York, NY, USA, 2000; ISBN 0824700244.
50. Ghosh, S.C.; Militz, H.; Mai, C. Natural weathering of scots pine (*Pinus sylvestris* L.) boards modified with functionalised commercial silicone emulsions. *BioResources* **2009**, *4*, 659–673. [CrossRef]
51. Scheffer, T.C. Natural resistance of wood to microbial deterioration. *Annu. Rev. Phytopathol.* **1966**, *4*, 147–168. [CrossRef]
52. Viitanen, H.A. Modelling the time factor in the development of mould fungi—The effect of critical humidity and temperature conditions on pine and spruce sapwood. *Holzforschung* **1997**, *51*, 6–14. [CrossRef]
53. Butler, M.J.; Day, A.W. Fungal melanins: A review. *Can. J. Microbiol.* **1998**, *44*, 1115–1136. [CrossRef]
54. El-Bialy, H.A.; El-Gamal, M.S.; Elsayed, M.A.; Saudi, H.A.; Khalifa, M.A. Microbial melanin physiology under stress conditions and gamma radiation protection studies. *Radiat. Phys. Chem.* **2019**. [CrossRef]
55. Jiang, H.; Liu, N.N.; Liu, G.L.; Chi, Z.; Wang, J.M.; Zhang, L.L.; Chi, Z.M. Melanin production by a yeast strain XJ5-1 of aureobasidium melanogenum isolated from the Taklimakan desert and its role in the yeast survival in stress environments. *Extremophiles* **2016**, *20*, 567–577. [CrossRef]
56. Kumar, C.G.; Mongolla, P.; Pombala, S.; Kamle, A.; Joseph, J. Physicochemical characterization and antioxidant activity of melanin from a novel strain of Aspergillus bridgeri ICTF-201. *Lett. Appl. Microbiol.* **2011**, *53*, 350–358. [CrossRef]
57. Hernandez, V.A.; Evans, P.D. Technical note: Melanization of the wood-staining fungus aureobasidium pullulans in response to UV radiation. *Wood Fiber Sci* **2015**, *47*, 120–124.
58. Gniewosz, M.; Duszkiewicz-Reinhard, W. Comparative studies on pullulan synthesis, melanin synthesis and morphology of white mutant aureobasidium pullulans B-1 and parent strain A.p.-3. *Carbohydr. Polym.* **2008**, *72*, 431–438. [CrossRef]
59. Lingappa, Y.; Sussman, A.S.; Bernstein, I.A. Effect of light and media upon growth and melanin formation in aureobasidium pullulans (DE BY.) ARN. (=Pullularia pullulans). *Mycopathol. Mycol. Appl.* **1960**, *20*, 109–128. [CrossRef]
60. Kucuktuvek, M.; Baysal, E.; Turkoglu, T.; Peker, H.; Gunduz, A.; Toker, H. Surface characteristics of scots pine wood heated at high temperatures after weathering. *Wood Res.* **2017**, *62*, 905–917.
61. Turkoglu, T.; Toker, H.; Baysal, E.; Kart, S.; Yuksel, M.; Ergun, M.E. Some surface properties of heat treated and natural weathered oriental beech. *Wood Res.* **2015**, *60*, 881–890.
62. Oberhofnerová, E.; Pánek, M. Surface wetting of selected wood species by water during initial stages of weathering. *Wood Res.* **2016**, *61*, 545–552.
63. Kalnins, M.A.; Feist, W.C. Increase in wettability of wood with weathering. *For. Prod. J.* **1993**, *43*, 55–57.
64. Kishino, M.; Nakano, T. Artificial weathering of tropical woods. Part 2: Color change. *Holzforschung* **2004**, *58*, 558–565. [CrossRef]
65. Gonzalezde, C.P.H.; Missio, A.L.; Mattos, B.D.; Gatto, D.A. Natural weathering performance of three fast-growing eucalypt woods. *Maderas Cienc. Y Tecnol.* **2016**, *17*, 799–808. [CrossRef]
66. Williams, R.R. Weathering of Wood. In *Handbook of Wood Chemistry and Wood Composites*; Rowel, M.R., Ed.; CRC Press: Boca Raton, FL, USA, 2005; pp. 139–185, ISBN 0849315883.
67. Goodell, B.; Winandy, J.E.; Morrell, J.J. Fungal degradation of wood: Emerging data, new insights and changing perceptions. *Coatings* **2020**, *10*, 1210. [CrossRef]
68. Di Francesco, A.; Ugolini, L.; D'Aquino, S.; Pagnotta, E.; Mari, M. Biocontrol of monilinia laxa by aureobasidium pullulans strains: Insights on competition for nutrients and space. *Int. J. Food Microbiol.* **2017**, *248*, 32–38. [CrossRef]
69. Bozoudi, D.; Tsaltas, D. The multiple and versatile roles of aureobasidium pullulans in the vitivinicultural sector. *Fermentation* **2018**, *4*, 85. [CrossRef]

Article

Physical, Optical, and Visual Performance of Coated Cross-Laminated Timber during Natural and Artificial Weathering

Gabrielly dos Santos Bobadilha [1,*], C. Elizabeth Stokes [1], Katie M. Ohno [2], Grant Kirker [2], Dercilio Junior Verly Lopes [1] and Mojgan Nejad [3]

1 Department of Sustainable Bioproducts, Forest and Wildlife Research Center (FWRC), Mississippi State University, Starkville, MS 39762, USA; ces8@msstate.edu (C.E.S.); dvl23@msstate.edu (D.J.V.L.)
2 Forest Products Laboratory-USDA, Madison, WI 53726, USA; katie.m.ohno@usda.gov (K.M.O.); grant.kirker@usda.gov (G.K.)
3 Department of Forestry, College of Agriculture & Natural Resources, Michigan State University, East Lansing, MI 48824, USA; nejad@msu.edu
* Correspondence: gd450@msstate.edu

Abstract: Cross-laminated timber (CLT) market demand is on the rise in the United States. Adequate protective measures have not been extensively studied. The objective of this study was to investigate the weathering performance of exterior wood coatings. We evaluated coated CLT sample surfaces based on visual appearance, color change (CIE*L*a*b), gloss changes, and water intrusion. From the five exterior wood coatings evaluated, only two showed adequate performance after twelve months field exposure. Based on visual ratings following the ASTM procedures, coating failure occurs more quickly in Mississippi than in Wisconsin, due to its greater decay zone. Both location and coating type impacted the aging of the samples. Artificial weathering results were consistent with natural weathering indicating the two adequate coatings were the most resistant to failure, color, and gloss change. For future studies, new coatings designed for the protection of end-grain in CLT panels should be a target of research and development.

Keywords: mass-timber; surface protection; outdoor exposure; wood degradation; cross-laminated timber

Citation: Bobadilha, G.d.S.; Stokes, C.E.; Ohno, K.M.; Kirker, G.; Lopes, D.J.V.; Nejad, M. Physical, Optical, and Visual Performance of Coated Cross-Laminated Timber during Natural and Artificial Weathering. *Coatings* **2021**, *11*, 252. https://doi.org/10.3390/coatings11020252

Academic Editor: Anna Sandak

Received: 27 January 2021
Accepted: 18 February 2021
Published: 20 February 2021

1. Introduction

The use of cross-laminated timber (CLT) in construction has expanded across international markets because of its structural performance, seismic behavior, and sustainability [1–3]. CLT was introduced in North America in the early 2000 s, and since then research have been developed to expand its use across the building construction sector. In spite of recent advances, limited methods and technologies are readily available to prolong the durability of buildings composed of such material. Due to aesthetics, exterior CLT structural materials have been left unprotected, where they are either semi- or fully exposed to weathering [4]. In such conditions, surface degradation of uncoated CLT is very likely.

Although wood can be used exteriorly and interiorly, when uncoated and exposed to weather, wood surfaces deteriorate over time [5]. The natural properties and appearance of wood are compromised due to water and sun exposure. Repetitive cycles of wetting and drying cause alteration of chemical bonds and oxidation [6,7]. In addition to wetting, sun exposure can seriously damage the surface of wood materials since surface photo-oxidation is catalyzed by ultraviolet radiation [8,9]. The combined effects of moisture and solar exposure are primarily responsible for checking, splitting, surface erosion, and degradation caused by microorganisms [10].

The occurrence and development of mold, decay, and stains by fungi are related to climate conditions, such temperature and moisture [11]. In the Southern US, where temperature and humidity are high compared to other regions in the country, the color and appearance of wood are rapidly modified by weathering, mold, and stain fungi [12]. A study comparing climate index for decay in wood structures in three different states (MS, OR, and WI) showed that highest decay susceptibility was found in Mississippi [13].

Outdoor performance of CLT is still unknown in many parts of the US. However, the literature reports that its abundant storage capacity may lead to biodeterioration [14,15]. To expand the use of CLT in mid- and high-rise construction markets, more research is needed to implement proper codes for managing moisture and weathering degradation [16]. Currently, there is a lack of studies on surface protection of mass timber exposed to outdoor environments. One of the reasons is that CLT was not primarily designed to be exteriorly exposed [16]. However, as its use increases throughout the country, more projects are designed with either full or partial external exposure. This study was designed to investigate the potential of readily available coatings in North America to protect CLT against weathering degradation.

Surface protection is generally used to maintain and prolong the aesthetics and service life of materials that may be affected by weathering factors during any stage from manufacturing to its end use. Paints, varnishes, stains, or water repellent coatings are commonly used to prevent weathering degradation [17]. They limit the passage of water in and out of wood, which may cause rapid dimensional changes [12]. In exterior applications, the use of penetrating semitransparent stains is recommended, since they do not crack or peel during exposure and are less influenced by dimensional changes caused by water differentials [18]. Proven hydrophobic coatings are recommended to be used without direct rainwater contact [19]. Consequently, the effectiveness of a finish is directly influenced by when and or where it is exposed. In a previous work, we investigated the use of ANN to predict color change based on visual grading on coated CLT naturally weathered in Mississippi [20]. In this study, our objective is to compare the performance of coated CLT exposed in Mississippi with the ones exposed in Madison. We also improve upon [20] by investigating gloss change and water uptake, which are properties that critically influence the correct deployment, preservation and maintenance of CLT.

2. Materials and Methods

2.1. Sample Preparation and Coating Systems

CLT blocks of hemlock-fir (*Tsuga sp.* and *Abies sp.*) were prepared from three-ply panels (SmartLam LLC; Whitefish, MT, USA) measuring 15 cm × 14 cm × 11 cm. In a previous study, 12 coatings were investigated (including paints, stains and water relents named from A to L) [21], and based on a series of preliminary tests on their water-repellency and anti-swelling efficiency, the five best-performing coatings were selected for a second study, reported here (Table 1).

Table 1. Coating treatment specifications.

Treatment	Coating or Surface Description	Resin Type	No. of Coats	No. of Reps
A (Alk/Acr, W[1])	Transparent penetrating wood finish	Alkyd/Acrylic	2	6
C (Alk/Acr, W)	Transparent, UV resistant	Alkyd/Acrylic	3	6
F (Acr, W)	Semi-transparent, water and UV resistant	Acrylic	2	6
I (Alk, S[2])	Transparent, mildew and water resistant	Alkyd	2	6
J (Alk, S)	Semitransparent and water repellent	Alkyd	1	6
Control	Uncoated sample	None	None	6

W[1] = water-based; S[2] = Solvent-based.

Coatings were applied on unprimed CLT samples with foam brush on the top surface and sides in accordance with manufacturer recommendation.

2.2. Natural and Artificial Weathering

CLT samples were exposed for 12 months (June 2019—June 2020) in Madison, Wisconsin (AWPA hazard zone 2) and in Starkville, Mississippi (AWPA hazard zone 4). Normally, exterior wood coatings exposed to natural weathering are tested on vertical positions with minimum length dimension of 152 mm [22]. Since the CLT samples were thicker, the samples contained different dimensions, and racks were designed to horizontally expose the samples. In Madison, the exposure site was located at the Forest Products Laboratory (FPL, Madison, WI, USA) Valley View field site. The CLT samples were atop racks that were constructed with garden mesh to avoid water trapping underneath the test samples. An overview of the climatic conditions during one year of natural weathering exposure is displayed in Table 2. One sample per treatment was left unexposed in an environmental chamber at 25 °C and 66% RH for each set of samples for further comparison.

Table 2. Weather conditions in Mississippi and Wisconsin during natural outdoor exposure of coated and uncoated CLT samples, adapted from [23,24].

		Mississippi *		
2019/2020		**Weather Conditions**		
Months	**Mean Temperature (°C)**	**Total Precipitation (mm)**	**Mean Radiation (kW-h/m^2)**	**Total Snow (mm)**
Jun (0)	26	211.6	6.3	-
Jul (1)	27.5	271.3	6.5	-
Aug (2)	27.5	140.5	6.4	-
Sept (3)	27.5	1	6.1	-
Oct (4)	19.2	278.4	4.5	-
Nov (5)	9.3	93.7	3.3	-
Dec (6)	9.9	172.5	2.7	-
Jan (7)	9.4	261.1	2.9	-
Feb (8)	9.3	373.4	3.7	-
Mar (9)	16.8	180.8	4.9	-
Apr (10)	16.1	287.3	6.1	-
May (11)	20.5	46.2	6.5	
Jun (12)	25.7	137.2	6.5	-
		Wisconsin		
Jun (0)	19.2	131.1	6.8	-
Jul (1)	24	146.6	6.7	-
Aug (2)	20.3	72.4	6	-
Sept (3)	18.9	172.7	4.9	-
Oct (4)	9.2	140.7	3.4	205.7
Nov (5)	−0.5	66.8	2.1	193.0
Dec (6)	−1.1	38.6	1.6	73.7
Jan (7)	−3.2	1.7	1.9	464.82
Feb (8)	−4.9	0.9	2.9	337.8
Mar (9)	3.6	88.1	4.1	71.2
Apr (10)	7.2	52.1	5.2	5.1
May (11)	13.5	137.7	6.2	-
Jun (12)	21.2	129.5	6.8	-

* Data from Mississippi were previously published [20].

The artificial weathering test was conducted in a weathering apparatus that consisted of a stainless-steel cabinet equipped with water spray, UV lamps, and temperature controlled forced air heating elements that simulated exterior conditions by alternating each cycle of irradiation and water. The unit was equipped with UV-A lamps (600 W·m^{-2} at 340 nm), maintaining an average temperature of 26 °C. The samples were exposed to weathering cycles of 12 h of UV-light irradiation and 12 h of water spray (0.36 Lpm) for an accelerated 15 days (360 h) and a longer term of 75 days (1800 h). The test panels were placed at a 45° angle and assignment of board location was randomized to eliminate potential position bias.

2.3. *Surface Appearance Analysis*

At the end of each month, superficial damage was evaluated according to the American Standard and Testing Materials [25–31]. Cracking, flaking, mildew growth and erosion were visually distinguishable after three months of exposure; Figure 1 shows the most common defects found on samples surface.

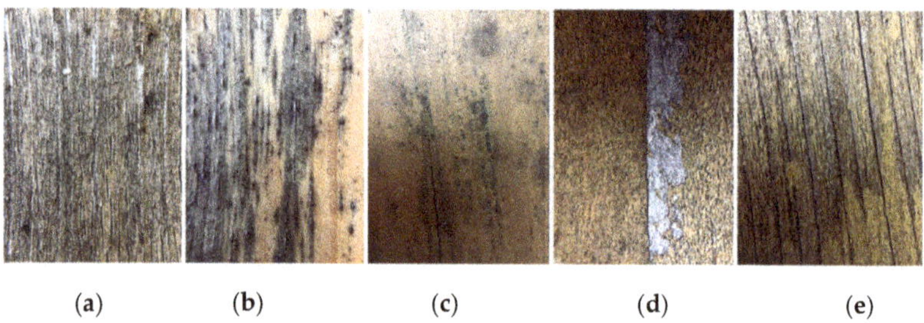

Figure 1. Types of coating failure after natural weathering exposure. (**a**) Several types of micro-checks and micro-cracks; (**b**) irregular micro-checks; (**c**) long line checks; (**d**) combination of checks, cracks, flaking, and erosion; (**e**) advanced coating erosion.

Stain and paint failures were evaluated by comparison with photographic standards. Degrees of cracking, flaking, and erosion range from 0 (complete failure) to 10 (no visible defect). Each procedure describes one type of failure. For example, cracking is manifested as breaking of paint or stain only where underlying surface is visible. Flaking is classified as detachment of coating from its substrate. Erosion describes paint removal that displays bare wood surface. Mold growth ratings followed the similar principle, where 0 described full coverage and 10 no visual growth (Figure 2).

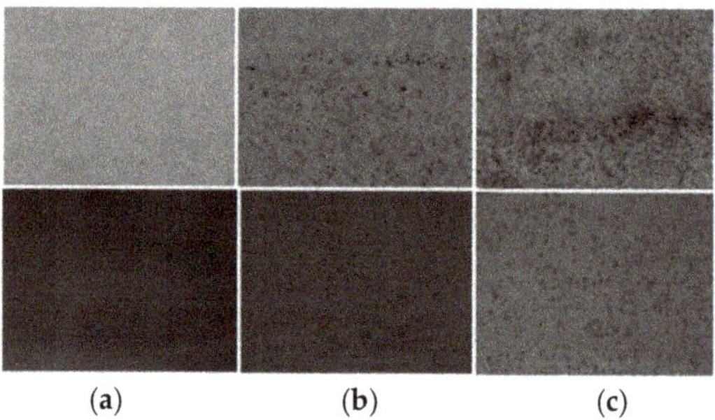

Figure 2. Mildew growth: (**a**) slight; (**b**) moderate, and; (**c**) severe mold pictures. First row corresponds to reference pictures [26], second row contains the mildew growing on CLT samples surface.

2.4. *Color and Gloss Measurements*

The color parameters of the tested blocks were measured using a hand-held spectrophotometer (CM-2300d, Konica Minolta, Osaka, Japan). Both natural and artificially weathered samples were evaluated, and five measurements were taken at the same location of each specimen following the Commission International de l'Eclairage [32] colorimetry method using color parameters *($L^*a^*b^*$)*. Where L^* represents lightness from 0 (black) to 100 (white), a^* chromaticity coordinates red (+) or green (-), and b^* chromaticity coordinates yellow (+) or blue (-). The color changes (ΔL^*, Δa^*, Δb^*) between the exposed period and initial state were determined. Color differences were calculated using Equation (1).

$$\Delta E^* = \sqrt{\Delta L^{*2} + \Delta a^{*2} + \Delta b^{*2}} \tag{1}$$

The surface luster of samples was measured using a glossmeter ETB-0686 following ISO 2813 [33]. Three measurements were made on each sample at a 60°. Alterations of the surface luster of coated and uncoated samples were assessed at the end of each month or at the end of the artificial weathering cycle. Results were based on a specular gloss value of 96 gloss units (GU), related to the perfect condition under identical illumination and view conditions of highly polished plain black glass surface. Color and gloss measurements were recorded at 0, 6, and 12 months after exposure.

2.5. Moisture Content during Exposure

During the manufacturing process, CLT is often only glued between layers and not on the sides. For this reason, all exposed surfaces contained an edge joint. At the end of each month for 6 months, and then after 12 months, all exposed samples were weighed to determine the mass. The initial weight was taken when the samples were already coated and at equilibrium moisture content (EMC = 12%) after 45 days in environmental chamber at 25 °C and 66% RH. Moisture content (MC) was calculated as displayed in Equation (2):

$$\text{MC} = \frac{W_2 - W_1}{W_1} \times 100 \tag{2}$$

where:

MC = moisture content after exposure in %,
W_2 = final weight gain in g,
W_1 = initial weight gain in g.

2.6. Statistical Analysis

Statistical Analysis of Variance (ANOVA) were performed for artificial weathering test variables. Color and gloss changes were analyzed as completely randomized design based on coating effect. Tests were performed at $\alpha = 0.05$, when the sources of variations were detected as significant by the Fisher-test ($p \leq 0.05$). Analyses were performed using Statistical Analysis System (SAS) software version 9.4 [34].

3. Results and Discussion

3.1. Natural Weathering

3.1.1. Visual Assessments

The types of failure present on coatings were mainly fungal growth, cracking, flaking, and erosion, consequently the samples were evaluated according to those characteristics (Table 3). Location had a great impact on coating degradation. For instance, after 3 months, all samples displayed mildew growth in Mississippi.

In Wisconsin, coating F Acr protected the samples against mildew until three months of exposure; however, after six months, mildew growth was visible. Coating C Alk/Acr as well did not show any mildew growth even after 6 months of exposure. Figure 3 shows surface degradation over time for all coatings tested in comparison to uncoated controls.

Table 3. Average visual rating for coatings after 3, 6 and 12 months of outdoor exposure in Mississippi and Wisconsin, reported as mean and standard deviation in parenthesis.

Coatings	Mississippi [1]			Wisconsin		
	Mildew Growth					
	3	6	12	3	6	12
A. Alk/Acr	4 (1.0)	3 (1.4)	7 (0.8)	7 (1.1)	5 (0.7)	5 (1.0)
C. AlkAcr	9 (0.5)	7 (1.1)	7 (1.1)	10 (0.5)	10 (0.9)	8 (0.8)
F. Acr	7 (1.1)	5 (0.8)	5 (0.0)	10 (0.4)	8 (0.5)	9 (0.5)
I. Alk	7 (1.3)	4 (0.9)	9 (0.5)	7 (0.8)	6 (1.3)	5 (0.9)
J. Alk	1 (0.6)	1 (0.0)	5 (1.6)	6 (0.5)	3 (0.4)	4 (0.4)
Control	5 (1.2)	5 (1.1)	9 (0.7)	7 (0.5)	7 (0.0)	9 (0.5)
	Checking					
	3	6	12	3	6	12
A. Alk/Acr	6 (0.9)	4 (2.3)	0 (1.1)	6 (1.4)	4 (0.9)	0 (0.0)
C. Alk/Acr	10 (0.0)	8 (1.8)	8 (2.2)	10 (0.0)	8 (1.1)	8 (1.7)
F. Acr	10 (0.9)	10 (0.9)	8 (1.9)	8 (0.9)	8 (0.0)	8 (0.0)
I. Alk	8 (1.1)	6 (1.1)	2 (3.6)	10 (0.0)	8 (0.0)	4 (1.1)
J. Alk	6 (1.7)	2 (1.1)	2 (1.7)	6 (1.7)	8 (1.1)	2 (0.0)
	Flaking					
	3	6	12	3	6	12
A. Alk/Acr	8 (1.4)	6 (1.1)	2 (1.1)	10 (0.0)	6 (1.7)	0 (0.0)
C. Alk/Acr	10 (0.0)	10 (0.0)	8 (2.2)	10 (0.0)	10 (0.0)	4 (1.1)
F. Acr	10 (0.0)	10 (0.0)	10 (0.0)	10 (0.0)	10 (0.0)	10 (0.0)
I. Alk	10 (1.8)	8 (0.9)	6 (1.7)	8 (0.0)	6 (1.1)	4 (1.1)
J. Alk	6 (1.1)	4 (0.9)	4 (2.3)	8 (1.1)	8 (1.1)	4 (0.0)
	Erosion					
	3	6	12	3	6	12
A. Alk/Acr	10 (0.0)	8 (1.1)	4 (1.1)	6 (1.7)	4 (0.0)	0 (0.0)
C. Alk/Acr	10 (0.0)	10 (0.0)	10 (0.0)	10 (0.0)	10 (0.0)	8 (0.9)
F. Acr	10 (0.0)	10 (0.0)	10 (0.0)	10 (0.0)	10 (0.0)	10 (1.1)
I. Alk	10 (0.9)	10 (0.9)	6 (2.0)	8 (0.0)	8 (1.1)	8 (0.9)
J. Alk	8 (1.4)	8 (1.1)	6 (1.7)	6 (2.3)	6 (0.0)	8 (0.0)

[1] Not applicable for uncoated samples. Mississippi results previously published [20].

In Wisconsin, mildew growth increased over time on the sample surfaces, except for the coating J Alk and the control. In Mississippi, a similar trend occurred after 3 and 6 months of exposure, yet after 12 months, mildew decreased. Coatings A and J had noticeably worse performance in Mississippi, most likely due to climate conditions such as higher temperature and humidity (see Table 1). In fact, the essential condition for mildew growth is a sporadic supply of free water. Coatings A Acr/Alk and J Alk were fairly eroded over time, and thus did not promote enough protection. Stirling [35] pointed out that semitransparent wood coatings (e.g., coating J Alk) frequently present signs of early discoloration caused by "black stain" fungi. Coatings C Acr/Alk and F Acr had better performance against mildew because of anti-microbial ingredients present in their composition. For instance, coating C Acr/Alk is composed of two antimicrobial agents (n-n diethylethanamine, DMEA and 3-iodo-2 propynyl butyl carbamate, IPBC), while coating F Acr only contains IPBC. As temperature increases, IPBC may degrade or evaporate [36].

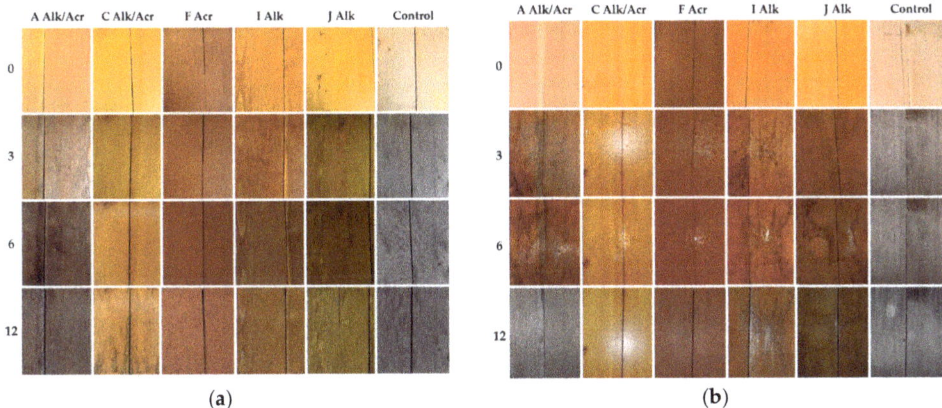

Figure 3. Visual changes of uncoated and coated CLT samples after twelve months of natural weathering in (**a**) Mississippi and (**b**) Wisconsin.

On coating degradation, we observed that most samples surfaces had more than one type of failure after twelve months of exposure; however, the most frequent were checking, flaking, and surface erosion. Most coating performance varied due to site location and time. However, checking assessments had low variations due to location. Overall, coating F Acr had the lowest degree of checking and flaking. Coatings I Alk presented checking earlier (three months of exposure) in Mississippi; yet in Wisconsin, checks were visible only after six months. Coating A Alk/Acr had the worst performance, with a high degree of degradation expressed with flaking and checking at both locations after 12 months of exposure. These failures advanced to surface erosion and a complete removal of the coating. Based on the manufacturer's information, coatings A Alk/Acr and J Alk are penetrating stains that, according to Roux et al. [37], are more likely to fail after a short period of outdoor exposure. Coating F Acr showed no visible sign of erosion in either location. CLT defects on the exposed surface influenced coatings efficiency. The minimum visual grade of longitudinal CLT layers is No. 2 [38], characterized by having knots and wane. Most samples had few defects on the exposed layer but when they did, coating failure was seen earlier.

CLT samples with higher proportions of latewood had early coating deterioration. For the purpose of coating, a high proportion of earlywood is desirable; however, in structural wood products such as CLT, it may result in tangential shrinkage and low stability [39]. As softwood life-cycles are becoming shorter, it is more common to find lumber with a high percentage of earlywood. According to [5], some species that have wide bands of earlywood and latewood do not absorb paint and coatings well. This variation creates stresses such as breaking adhesion between coating and wood resulting in checks and cracks. Penetrating coatings such as A Alk/Acr and J Alk were visually eroded after three months of exposure, whereas film forming coatings were not. Penetrating coatings perform better on porous surfaces. In our study, we exposed the tangential surface, and therefore coatings could not penetrate into the cells.

3.1.2. Color and Gloss Change

The color parameters (L^*, a^*, and b^*) were evaluated based on type of coating, site location, and duration of exposure (0, 3, 6, and 12 months of weathering). Coatings provided a degree of protection against ultraviolet degradation. However, they (except F Acr) expressed some type of discoloration as observed in changes in lightness (brightness) and color after irradiation in either location (Figure 4).

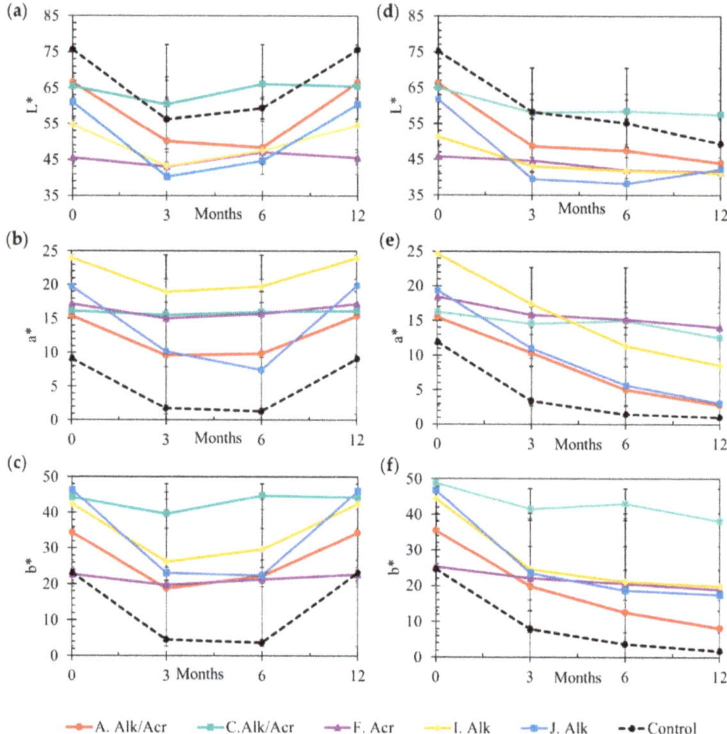

Figure 4. Describes CIELAB color parameters measured during 12 months of exposure. Graphs (**a**), (**b**), and (**c**) respectively display L^* lightness, a^* red to green, b^* yellow to blue variations in Wisconsin, and (**d**), (**e**), (**f**) respectively: L^*, a^*, and b* variations in Mississippi. Error bars represent standard deviation at 95% of confidence interval.

One year of weathering exposure in Mississippi and Wisconsin showed that different climate conditions impact performance of both coatings and wood. Samples exposed in Mississippi decreased in lightness and color as time progressed. In Wisconsin, there was no specific trend in color discoloration of samples, and alterations in lightness and color were expressed only in first three months and after one year. Rapid discoloration of surfaces can be associated with higher temperatures and UV light between the initiation of the experiment and the first data collection (June and September).

Overall, coatings F Acr and C Alk/Acr were less susceptible to color degradation over time. Even though literature reports low photo-degradation resistance of clear coatings, coating C Alk/Acr was the second most stable color. Coating J Alk was the most susceptible to both darkening and color change. As mentioned in Section 3.1.1, the surface of the J Alk samples was completely covered with mold. Fungal growth is the major cause of surface graying when moisture is present [40]. They usually produce dark colored spores and mycelia that may grow on either raw or coated wood surfaces that gives the weathered wood on color change overtime. Coating J Alk also degraded faster than the other finishes.

Initial gloss of samples before installation varied between was initially very low ranging between 2–10 GU (except for coating C Alk/Acr around 20 GU), due to both wood surface variation and coating type. Overall, the outdoor exposure did not affect the gloss of coatings A Alk/Acr, F Acr, and J Alk or the uncoated control samples (Figure 5).

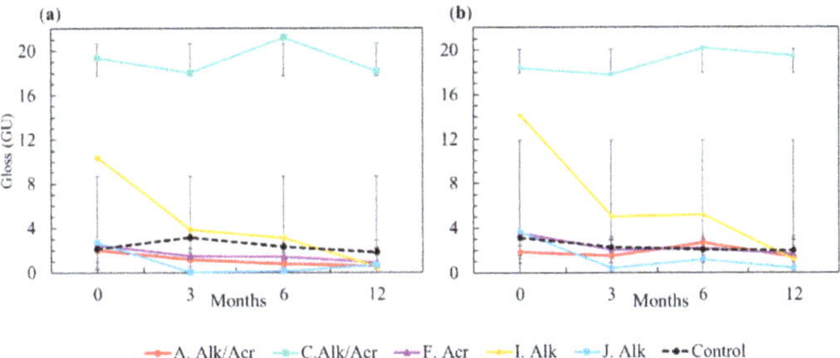

Figure 5. Gloss variation (GU) of tested coatings exposed to natural weathering in (**a**) Mississippi and (**b**) Wisconsin.

Coatings I Alk and C Alk/Acr were the least resistant to gloss change. Loss of gloss is an indicator of initial degradation and is caused by either non-chemical changes (e.g., cracking, checking) or by chemical changes located in the topmost portion of the coating [41]. Since some coatings had very low gloss values before exposure due to their opaque nature (e.g., coating J), alterations to their surface luster were not detected.

3.1.3. Moisture Content during Exposure

The differences in moisture content of samples exposed in Mississippi compared to Wisconsin were closely related to weathering factors such as temperature, precipitation, and radiation. Moisture content increase in samples exposed in Wisconsin was higher. Low temperatures were reported at the site, along with snow and ice (Figure 6).

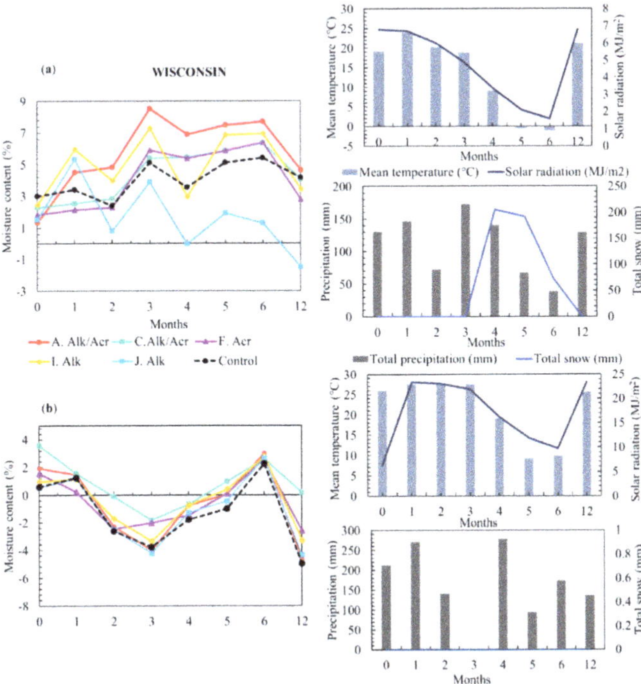

Figure 6. Moisture content of uncoated and coated samples exposed to 12 months of natural weathering in (**a**) Wisconsin and (**b**) Mississippi.

Even though the precipitation in Mississippi was higher, the intense solar radiation and high temperatures resulted in lower moisture content. Differences in temperature cause stress to any material due to gradients of thermal expansion [42]. Shrinking and swelling can result in coating and adhesive failure. In Mississippi, for example, three samples had adhesive failure because of dimensional changes resulted from differential in moisture and temperature (Figure 7).

Figure 7. Delamination of coated CLT (I Alk) after 12 months of outdoor exposure.

Another factor that may have impacted increase in moisture content is the end joint present in each sample. Many CLT panels manufactured in North America are not edge glued, i.e., they are only glued between each layer, and not on the sides. These spaces may trap water and weaken the bond between wood and adhesive. Water absorption in CLT during service raises concerns with respect to the dimensional stability and durability of wood. Polyurethanes are the most common adhesive in CLT production because of its considerable resistance to water and fire [43]. However, combinations of liquid water, shrinking, and swelling tend to break chemical bonds between wood and adhesive resulting in CLT delamination. High moisture content also contributes to mold and decay development [15].

3.2. *Artificial Weathering*

3.2.1. Visual Analysis

The transparent and semi-transparent coatings had different performance during artificial exposure. The coatings A Acr/Alk, J Alk, and I Alk presented some type of chalking that occasionally could result in surface erosion. It is important to mention that some studies describe degradation of clear coating as cracking or flaking. The type of degradation found for the opaque coatings used in this study (A Acr/Alk and J Alk) were best described as chalking due to their powdery appearance (Figure 8).

Although the short-term exposure resulted in no major visual change on most of the treatments, coating I exhibited decrease in brightness with some degree of bleaching. Similar results were found on commercial coatings after 1000 h of accelerated weathering [43]. The long exposure of 1800 h resulted in slight chalking of coatings A and J and moderate chalking of coating I.

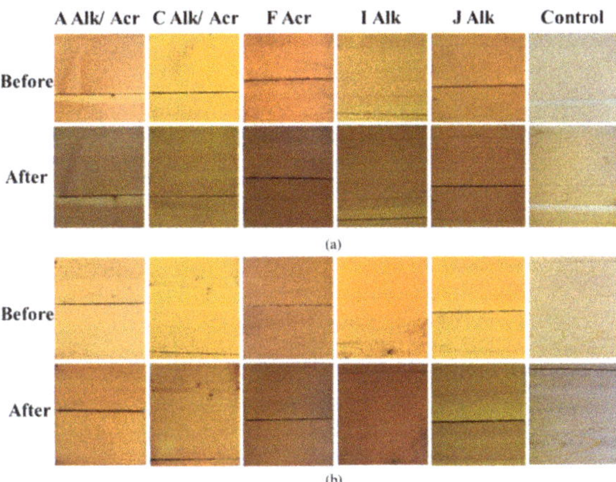

Figure 8. Surface change of selected tested samples. First and second rows correspond to before and after exposure, respectively. (**a**) After 360 h of exposure; and (**b**) after 1800 h of exposure.

3.2.2. Color and Gloss Changes

The color changes of coated and uncoated CLT samples exposed for 360 h and 1800 h are summarized in Table 4. Although short-term exposure (360 h) showed discrete changes, there was statistical difference between treatments ($\alpha = 0.05$). Overall, coating did not express great lightness degradation (ΔL^*) in the first accelerated weathering test except for coating J (−4.8 units). The lowest value of ΔL^* was reported for untreated samples (−10.4 units) that became darker after the test. This result was expected because wood chemical components, such as extractives, rapidly degrade with photo-radiation exposure leaving them darker [41].

Table 4. Color change values of artificial weathered CLT samples. Mean and (standard deviation).

Coating	360 h				1800 h			
	CIE Lab Coordinates							
	ΔL^*	Δa^*	Δb^*	ΔE^*	ΔL^*	Δa^*	Δb^*	ΔE^*
A. Alk/Acr	−2.6 (1.3)	−0.9 (0.5)	0.2 (0.7)	2.9 (1.1)	−12.9 (1.4)	3.2 (0.7)	−1.3 (0.8)	13.4 (1.5)
C. Alk/Acr	−1.2 (0.2)	0.0 (0.6)	−1.1 (0.7)	1.8 (0.7)	−7.6 (1.1)	3.0 (0.4)	−3.4 (0.2)	8.9 (1.0)
F. Acr	1.5 (2.2)	−0.2 (0.2)	−0.3 (0.3)	1.7 (2.1)	−1.7 (0.4)	−1.3 (1.3)	−1.6 (0.3)	2.9 (0.8)
I. Alk	−1.6 (1.7)	0.3 (1.3)	−3.5 (2.7)	4.2 (2.9)	−11 (2.2)	−0.6 (1.0)	−12.6 (3.9)	16.9 (4.2)
J. Alk	−4.8 (0.5)	−0.4 (1.0)	−2.4 (0.4)	5.5 (0.4)	−12 (0.6)	0.7 (1.7)	−6.7 (1.9)	13.9 (0.4)
Control	−10.4 (0.8)	2.0 (1.0)	7.3 (1.8)	12.6 (1.8)	−11.3 (0.2)	−0.8 (0.6)	−5.5 (1.9)	13 (0.9)

Color of samples were not degraded during short-term exposure. The highest change for coated wood was found on coating I Alk ($\Delta b^* = -3.5$ units) and uncoated samples ($\Delta b = 7.3$).

Samples coated with either coating I or J were less stable (ΔE^*). The acrylic water-based coatings C Alk/Acr and F Acr had better performance at the beginning of the test.

The overall ranking associated with resistance to color change was: C Alk/Acr > F Acr > A Alk/Acr > I Alk > J Alk > Control.

The color changes after 1800 h of artificial weathering were statistically different among treatments ($\alpha = 0.05$). Long-term exposure resulted in low resistance to darkening of coated and uncoated samples. The coatings A Alk/Acr, J Alk, and I Alk showed high sensitivity to light degradation (−12.9, −12.0, 11.0 respectively). This finding corroborates [44], which found effects of aging much earlier in alkyd coatings.

Overall, coatings did not show instability to changes in the Δa^* spectrum. The highest values were found for coatings A Alk/Acr and C (3.23 and 3.00, respectively). The major change in Δb^* was measured for coating I Alk (−12.6 units) followed by coating J Alk and control samples (−6.7 and 5.5 respectively). The higher color change after 1800 h of accelerated weathering may be related to the degradation of the protective coatings and the leaching of wood surface components (extractives and lignin). Coating F Acr was the most stable color treatment, which is consistent with the results of other research [44–46] that reported pigmented coatings to be more resistant to photo-degradation than clear coatings.

The gloss of coated and uncoated CLT significantly changed after artificial weathering exposure. Based on the initial surface luster of the samples, the oil-based coatings were affected more after exposure than water-based coatings for either exposure time (Table 5). Oil and alkyd finishes are less permeable and are more likely to decompose as time progresses [45,46]. If the coating is transparent, it is even more susceptible and sensitive to UV-degradation.

Table 5. Gloss change (ΔG) of coated and uncoated CLT after 360 h and 1800 h of artificial weathering (Standard deviation).

Treatment	ΔG_{360}	ΔG_{1800}
A. Alk/Acr	−0.5 (0.4)	−0.7 (0.1)
C. Alk/Acr	0.2 (0.4)	4.2 (0.9)
F. Acr	−0.3 (0.2)	−0.2 (0.4)
I. Alk	−0.9 (0.7)	−0.9 (0.8)
J. Alk	−0.9 (0.7)	−1.2 (0.2)
Control	−0.8 (0.3)	1.0 (0.5)

Similar results were found by [45], who reported gloss degradation on an oil-based coating after three weeks of artificial weathering exposure. The loss of gloss indicates that degradation is occurring due to non-chemical changes (surface wrinkling) or chemical changes located in the topmost portion of the coating [41].

4. Conclusions

Visual rankings and degrees of color change reported for samples exposed to outdoor weathering were highly consistent. In both site locations, coatings C Alk/Acr and F Acr were the most resistant. A reason for their superior performance is likely the inclusion of anti-microbial ingredients in their composition. Coatings A Alk/Acr and J Alk failing to protect the CLT surface coincided with increased mold growth, chalking, erosion, and color change over other treated samples. Gloss changed over time, specifically for coatings I Alk and C Alk/Acr, while other variations were not reported due to low values during initial exposure. Water uptake is influenced by substrate variations (defects, type of grain, earlywood/latewood, and end-joint) and climatic conditions. For these reasons, the effect of coatings on moisture content during exposure was not significant. Combinations of water, temperature, and solar radiation impacted coating performance. Even when the wood surface is protected, variations in the CLT panels such as end-joint, cracks, and checks can facilitate water uptake that eventually will result in coating failure, delamination, and fungal attack.

Artificial weathering results were similar to the natural weathering. Coatings A Alk/Acr, I Alk, and J Alk had slight to moderate chalking after long-term exposure. These same coatings were the most sensitive to changes in lightness, color, and gloss. Therefore, an artificial weathering test of 1800 h or greater may screen potential durable coatings for CLT. However, it is important to consider that in artificial weathering, biological agents such as fungi and bacteria are not present. Once biological factors are added, the service life of coatings will be diminished.

Author Contributions: Conceptualization, M.N., K.M.O., and C.E.S.; methodology, C.E.S., G.d.S.B., M.N., and G.K.; formal analysis, G.d.S.B., D.J.V.L., and C.E.S.; investigation, G.d.S.B., C.E.S., G.K., and K.M.O. writing—original draft preparation, G.d.S.B.; writing—review and editing, C.E.S., D.J.V.L., M.N., K.M.O., and G.d.S.B.; supervision G.d.S.B., and C.E.S.; project administration C.E.S.; funding acquisition, M.N. All authors have read and agreed to the published version of the manuscript.

Funding: This research was funded by USDA Forest Service, grant number FPL #16-JV-11111136-048. This publication is a contribution of the Forest and Wildlife Research Center, Mississippi State University.

Institutional Review Board Statement: Not applicable.

Informed Consent Statement: Not applicable.

Data Availability Statement: The data presented in this study are available on request from the corresponding author. The data are not publicly available due to ethical restrictions.

Acknowledgments: The authors wish to acknowledge the support of U.S. Department of Agriculture (USDA), Forest Service. Any opinions, findings, conclusion, or recommendations expressed in this publication are those of the author(s) and do not necessarily reflect the view of the U.S. Department of Agriculture.

Conflicts of Interest: The authors declare no conflict of interest.

References

1. Mallo, M.F.L.; Espinoza, O. Outlook for cross-laminated timber in the United States. *BioResources* **2014**, *9*, 7427–7443.
2. Zumbrunnen, P.; Fovargue, J. Mid-rise CLT buildings, the UK's experience and potential for Australia and New Zealand. In Proceedings of the 12th World Conference on Timber Engineering, Auckland, New Zealand, 15–19 July 2012; pp. 91–98.
3. Pei, S.; Rammer, D.; Popovski, M.; Williamson, T.; Line, P.; van de Lindt, J.W. An Overview of CLT Research and Implementation in North America. In Proceedings of the 16th World Conference on Timber Engineering (WCTE), Vienna, Austria, 15–22 August 2016.
4. Morris, P. Optimizing the Design of Mass Timber in Exterior Applications. 2015. Available online: www.durable-wood.com (accessed on 12 December 2020).
5. Williams, R.S. Finishing of Wood. In *Wood Handbook: Wood as an Engineering Material*; Ross, R.J., Ed.; General Technical Report FPL: Madison, WI, USA, 1999; pp. 15–37.
6. Cappellazzi, J.; Konkler, M.J.; Sinha, A.; Morrell, J.J. Potential for decay in mass timber elements: A review of the risks and identifying possible solutions. *Wood Mat. Sci. Eng.* **2020**, *15*, 351–360. [CrossRef]
7. Joshi, A.A.; Pagni, P.J. Fire induced thermal fields in window glass. II- Experiments. *Fire Saf. J.* **1994**, *22*, 45–65. [CrossRef]
8. Clausen, C.A. Biodeterioration of Wood. In *Wood Handbook: Wood as an Engineering Material*; Ross, R.J., Ed.; General Technical Report FPL: Madison, WI, USA, 2010; pp. 1–16.
9. Kataoka, Y.; Kiguchi, M.; Williams, R.S.; Evans, P.D. Violet light causes photodegradation of wood beyond the zone affected by ultraviolet region. *Holzforschung* **2007**, *61*, 23–27. [CrossRef]
10. Shupe, T.; Lebow, S.; Ring, D. Wood Decay, Degradation & Stain. In *Causes and Control of Wood Decay, Degradation & Stain*; Louisiana Cooperative Extension Service: Baton Rouge, LA, USA; Louisiana State University Agricultural Center: Baton Rouge, LA, USA, 2008; p. 26.
11. Isaksson, T.; Thelandersson, S. Experimental investigation on the effect of detail design on wood moisture content in outdoor above ground applications. *Build. Environ.* **2013**, *59*, 239–249. [CrossRef]
12. Koch, P. Finishing. In *Utilization of the Southern Pines*; U.S. Southern Forest Experiment Station, USDA: Washington, DC, USA, 1972; pp. 1327–1366.
13. Scheffer, T.C. A climate index for estimating potential for decay in wood structures above ground. *For. Prod. J.* **1971**, *21*, 25–31.

14. APA—The Engineered Wood Association. *Cross Laminated Timber Standards*; American National Standards Institute: Tacoma, WA, USA, 2018.
15. Lepage, R.; Higgins, J.; Finch, G. Moisture uptake test for CLT floor panels in a tall wood building in Vancouver. In Proceedings of the 15th Canadian Conference on Building Science and Technology, British Columbia Building Envelope Council, Vancouver, BC, Canada, 6–8 November 2017; p. 17.
16. Crespell, P.; Gagnon, S. Cross Laminated Timber: A Primer, FPInnovations, Special Publication (52). Vancouver, British Columbia. 2010. Available online: https://fpinnovations.ca/media/factsheets/Documents/cross-laminated-timber-the-boook.pdf (accessed on 25 November 2020).
17. de Meijer, M.; Thurich, K.; Militz, H. Quantitative measurements of capillary coating penetration in relation to wood and coating properties. *Holz Roh Werkst* **2001**, *59*, 35–45. [CrossRef]
18. Bonura, T.; Bussjeager, S.; Christensen, L.; Daisey, G.; Daniels, T.; Hirsch, M.; Jourdain, C.J.; Mall, D.D.; Springate, B.; Wagner, L.E.; et al. Finishes checklist: A guide to achieving optimum coating performance on exterior wood surfaces. *Coat. Technol.* **2004**, *1*, 36–49.
19. Pánek, M.; Oberhofnerová, E.; Zeidler, A.; Šedivka, P. Efficacy of hydrophobic coatings in protecting oak wood surfaces during accelerated weathering. *Coatings* **2018**, *7*, 172. [CrossRef]
20. Bobadilha, G.S.; Stokes, C.E.; Verly Lopes, D.J. Artificial neural networks modelling based on visual analysis of coated cross laminated timber (CLT) to predict color change during outdoor exposure. *Holzforschung* **2020**. published online ahead of print 2020. [CrossRef]
21. Bobadilha, G.S.; Stokes, C.E.; Kirker, G.; Ahmed, S.A.; Ohno, K.M.; Lopes, D.J.V. Effect of exterior wood coatings on the durability of cross-laminated timber against mold and decay fungi. *BioRes* **2020**, *15*, 8420–8433.
22. ASTM D1006 / D1006M-13. *Standard Practice for Conducting Exterior Exposure Tests of Paints on Wood*; American Society for Testing Material: West Conshohoken, PA, USA, 2013.
23. National Oceanic and Atmospheric Administration—NOAA. Global Summary of the Month Station Details. Mississippi State University Station, 2020. Available online: https://www.ncdc.noaa.gov/cdo-web/datasets/GSOM/stations/GHCND:USC00228374/detail (accessed on 29 September 2020).
24. National Oceanic and Atmospheric Administration—NOAA. Global Summary of the Month Station Details. Madison Dane CO Regional Airport Station, 2020. Available online: https://www.ncdc.noaa.gov/cdo-web/datasets/GSOM/stations/GHCND:USW00014837/detail (accessed on 29 September 2020).
25. ASTM D772. *Standard Test Method for Evaluating Degree Flaking (Scaling) of Exterior Paints*; American Society for Testing Material: West Conshohoken, PA, USA, 2018.
26. ASTM D3274-09. *Standard Test Method for Evaluating Degree of Surface Disfigurement of Paint Films by Fungal or Algal Growth, or Soil and Dirt Accumulation*; American Society for Transforming Material: West Conshohocken, PA, USA, 2017.
27. ASTM D714. *Standard Test Method for Evaluating Degree of Blistering of Paints*; American Society for Testing Material: West Conshohoken, PA, USA, 2015.
28. ASTM D4214. *Standard Test Method for Evaluating Degree of Chalking of Exterior Paint Films*; American Society for Testing Material: West Conshohoken, PA, USA, 2015.
29. ASTM D660-93. *Standard Test Method for Evaluating Degree of Checking of Exterior Paints*; American Society for Testing and Material: West Conshohocken, PA, USA, 2011.
30. ASTM D661-93. *Standard Test Method for Evaluating Degree of Cracking of Exterior Paints*; American Society for Testing Material: West Conshohocken, PA, USA, 2011.
31. ASTM D662-93. *Standard Test Method for Evaluating Degree of Erosion of Exterior Paints*; American Society for Testing Material: West Conshohocken, PA, USA, 2011.
32. ISO 11664-4. *Colorimetry—Part 4: CIE 1976 L*a*b* Colour Space*; International Organization for Standardization: Geneva, Switzerland, 2008.
33. ISO 2813. *Paints and Varnishes–Determination of Gloss Value at 20 Degrees, 60 Degrees and 85 Degrees*; International Organization for Standardization: Geneva, Switzerland, 2014.
34. SAS Institute Inc. *SAS/ACCESS®9.4 Interface to ADABAS: Reference*; SAS Institute Inc: Cary, NC, USA, 2013.
35. Stirling, R.; Uzunovic, A.; Morris, P.I. Control of black stain fungi with biocides in semitransparent wood coatings. *For. Prod. J.* **2011**, *61*, 359–364. [CrossRef]
36. Schultz, T.P.; Militz, H.; Freeman, M.H.; Goodell, B.; Nicholas, D.D. *Development of Commercial Wood Preservatives*; American Chemical Society Symposium Series 982; Oxford University Press: Washington, DC, USA, 2008; 655p.
37. Roux, M.I.; Wozniak, E.; Miller, E.R.; Boxall, J.; Botcher, P.; Kropf, F.; Sell, J. Natural weathering of various surface coatings on five species at four European sites. *Holz Roh Werkst* **1988**, *46*, 165–170. [CrossRef]
38. ANSI/APA—American National Standard. *Standard for Performance-Rated Cross-Laminated Timber*; Form No. PRG 320-2019; APA: Tacoma, WA, USA, 2020.
39. Han, Y.; Park, Y.; Yang, S.-Y.; Chung, H.; Chang, Y.-S.; Yeo, H. Dimensional changes of cross-laminated specimens produced under different conditions due to humidity variation. *BioRes* **2019**, *14*, 4035–4046.
40. Feist, W.C.; Hon, D.N.S. Chemistry of weathering and protection. In *The Chemistry of Solid Wood*; Rowell, R.M., Ed.; Advances in Chemistry Series 20; American Chemical Society: Washington, DC, USA, 1984; pp. 401–454.

41. Wood, K.A.; Cypcar, C.; Hedhli, L. Predicting the exterior durability of new fluoropolymer coatings. *J. Fluor. Chem.* **2000**, *104*, 63–71. [CrossRef]
42. Wang, J.Y.; Stirling, R.; Morris, P.I.; Taylor, A.; Lloyd, J.; Kirker, G.; Lebow, S.; Mankowski, M.; Barnes, H.M.; Morrel, J.J. Durability of mass timber structures: A review of the biological risks. *Wood Fiber. Sci.* **2018**, *50*, 110–127. [CrossRef]
43. Grigsby, W.; Steward, D. Applying the protective role of condensed tannins to acrylic-based surface coatings exposed to accelerated weathering. *J. Pol. Environ.* **2018**, *26*, 895–905. [CrossRef]
44. Feist, W.C. Weathering performance of painted wood pretreated with water-repellent preservatives. *For. Prod. J.* **1990**, *40*, 21–26.
45. Berdahl, P.; Akbari, H.; Levinson, R.; Miller, W.A. Weathering of roofing materials–an overview. *Constr. Build. Mater.* **2008**, *22*, 423–433. [CrossRef]
46. Grüneberger, F.; Künniger, T.; Zimmermann, T.; Arnold, M. Nanofibrillated Cellulose in Wood Coatings: Mechanical Properties of Free Composite Films. *J. Mater. Sci.* **2014**, *49*, 6437–6448. [CrossRef]

Article

The Potential Use of Seaweed (*Posidonia oceanica*) as an Alternative Lignocellulosic Raw Material for Wood Composites Manufacture

Ekaterini Rammou [1], Andromachi Mitani [1], George Ntalos [1], Dimitrios Koutsianitis [1], Hamid R. Taghiyari [2] and Antonios N. Papadopoulos [3,*]

[1] Department of Forestry, Wood Science and Design, University of Thessaly, GR-431 00 Karditsa, Greece; rammou@uth.gr (E.R.); amitani@uth.gr (A.M.); gntalos@uth.gr (G.N.); dkoutsianitis@uth.gr (D.K.)

[2] Wood Science and Technology Department, Faculty of Materials Engineering & New Technologies, Shahid Rajaee Teacher Training University, Tehran 16788-15811, Iran; htaghiyari@sru.ac.ir

[3] Laboratory of Wood Chemistry and Technology, Department of Forestry and Natural Environment, International Hellenic University, GR-661 00 Drama, Greece

* Correspondence: antpap@for.ihu.gr

Abstract: A big challenge in the composites industry is the availability of cheap raw lignocellulosic materials, potential candidates to replace slow growing trees, in order to minimize the production cost. Therefore, a variety of plants were studied and tested worldwide in composites manufacturing. The objective of this study was to investigate the technical feasibility of manufacturing particleboards from seaweed leaves (*Possidonia oceanica*—PO). The use of such a material may benefit both socioeconomic and environmental development since these leaves settle on seashores and decay. The results showed that an incorporation of up to 10% PO leaves did not significantly affect the mechanical properties of the board. Internal bond strength was more severely affected than the other mechanical properties. The incorporation of PO leaves up to 25% did not significantly improve the dimensional stability of the boards. Markedly, boards made from 50% wood particles and 50% PO leaves showed the best thickness swelling values. It is suggested that higher resin dosage and an alternative resin system, such as isocyanates, may improve the panel properties.

Keywords: seaweeds; *Posidonia oceanica*; lignocellulosic materials; wood composite panels

Citation: Rammou, E.; Mitani, A.; Ntalos, G.; Koutsianitis, D.; Taghiyari, H.R.; Papadopoulos, A.N. The Potential Use of Seaweed (*Posidonia oceanica*) as an Alternative Lignocellulosic Raw Material for Wood Composites Manufacture. *Coatings* **2021**, *11*, 69. https://doi.org/10.3390/coatings11010069

Received: 22 December 2020
Accepted: 7 January 2021
Published: 8 January 2021

1. Introduction

Seaweeds or marine macroalgae are plant-like organisms which are in general live attached to rocks or other substrata in coastal territories. This aquatic flowering plant grows at the bottom of the sea and consists of about 60 species. A total of 42 countries worldwide are involved in the commercialization of seaweeds and it is reported that the entire area coverage of this plant approaches a value of about 177,000 km^2 [1]. Indonesia produces 800,000 tons/year dried seaweed, which corresponds to almost the half of the world's production, and approximately the 85% of that figure is exported [2]. Seaweed is widely used for industrial purposes such as in cosmetics, medicine, food and beverages, ink and paper [3,4]. Seaweed can also be manufactured artificially with different levels of viscosity and can be used as industrial adhesives, known as hydrocolloids [5]. Seaweeds are important habitats for various microorganisms living in the sea, however, they are considered to be a waste material, since many leaves break away after their growing season, settle on the sea shores and decay; furthermore, their appearance becomes an eyesore. It is reported that a moderately wide belt of *Possidonia oceanica* seagrass may deliver more than 125 kg of dry material per square meter of the coastline annually [6,7].

Possidonia oceanica is a lignocellulosic material that can be found on the shores of the Mediterranean Sea, covering approximately 40,000 km^2 of the seabed, and can be found in the form of seagrass balls and leaves [8]. The former has a fibrous form and comes from the rhizome of the plant and the latter comes from the living leaves. Seagrass has been investigated mainly because it was considered as a potential insulation material for buildings [6,7]. In addition, seagrass balls have been incorporated as a reinforcing agent in the manufacture of polyethylene composites [9,10]. Bettaieb et al. [11,12] examined the chemical and morphological characteristics of *Possidonia oceanica* leaves and concluded that they exhibited encouraging perspectives as nano-fillers for polymer matrices. Garcia et al. [13] studied the physical and mechanical properties of fiberboards made from *Possidonia oceanica* fibers and concluded that can be considered as an alternative to the conventional fiberboards. Similar observations were reported by Alamsjah et al. [14].

Although there is intense research into *Possidonia oceanica* fibers, the leaves have received far less attention. Saval et al. [15] manufactured cement-bonded particleboards from *Possidonia oceanica* leaves and outlined the possibility of their application in construction. Liew et al. [16] studied the physico-mechanical properties of particleboard made from seaweed adhesive and tapioca starch flour. They found that increasing the amount of tapioca starch flour in the seaweed adhesive resulted in improved mechanical properties. Kuqo et al. [17] made particleboards from *Possidonia oceanica* leaves and used isocyanate resin as a binder. They reported that seagrass leaves are propitious for application in construction and furniture industries.

A big challenge in composites industry is the availability of cheap raw lignocellulosic materials, potential candidates to replace slow growing trees, in order to minimize the production cost. A variety of plants were studied and tested worldwide in composites manufacturing, including vine stalks [18], topinambur stalks [19], cotton stalks [20,21], bamboo chips [22], canola straws [23], reed stem [24], date palms [25], oil palms and poppy husks [26], rice and wheat straw [27,28], stalks from cotton [29], camelthorn [30], and even chicken feathers [31,32]. This laboratory has extensive experience in the utilization of various lignocellulosic materials for composites manufacture, including vine prunings [33], castor stalks [34], bamboo and coconut chips [35,36], flax and banana chips [37,38] and cotton stalks [39]. As a consequence, the objective of this paper was to investigate the technical feasibility of manufacturing particleboards from seaweed leaves (*Possidonia oceanica*). The use of such materials may benefit both socioeconomic and environmental development since these leaves settle on the seashores and decay.

2. Materials and Methods

2.1. Raw Material

Possidonia oceanica (PO) leaves were collected from the coastline of Volos, central Greece. Their size varied from 8 to 10 mm in width and 50 to 150 mm in length. The leaves were washed and rinsed with distilled water in order to eliminate sand and other contaminations. After that, they were dried at room temperature for about two months. In their dried form, they have a brown appearance (Figure 1). Their moisture content, absolute density and pH were 118%, 0.35 kg/m^3 and 8.2, respectively. The leaves were dried at 105 °C to 6.5–7% moisture content. Industrially produced wood chips comprising of predominantly mixed softwoods were supplied by a local plant. The wood chips were first screened through a mesh with 5 mm apertures to remove oversized particles, and they were then put through a mesh with 1 mm apertures to remove undersized (dust) particles.

After screening, the chips were dried to 6.5–7% moisture content. The boards were manufactured from particles dried to this moisture content.

2.2. Board Manufacture and Testing

A urea-formaldehyde resin (UF) (200–400 cP in viscosity, 47 s of gel time, and 1.277 kg/m^3 in density), 7% as a percentage of the oven dry weight of raw material, was applied for single layer board manufacture. Mattresses (50 × 50 cm^2) were hot pressed at 180 °C for 6 min. The specific pressure of the plates was 24 kg/cm^2 (with 200 kgf as the total nominal pressure). The target board thickness was 16 mm and the target density was 0.55 kg/m^3. A 2% aqueous solution of ammonium chloride (20% solids content), based on resin solids, was added to the UF as a hardener before spraying. No water-repelling agent was used in this study. Four types of panel were made, consisting of varying mixtures of wood chips and PO leaves (the percentages of wood to PO leaves were 90:10, 75:25 and 50:50, respectively) and control boards with no PO content were made. Three replicates were made for each board type. The flow diagram of the experimental procedure is depicted in Figure 2.

Figure 1. *Possidonia oceanica* (PO) leaves as collected (**a**) and after drying (**b**).

Figure 2. The flow diagram of the experimental procedure.

2.3. Board Testing

The boards were conditioned at 20 °C and 65% relative humidity prior to testing mechanical properties—internal bond strength (IBS), modulus of rupture (MOR) and modulus of elasticity (MOE), resistance to axial withdrawal of screws and physical properties—

thickness swelling after 24 h immersion in water [40–43]. In addition, thickness swelling was also determined after 48 h immersion in water.

2.3.1. Internal Bond Strength

The wide faces of the 50 by 50 mm samples were glued to slotted aluminum blocks that were then pulled apart on a universal Zwick-Roell Z020 universal testing machine (Zwick-Roell, Kennesaw, GA, USA) and the load required to achieve separation was recorded.

2.3.2. Flexural Tests

The 50 by 350 mm long beams were tested in third-point loading at a span of 320 mm at a loading rate of 3 mm per minute. The load and deflection were continuously recorded, and the resulting data were used to calculate modulus of rupture (MOR) and modulus of elasticity (MOE).

2.3.3. Screw Withdrawal Test

The screw withdrawal tests were performed on the faces and edges of 75 mm square sections using 4.25 mm diameter MDF screws at a withdrawal speed of 2.5 mm/min. The tests were conducted with a 10 kN capacity INSTRON-4486 test machine. A 2 mm pilot hole was drilled prior to inserting the screws to a depth of 17 mm in the panels, leaving 1 mm of the screw above the panel surface for testing. Six replicate specimens were tested for each panel type.

2.3.4. Thickness Swelling

Samples of 50 by 50 mm were weighed and their dimensions were measured with digital calipers (to the nearest 0.01 mm) before being immersed in distilled water. Differences in dimensions were measured after 24 and 48 h of immersion and changes were used to calculate % thickness swell.

2.4. Statistical Analysis

Statistical analysis was conducted using SPSS software program, version 24.0 (IBM, Armonk, NY, USA, 2018). One-way ANOVA was performed to identify significant differences at the 95% level of confidence, with Duncan's multiple range test grouping.

3. Results and Discussion

The mechanical properties of the single layer particleboards made from various wood/PO leaves combinations are shown in Table 1. It can be seen that using higher levels of PO leaves resulted in inferior board properties. An incorporation of up to 10% PO leaves did not significantly affect the mechanical properties of the board. In this regard, cluster analysis based on mechanical properties demonstrated close clustering of boards with only 10% PO with the control boards (that is, boards with no PO content) (Figure 3A). Internal bond strength (IBS) was more severely affected in comparison to the bending properties and the screw withdrawal resistance. Similar observations were also made by Grigoriou [44] with straw-based panels, by Papadopoulos and Hague [37] with flax-based panels, and by Hague et al. [45]. The significant reduction in IBS, especially in boards that contained 50% PO leaves, can be attributed to the fact that PO leaf chips are mainly comprised of relatively thin, short walled and weak cells [5]. Consequently, PO leaves are relatively weak and vulnerable to critical defects inside the panel structure, and therefore a rapid decrease in the IBS of the panel is observed as the PO leaves content increases. In addition, it must be pointed out that in boards made with 50% PO leaves, visible checks and cracks (internal blows) occurred in the core section of the mat, as clearly highlighted in Figure 4.

Table 1. The mechanical properties of various board types. Standard deviations are given in parentheses. The different letters show which values are statistically different at the 5% level.

Board Type (Wood Particles: PO Leaves)	Density (Kg/m^3)	IBS (N/mm^2)	MOR (N/mm^2)	MOE (N/mm^2)	Screw Withdrawal Resistance * (N)	
					⊥	//
100:0	0.5 A (0.03)	0.12 A (0.04)	3.35 A (0.50)	648.11 A (70.35)	1652.25 A (267.27)	768.62 A (113.71)
90:10	0.55 A (0.05)	0.12 A (0.04)	2.83 A (0.29)	583.29 A (167.50)	1488.88 A (181.22)	669.57 A (69.58)
75:25	0.53 A (0.03)	0.07 B (0.04)	2.17 B (0.45)	461.37 B (105)	1148.01 B (187.68)	508.20 B (61.78)
50:50	0.53 A (0.03)	0.03 B (0.01)	1.26 C (0.29)	279.23 B (60.27)	802.49 C (68.11)	198.42 C (42.52)

*, ⊥ vertical to the surface, // parallel to the surface.

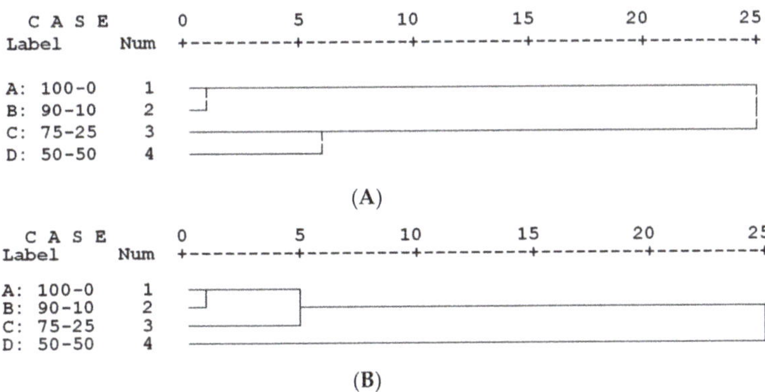

Figure 3. Cluster analyses of the four board types based on the mechanical properties (**A**), and based on all physical and mechanical properties (**B**).

Figure 4. Cracks (blows) in the core layer of boards made with 50% PO leaves, (**a**) after hot pressing; (**b**) after trimming.

Cluster analysis based on mechanical properties also illustrated distinct difference clustering of boards with higher PO contents (25% and 50%) with those containing lower PO contents (0% and 10%) (Figure 3A). The contour and surface plots demonstrated a close relationship between the mechanical properties, showing the nearly uniform effect of the increase in PO content on all properties studied here (Figure 5A,B).

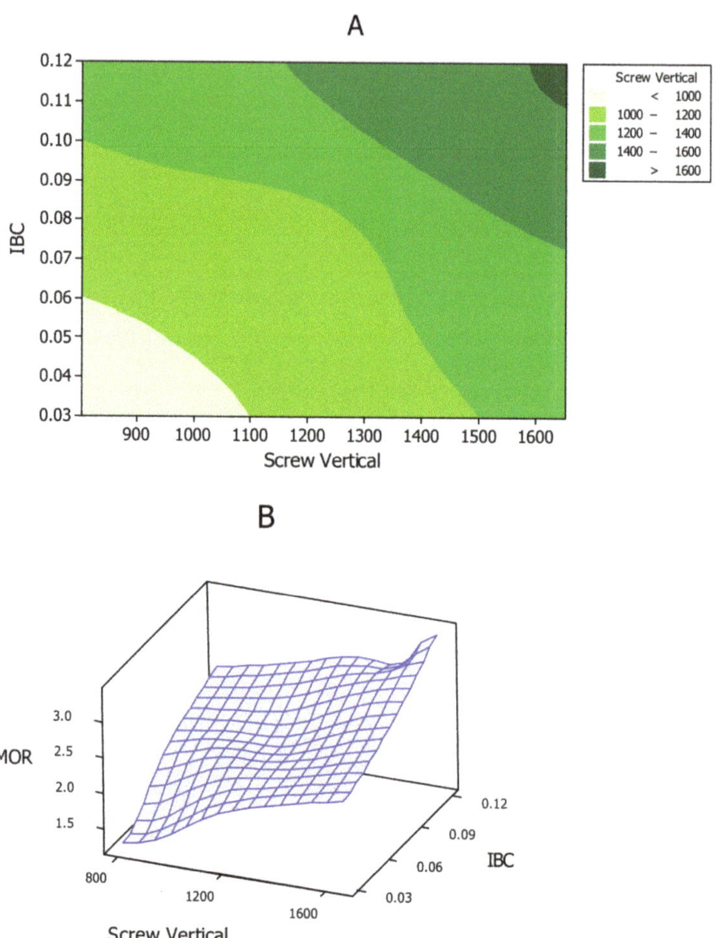

Figure 5. Contour (**A**) and surface (**B**) plots, based on mechanical properties of the four board types.

The thickness swell values after 24 h immersion in water are summarised in Table 2. The results showed that the incorporation of PO leaves up to 25% did not significantly affect the dimensional stability of the boards. It is to be mentioned that no water-repelling agent was used in this study. What can be deducted from the data presented in Table 2 is that boards made from 50% wood particles and 50% PO leaves showed the best thickness swelling values. In fact, this value is significantly different from the corresponding value of boards made from pure wood chips. This tendency remained the same after 48 h immersion in water. Cluster analysis clearly demonstrated distinct different clustering of boards containing 50% PO with the other three board types, indicating the significant effect of the increase in PO content on the overall properties of the boards (Figure 3B). Two possible explanations can be offered for this behaviour; firstly, the great resistance to water that PO leaves have as an aquatic plant contributed in a decrease in thickness swelling, and secondly, their flat shape served as a coating layer, which in turn protected the internal part of the board.

Table 2. The physical properties of various board types. Standard deviations are given in parentheses. The different letters show which values are statistically different at the 5% level.

Board Type (Wood Particles: PO Leaves)	Moisture Content (%)	Thickness Swelling (%)	
		24 h	48 h
100:0	9.90 A (0.21)	63.31 A (5.75)	66.77 A (7.80)
90:10	9.93 A (0.31)	64.99 A (6.25)	66.70 A (7.01)
75:25	10.31 A (0.11)	66.44 A (6.67)	69.21 A (5.68)
50:50	11.41 A (0.20)	56.07 B (9.45)	57.42 B (9.44)

4. Conclusions

This study made an approach to investigate the technical feasibility of manufacturing particleboards from seaweed leaves (*Possidonia oceanica*—PO). The use of such a material may benefit both socioeconomic and environmental development since these leaves settle on seashores and decay, and therefore they are generally considered to be a waste material of no industrial value. An incorporation of up to 10% PO leaves did not significantly affect the mechanical properties of the board. Internal bond strength was more severely affected than the other mechanical properties. The results showed that the incorporation of PO leaves up to 25% did not significantly improve the dimensional stability of the boards. Markedly, boards made from 50% wood particles and 50% PO leaves showed the best thickness swelling values. It is suggested that a higher resin dosage and an alternative resin system, such as isocyanates, may improve the panel's properties and could allow a higher content of PO leaves to be incorporated in the panel. Such strategies have been successfully employed in the commercial manufacture of panels from cereal straws.

Author Contributions: Methodology, E.R. and G.N.; Validation, E.R. and G.N.; Investigation, E.R., A.M., and D.K.; Writing—Original Draft Preparation, E.R., A.M., and D.K.; Writing—Review and Editing, G.N., H.R.T., and A.N.P.; Visualization, E.R., A.M.; D.K., and G.N.; Supervision, G.N. and A.N.P. All authors have read and agreed to the published version of the manuscript.

Funding: This research received no external funding.

Data Availability Statement: The data presented in this study are available on request from the corresponding author.

Conflicts of Interest: The authors declare no conflict of interest.

References

1. Green, E.; Short, F. *World Atlas of Seagrasses*; University of California Press: Berkeley, CA, USA, 2003; p. 40.
2. Alamsjah, M.A.; Subekti, S.; Lamid, M.; Pujiaastuti, D.Y.; Kurnia, H.; Rifadi, R.R. Porosity structure of green polybag of medium density fiberboard from seaweed waste. *IOP Conf. Ser. Earth Environ. Sci.* **2018**, *137*, 012084. [CrossRef]
3. Duarte, C.; Marbà, N.; Agawin, N.; Cebrián, J.; Enríquez, S.; Fortes, D.; Gallegos, M.; Merino, M.; Olesen, B.; Sand-Jensen, K.; et al. Reconstruction of seagrass dynamics: Age determinations and associated tools for the seagrass ecologist. *Mar. Ecol. Prog. Ser.* **1994**, *107*, 195–209. [CrossRef]
4. Kurnia, H.; Rifaldi, R.R.; Agustono, M.; Amin, N.G.; Sudjarwo, S.A.; Alamsjah, M.A. The potential of seaweed wastes as a medium density fiberboard-based pot material for better water use efficiency in tomato plants. *IOP Conf. Ser. Earth Environ. Sci.* **2019**, *236*, 012110. [CrossRef]
5. Shakel, A.; Chaudhery, M.H. *Green and Substainable Advanced Materials*; Scrivener Publishing LLC: Beverly, MA, USA, 2018; Volume 1, pp. 191–212.
6. Cebrian, J.; Duarte, C.M. Detrital stocks and dynamics of the seagrass *Posidonia oceanica* (L.) Delile in the Spanish Mediterranean. *Aquat. Bot.* **2001**, *70*, 295–309. [CrossRef]
7. Cocozza, C.; Parente, A.; Zaccone, C. Chemical, physical characterization of *Posidonia oceanica* (L.) Del. residue and their possible recycle. *Biomass Bioenergy* **2011**, *35*, 799–807. [CrossRef]

8. Duarte, C.M. How can beaches be managed with respect to seagrass litter? In *European Seagrasses: An introduction to Monitoring and Management*; Borum, J., Duarte, C.M., Krause-Jensen, D., Eds.; EU project Monitoring and Management of European Seagrass Beds: Copenhagen, Denmark, 2004; pp. 83–87.
9. Ferrero, B.; Boronat, T.; Moriana, F.; Fenolar, O.; Balart, R. Green composites based on wheat gluten matrix and *Posidonia oceanica* waste fibers as reinforcements. *Polym. Compos.* **2013**, *34*, 1663–1669. [CrossRef]
10. Puglia, D.; Petruci, R.; Luzi, F.; Kenny, J.M.; Torre, L. Revalorisation of *Posidonia oceanica* as reinforcement in polyethylene/maleic anhydride grafted polyethylene composites. *J. Renew. Mater.* **2014**, *2*, 66–76. [CrossRef]
11. Bettaieb, F.; Khiari, R.; Dufrence, A.; Mhenni, M.F.; Putaux, J.L.; Boufi, S. Nanofibrillar cellulose from *Posidonia oceanica*: Properties and morphological features. *Ind. Crop. Prod.* **2015**, *72*, 97–106. [CrossRef]
12. Bettaieb, F.; Khiari, R.; Hassan, M.; Belgacem, M.N.; Bras, J.; Dufresne, A.; Mhenni, M.F. Preparation and characterization of new cellulose nanocrystals from marine biomass *Posidonia oceanica*. *Ind. Crop. Prod.* **2014**, *72*, 175–182. [CrossRef]
13. Garcia, D.G.; Carrillo, L.Q.; Montanes, N.; Fombuena, V.; Balart, R. Manufacturing and characterization of composite fibreboards with *Posidonia oceanica* wastes with an environmentally-friendly binder from epoxy resin. *Materials* **2017**, *11*, 35. [CrossRef]
14. Alamsjah, M.A.; Sulmartiwi, S.; Pursetyo, K.T.; Amin, A.N.G.; Wardani, K.A.; Arifianto, M.D. Modifying bioproduct technology of medium density fibreboard from the seaweed waste *Kappaphycus alvarezii* and *Gracilaria verrucosa*. *J. Indian Acad. Wood Sci.* **2017**, *14*, 32–45. [CrossRef]
15. Saval, J.M.; Lapuente, R.; Navarro, V.; Tenza-Abril, A.J. Fire-resistance, physical, and mechanical characterization of particleboard containing *Oceanica Posidonia* waste. *Mater. Constr.* **2014**, *64*, e019. [CrossRef]
16. Liew, K.C.; Ting, P.B.; Tan, Y.F. Physico-mechanical properties of particleboard made from seaweed adhesive and tapioca starch flour. *J. Indian Acad. Wood Sci.* **2018**, *15*, 199–203. [CrossRef]
17. Kuqo, A.; Korpa, A.; Dhamo, N. Posidonia oceanic leaves for processing of PMDI composite boards. *J. Compos. Mater.* **2019**, *53*, 1697–1703. [CrossRef]
18. Yeniocak, M.; Göktas, O.; Erdil, Y.Z.; Özen, E. Investigating the use of vine pruning stalks (*Vitis Vinifera,* L. *CV. Sultani*) as raw material for particleboard manufacturing. *Wood Res.* **2014**, *59*, 167–176.
19. Klimek, P.; Meinlschimidt, P.; Wimmer, R.; Plinke, B. Using sunflower (*Helianthus annuus* L.), topinambour (*Helianthus tuberosus* L.) and cup-plant (*Silphium perfoliatum* L.) stalks as alternative raw materials for particleboards. *Ind. Crop. Prod.* **2016**, *92*, 157–164. [CrossRef]
20. Gansberger, M.; Montgomery, L.F.R.; Liebhard, P. Botanical characteristics, crop management and potential of *Silphium perfoliatum* L. as a renewable resource for biogas production: A review. *Ind. Crop. Prod.* **2015**, *63*, 362–372. [CrossRef]
21. Nazerian, M.; Beyki, Z.; Gargari, R.; Kool, F. The effect of some technological production variables on mechanical and physical properties of particleboard manufactured from cotton stalks. *Maderas Cienc. Technol.* **2016**, *18*, 167–178. [CrossRef]
22. De Araujo, P.; Arruda, L.; Menezzi, C.; Texeira, D.; de Souza, M. Lignocellulosic composites from Brazilian giant bamboo. Part 2: Properties of cement and gypsum bonded particleboard. *Maderas Cienc. Technol.* **2011**, *13*, 297–306. [CrossRef]
23. Kord, B.; Zare, H.; Hosseinzabeh, A. Evaluation of the mechanical and physical properties of particleboard manufactured from canola straws. *Maderas Cienc. Technol.* **2016**, *18*, 9–18. [CrossRef]
24. Kord, B.; Roohani, M.; Kord, B. Characterisation and utilization of reed stems as a lignocellulosic residue for particleboard production. *Maderas Cienc. Technol.* **2015**, *17*, 517–524.
25. Amirou, S.; Zerizer, A.; Pizzi, A.; Haddadou, I.; Zhou, X. Particleboards production from date palm biomass. *Eur. J. Wood Prod.* **2013**, *71*, 717–723. [CrossRef]
26. K#xFC;çüktüvek, M.; Kasal, A.; Kuşkun, T.; Erdil, Y. Utilizing poppy husk-based particleboards as an alternative material in case furniture construction. *BioResources* **2017**, *12*, 839–852.
27. Li, X.; Cai, Z.; Winandy, J.; Basta, A. Selected properties of particleboard panels manufactured from rice straws of different geometries. *Biores. Technol.* **2010**, *101*, 4662–4666. [CrossRef] [PubMed]
28. Mo, X.; Cheng, E.; Wang, D.; Sun, X. Physical properties of medium-density wheat straw particleboard using different adhesives. *Ind. Crop. Prod.* **2003**, *18*, 47–53. [CrossRef]
29. Guler, C.; Ozen, R. Some properties of particleboards made from cotton stalks (*Gossypium hirsitum* L.). *Holz als Roh-und Werkstoff* **2004**, *62*, 40–43. [CrossRef]
30. Esmailpour, A.; Taghiyari, H.R.; Majidi, R.; Babaali., S.; Morrell, J.J.; Mohammadpanah, B. Effects of adsorption energy on air and liquid permeability of nanowollastonite-treated medium-density fiberboard. *IEEE Trans. Instrum. Meas.* **2020**. [CrossRef]
31. Taghiyari, H.R.; Bari, E.; Schmidt, O.; Tajick Ghanbary, M.A.; Karimi, A.; Tahir, P.M.D. Effects of nanowollastonite on biological resistance of particleboard made from wood chips and chicken feather against *Antroia vaillantii*. *Int. Biodeterior. Biodegrad.* **2014**, *90*, 93–98. [CrossRef]
32. Taghiyari, H.R.; Majidi, R.; Esmailpour, A.; Sarvari Samadi, Y.; Jahangiri, A.; Papadopoulos, A.N. Engineering composites made from wood and chicken feather bonded with UF resin fortified with wollastonite: A novel approach. *Polymers* **2020**, *12*, 857. [CrossRef]
33. Ntalos, G.A.; Grigoriou, A.H. Characterization and utilisation of vine prunings as a wood substitute for particleboard production. *Ind. Crop. Prod.* **2002**, *16*, 59–68. [CrossRef]
34. Grigoriou, A.H.; Ntalos, G.A. The potential use of *Ricinus communis* L. (Castor) stalks as a lignocellulosic resource for particleboards. *Ind. Crop. Prod.* **2001**, *13*, 209–218. [CrossRef]

35. Papadopoulos, A.N.; Hill, C.A.S.; Gkaraveli, A.; Ntalos, G.; Karastergiou, S. Bamboo chips (*Bambusa vulgaris*) as an alternative lignocellulosic raw material for particleboard manufacture. *Holz als Roh-und Werkstoff* **2004**, *62*, 36–39. [CrossRef]
36. Papadopoulos, A.N.; Traboulay, E.; Hill, C.A.S. One layer Experimental Particleboard from Coconut Chips (*Cocos nucifera* L.). *Holz als Roh-und Werkstoff* **2002**, *60*, 394–396. [CrossRef]
37. Papadopoulos, A.N.; Hague, J.R.B. The potential use of *Linum usitatissimun* (flax) chips as a raw lignocellulosic material for particleboards. *Ind. Crop. Prod.* **2003**, *17*, 143–147. [CrossRef]
38. Papadopoulos, A.N. Banana chips (*Musa acuminate*) as an alternative lignocellulosic raw material for particleboard manufacture. *Maderas Cienc. Technol.* **2018**, *20*, 395–402.
39. Papadopoulos, A.N.; Kyzas, G.Z.; Mitropoulos, A.C. Lignocellulosic composites from acetylated sunflower stalks. *Appl. Sci.* **2019**, *9*, 646. [CrossRef]
40. EN 310. *Wood Based Panels-Determination of Modulus of Elasticity in Bending and of Bending Strength*; Comite Europeen de Normalisation: Brussels, Belgium, 1993.
41. EN 319. *Particleboards and Fiberboards-Determination of Tensile Strength Perpendicular to the Plane of The Board*; Comite Europeen de Normalisation: Brussels, Belgium, 1993.
42. EN 317. *Particleboards and Fiberboards-Determination of Swelling in Thickness after Immersion in Water*; Comite Europeen de Normalisation: Brussels, Belgium, 1993.
43. EN 320. *Particleboards and Fiberboards-Determination of Resistance to Axial Withdrawal of Screws*; Comite Europeen de Normalisation: Brussels, Belgium, 2011.
44. Grigoriou, A.H. Straw-wood composites bonded with various adhesive systems. *Wood Sci. Technol.* **2000**, *34*, 355–365. [CrossRef]
45. Hague, J.R.B.; McLauchlin, A.; Quinney, R. Agri-materials for panel products: A technical assessment of their viability. In Proceedings of the 32nd International Particleboard/Composite Symposium, Washington State University, Pullman, WA, USA, 31 March–2 April 1998; pp. 151–159.

Review

Fungal Degradation of Wood: Emerging Data, New Insights and Changing Perceptions

Barry Goodell [1], Jerrold E. Winandy [2] and Jeffrey J. Morrell [3,*]

1 Microbiology Department, University of Massachusetts, Amherst, MA 01003, USA; bgoodell@umass.edu
2 Bioproducts and Biosystems Engineering Department, University of Minnesota, St. Paul, MN 55108, USA; jwinandy@umn.edu
3 Centre for Timber Durability and Design Life, University of the Sunshine Coast, Brisbane 4102, Australia
* Correspondence: jmorrell@usc.edu.au

Received: 30 October 2020; Accepted: 8 December 2020; Published: 11 December 2020

Abstract: Wood durability researchers have long described fungal decay of timber using the starkly simple terms of white, brown and soft rot, along with the less destructive mold and stain fungi. These terms have taken on an almost iconic meaning but are only based upon the outward appearance of the damaged timber. Long-term deterioration studies, as well as the emerging genetic tools, are showing the fallacy of simplifying the decay process into such broad groups. This paper briefly reviews the fundamentals of fungal decay, staining and mold processes, then uses these fundamentals as the basis for a discussion of fungal attack of wood in light of current knowledge about these processes. Biotechnological applications of decay fungi are reviewed, and an overview is presented on how fungi surmount the protective barriers that coatings provide on surfaces. Advances in biochemical analyses have, in some cases, radically altered our perceptions of how wood is degraded, and even the relationships between fungal species, while other new findings have reinforced traditional perspectives. Suggestions for future research needs in the coatings field relative to enhanced fungal and environmental protection are presented.

Keywords: wood degradation; brown rot; white rot; soft rot; mechanisms of decay; coating performance; bioproducts; bio-coatings

1. Introduction

Wood is among the most durable cellulosic materials, but it can be degraded by a number of biotic and abiotic agents. These agents often act simultaneously making it difficult to completely separate causal agents. This review will concentrate on the role of fungi in degradation but will also discuss other agents as they relate to the overall process. Our discussion will focus on those fungi capable of degrading the primary cell wall polymers of cellulose, hemicellulose and lignin, but will also touch on the many mold and stain fungi that not only cause aesthetic concerns and disfigure coatings but can digest the stored compounds in the parenchyma cells and the pit membranes while promoting more limited damage to the structural elements of wood.

Wood poses a major challenge to organisms seeking to extract the energy from its polymeric structure. While the stored compounds in parenchyma cells are digestible by many organisms, accessing the more complex polymers is a key challenge. The chemistry and arrangement of the cellulose and lignin polymers in the wood cell wall sharply reduces the number of agents capable of causing damage. Many fungi are cellulolytic but are unable to unlock the chemistry of the lignin polymer that both enrobes and is interspersed with the cellulosic components of the lignocellulosic matrix. Only those fungi that have developed strategies to surmount the recalcitrance of lignin are able to fully extract the embodied energy of the lignocellulose cell wall.

2. Wood as a Polymeric Material

Cellulose represents 40–44% of most wood and endows wood with many of its unique material properties [1]. At the nanoscale, cellulose is arranged in discrete units known as elementary fibrils, with the inner core of those fibrils being a tightly packed crystalline material and the outer layers being more loosely packed amorphous cellulose. The crystalline nature of the inner core of the elementary cellulose fibrils gives wood its exceptional tensile strength—comparable to steel and aramid (Kevlar) at the nanoscale [2]. Crystallinity also renders the inner portion of the cellulose elementary fibril more resistant to degradation than the other more amorphous carbohydrate polymers also present in the wood cell wall.

Lignin represents 18 to 35% of the wood cell wall and is a heteropolymer consisting of repeating units of phenyl propane with a diverse array of bonds between three variations of the monomer form. This diversity in bonding pattern makes lignin extremely resistant to degradation and few microorganisms other than select wood degrading fungi have been able to unravel this system. Furthermore, the higher levels of lignin in wood compared to other plant materials, and manner in which lignin is intimately integrated with the holocellulose components, are major reasons for wood's resistance to degradation.

Hemicellulose is a branching heteropolymer of pentose and hexose sugar monomers representing 15–32% of the wood cell wall. Hemicellulose is a more amorphous polymer compared to cellulose, and it is considered to be the link between cellulose and lignin that allows the three polymers to behave as an integrated matrix. Hemicellulose is also more susceptible to degradation and the many forms of the hemicellulose polymer can be degraded by excessive heat such as that occurring in thermal modification processes used in industrial wood protection [3,4], or by relatively dilute acids, as was observed with significant mechanical property losses following treatment by some early-industrial acidic fire-retardant and preservative formulations [5]. These forms of degradation are discussed more completely in the section on Abiotic Degradation below. Important relative to this chapter are several types of fungi that attack hemicellulose in early stages of decay leading to mechanical property loss [6]. Although hemicellulose is an integral part of the wood structure, it is often viewed as the weak link in the armor of wood as a material, since its degradation profoundly affects the ability of lignocellulose to behave as a rigid structural system. Hemicellulose is often the first major component of wood to be attacked and deconstructed by fungi [7,8].

In addition to the cell wall polymers, other components including proteins, fatty acids, pectin, starch and other polysaccharides and sugars are also present at levels ranging from 2 to 15% (Average ~6%) of the wood mass [1]. While these materials can have dramatic effects on resistance to microbial attack, this review will concentrate on the roles of the three primary cell wall polymers in relation to fungal attack.

3. The Importance of the Wood–Water Relationship

Water is an important constituent of wood materials and biomass in the natural environment and can affect wood properties both directly and indirectly. The uptake of water is an integral factor associated with wood's ability to be attacked by fungi and to decay, but moisture also impacts wood by moving into the cellular structure to associate with the cellulose, causing dimensional changes. Even though water is held in wood primarily by hydrogen bonding at the molecular level, the collective force of the hydrogen bonds formed is sufficient to swell the wood to the extent that, at a practical level, is capable of splitting rocks, tearing thin metal plates or forcing structural walls apart. Shrinkage and swelling of wood with changes of humidity in the environment creates substantial challenges for coatings. Polymeric coatings typically are unable to prevent moisture vapor from diffusing in and out of wood as humidity changes. This results in repeated swelling and shrinkage of wood, particularly in exterior environments. Few polymeric coatings can withstand the recurring cycles of expansion and contraction that wood surfaces undergo with significant humidity changes.

The stress on wood coatings due to moisture changes, combined with other factors such as mechanical action and abrasion, creates an environment where coatings are subject to rupture at the microscopic level. Wood surfaces are subjected to a continual dusting of fungal propagules that, once moisture levels and environmental conditions are correct, can germinate and exploit the micro-cracks that can develop in coatings. This will be addressed in the section on future research needs for bio-based coatings.

4. Abiotic Degradation

Non-living agents can cause different types of wood degradation that are often confused with fungal degradation. For this reason, it is important to review some of the most important abiotic (non-biological) degradation agents which are often confused with fungal attack of wood.

Wood weathering is the most common type of non-biological degradation occurring in exterior-use wood [9]. The low energy levels in sunlight can be quite disruptive to the polymeric components of wood at the molecular level. During UV (sunlight) exposure, the primary energetic components which can attack lignocellulose are reactive oxygen species (ROS or oxygen radicals) that randomly attack the cell wall polymers, especially lignin. This energy transforms lignin into a variety of radical forms that transfer the energy to hemicellulose and cellulose which are then readily depolymerized [10,11]. This process begins almost immediately after timber is exposed to sunlight, but the effect on the timber is shallow until the outer layers of UV-degraded wood are sloughed off. Weathering related to UV exposure is typically slow but can have substantial effects on adhesion of coatings as well as declines in appearance leading to premature replacement. For example, a preweathering exterior exposure of Western redcedar siding in direct sunlight for 1, 2, 4, 8 or 16 weeks prior to being painted reduced paint adhesion by up to 50% [12,13]. Increased preweathered was associated with shorter paint service life: paint on panels preweathered for 16 weeks began to flake and peel after 4 to 5 years; preweathering for 8, 4, 2 or 1 weeks resulted in paint failures after 7 to 8 years, 9 to 10 years, 10 to 11 years or 13 years, respectively. Coatings on non-preweathered controls were still in almost perfect condition after 17 years of exposure.

Another common abiotic source of wood degradation that can be confused with fungal attack is heat- or thermal degradation. Wood has a well-known propensity to burn, but it also experiences more subtle changes upon extended exposure to temperatures between 65 °C and the ignition point [5,14]. Heating affects each polymer differently, with hemicelluloses being the most susceptible, followed by cellulose and finally, lignin [15]. Thermal-related modification can have negative effects on wood, particular if the wood has been modified with other treatments such as some types of fire-retardant treatment. In the 1980s, most of the fire retardants used in North America were acidic. During that period, issues gradually emerged where fire-retardant treated wood in structures began experiencing extensive degradation [5,16]. Subsequent investigations showed that the acidic formulations tended to react with, and degrade, wood maintained at temperatures typically found in attics and enclosed roof spaces during the summer. Further study revealed that hemicelluloses were the first polymers affected, again illustrating the critical role played by these polymers in lignocellulose performance [14]. Disruption of the biological degradation process by thermally modifying hemicellulose has been studied since the 1930s [17]. This research was initially directed at modifying colors to render lighter, less valuable woods darker, but it was also shown to affect wood/moisture relationships. The process was never fully exploited, but was revisited in the 1980s as European researchers sought non-biocidal methods for improving timber durability [4]. Extensive subsequent research has shown that thermal modification alters the hemicelluloses, thereby reducing moisture uptake which in turn reduces, but does not completely eliminate, the risk of decay.

5. Biotic Deterioration—Wood Decay and Requirements for Fungal Attack

Wood decay is largely caused by fungi that fall into categories depending on the appearance of the degraded wood which is, in turn, related to polymeric materials that are degraded. Brown rot decay is an informal name for the most common type of decay occurring in timber products. Fungi that cause brown rot depolymerize cellulose and hemicellulose (holocellulose) for digestion, while lignin is also depolymerized and modified before being rapidly repolymerized. The general categories of white rot fungi and soft rot fungi are the other major types of decay, and these are covered later in this review, as these fungal decays can be quite important in certain environments. Fungi are Eukaryotic organisms that are in the same Domain in the Tree of Life as plants and animals [18]. Species within other Domains in the tree of life comprising the Bacteria and Archaea can also live in wood. Some species of Bacteria have been shown to cause limited wood deterioration over long periods of time (several centuries), resulting in mechanical property loss in wood. Because of their minor importance in deterioration of structures, these microorganisms will not be considered further in this chapter, although we recognize that they are almost always present and have been suggested to play supporting roles in the degradation process including pre-conditioning of wood, and extractives detoxification. Bacteria are also active in long-term degradation of submerged wooden foundation piling, which typically occurs over many centuries [19]. Insects and some types of marine boring animals also can cause significant biodeterioration of wood under some circumstances, but deterioration by these animals is reviewed elsewhere [20].

For all fungi, spores (microscopic seeds) or other small fragments of the fungi must be produced and be transported either in the air, water or on other organisms (such as insects) to other pieces of wood where a new fungal colonization can initiate. Several requirements must be met for colonization to occur. In addition to the wood substrate itself, these include:

5.1. *Water and Air*

Typically, wood must be at or near the point where the wood cell wall is saturated with "bound" water, known as the fiber saturation point (FSP) for the fungal spores or fragments to germinate and initiate new fungal colonies. There is considerable debate about the minimum moisture level required for fungi to colonize and decay wood exposed out of soil contact [21]; however, because the filamentous strands of fungi (hyphae-see Fungal Physiology/Anatomy below) are required to be surrounded by a watery extracellular polysaccharide matrix (ECM) when wood is being attacked, some level of free moisture in the wood cell lumen is required to support fungal growth. In most cases, fungal decay can begin at approximately 30% moisture content (oven dry basis), reaching an optimum between 40 and 80%, then declining with increasing moisture levels above 100% as cell lumens begin filling with water and oxygen becomes limiting. Water is often the most important limiting factor in decay, and some paradigm-shifting current literature [22] focuses on the critical moisture content that allows fungal metabolites responsible for depolymerization of the wood polymers to diffuse within the wood cell wall. However, fungi require liquid water to initiate the secretion of metabolites required for decay, even in the absence of wood or other suitable substrates. Because the biosynthesis of these metabolites is a prerequisite for their diffusion within the cell wall, it is important for wood to first attain a moisture content where fungi can synthesize metabolites as well as one that supports robust ECM production surrounding and attaching the fungal cells to the wood cell wall to allow compounds to diffuse between fungus and the wood cell wall. In this regard, appropriate moisture conditions for fungal activity are required even before diffusional aspects of metabolites within the wood can be considered, and inhibition of fungal ECM production represents a fruitful area for future research in controlling fungal growth.

Wood in contact with the ground is frequently above the FSP and, in addition to having a fully saturated cell wall, contains liquid water in the lumens of the fibers. For fungal growth to occur in wood, the moisture content of wood must also not be too high to preclude adequate oxygen levels. Although fungi do not require as much oxygen as humans, they are aerobic organisms and wood decay fungi typically will not grow on wood that is completely saturated or submerged. Oxygen is rarely limiting for fungal attack, although complete saturation of logs by "ponding" (submersion in natural or artificial bodies of water) has been used to limit decay for periods of several months prior to processing in mills. In some cases, logs that sank during freshwater storage have been retrieved decades and sometimes even centuries later and processed with little deterioration noted. Many of these logs remain on the bottom of fresh water bodies and are in pristine condition because of the lack of oxygen, and also because relatively cold temperatures limited anaerobic bacterial deterioration.

5.2. Temperature

Temperature is critical for most physiologic reactions. Fungi may continue to grow as temperatures decline to levels near freezing [23], but reaction rates involved in both fungal metabolism and in the chemistry/biochemistry of wood depolymerization decline, and the decay process ultimately stops at freezing temperatures. Metabolic reactions increase with increasing temperature with most fungi having growth optima between 24 and 32 °C [20]. This is well within the temperature range inside most inhabited structures but specialized fungi can grow well outside this range. As a result, temperature is generally not a limiting factor in decay. For most decay fungi, fungal metabolism becomes more constrained as temperatures exceed 39–40 °C; however, some thermophilic fungi survive and have been observed to be active at temperatures exceeding 50 °C in specialized environments such as pulp chip piles [24,25]. Exposure to temperatures about 56 °C results in permanent denaturing of proteins and DNA, effectively killing most non-thermophilic organisms.

5.3. Other Essential Components for Fungal Growth and Decay Potential

As noted above, a primary requirement for decay is a nutrient source which is typically the timber itself. This wood or surrounding supporting materials/soils must also contain various micronutrients and microelements, including nitrogen, that are essential to fungal growth and the decay process. There are vast volumes written about the relationships between decay and wood properties, e.g., [26]. Relatively few fungi have evolved the ability to degrade and utilize the three primary cell wall polymers, and the mechanisms by which they accomplish this task will be the subject of the remainder of this paper.

6. The Decay Environment

Many microorganisms are ubiquitous and will colonize almost any substrate meeting the basic requirements for growth of oxygen, food, moisture & temperature. However, the nature of the wood substrate affects the types of wood degrading fungi that attack it, the colonization patterns, and the rates at which wood is utilized. For this discussion, we will ignore aquatic environments and concentrate on the terrestrial decay process. Decay can be categorized as either in ground or above ground, recognizing that these are arbitrary terms that overlap at their margins. In-ground or ground-contact decay is typically more rapid, reflecting the presence of adequate moisture along with an abundance of organisms with differing wood-attacking or wood-inhabiting capacities that are always present and in competition with each other. The above-ground environment is more challenging with fewer organisms present, and a greater potential for drying that can slow the decay process. There is also more limited potential for the transfer of moisture and exogenous nutrient resources into the wood and into the fungal thallus. These differences result in dramatic differences in the rates of degradation as well as substantial changes in the organisms involved.

7. Basics of Fungal Anatomy/Physiology and Evolution

7.1. Biology and Evolution

Within the Eukarya, the Kingdom of Fungi are classified into Subkingdoms and further into Orders/Divisions. The subkingdom of Dikarya, which are fungi that largely produce filamentous cells known as hyphae, are of particular interest with regard to coatings [27], and are described in more detail below. Two Orders are encompassed by Dikarya, the Basidiomycota and the Ascomycota, and both can play significant roles in wood and/or coating degradation. Ongoing research suggests that both Basidiomycota and Ascomycota evolved from earlier progenitor fungi between 400 and 500 million years ago [28–31], with species of white rot fungi that had the ability to decay lignocellulose evolving only perhaps 280 million years ago within a Class of fungi known as the Agaricomycetes [31]. Present day Agaricomycetes include many well-known mushroom-producing species, but many of the Agaricomycetes are rather nondescript in their macroscopic features, yet they have the ability to aggressively attack wood and disrupt any overlying coating materials.

It is interesting to note that plants initially evolved the ability to produce the cellulosic components of their cell walls long before they evolved the ability to produce lignin, which only later started to appear as a cell wall component in some types of woody plants [32]. Lignin "stiffened" the cellulose into a rigid polymer-matrix composite, allowing the precursors of modern-day trees to grow much taller than the grasses and sedges which existed prior to the Devonian period. The first lignin-producing plants evolved about 420 million years ago [33], long before fungi evolved enzymes and non-enzymatic systems with the capacity to depolymerize and further degrade lignin. Precursors to white- and brown rot fungi evolved lignin-degrading enzymes only about 295 million years ago [32], meaning that woody biomass was unable to be efficiently degraded by fungi for approximately 125 million years, and it has been proposed that inability to decay wood over this period of time led to the formation of coal seams in many areas of the world. However, some geology experts [33] examining fossilized woody tissue from the Carboniferous period dispute this premise and report that fungi were present in the fossilized wood cells showing "evidence of decay that is pervasive", suggesting that the fungi of that period were still somehow capable of decaying the woody tissues. However, the authors do not show convincing evidence of fungal attack of wood cell walls, and the images of fossilized fungal hyphae they present for interpretation are larger in diameter, and more consistent with non-decay Ascomycota species rather than Basidiomycota decay fungi. This suggests that there is still much more to be learned about the how wood decay fungi arose on earth, and their impacts, including the impact they have in decaying current day wood products and disrupting the coatings on those products.

7.2. Growth and Infection of Wood

Wood decay fungi primarily initiate as fungal spores or mycelial fragments. If conditions are favorable, spores germinate to produce fine hair-like structures known as fungal hyphae, which are elongated cells of the fungus which grow end-to-end. Hyphal fragments landing on wood can also in many cases initiate growth leading to broader colonization of the wood. Single-celled fungal growth is not common in wood degrading fungi except during spore formation. As the fungal hyphae grow along the surface of materials, some species can form a mat consisting of multiple layers of interwoven hyphae known as mycelium or a mycelial mat. The tips of the fungal hyphae initially seek out relatively simple pathways through the microstructure of wood, exploiting interconnecting cell wall pits (interconnecting channels between wood cells) to extend from one fiber to the next, and ramifying through the wood in this manner. During this initial growth phase, all wood-inhabiting fungi seek out stored products in the parenchyma as a ready nutrient source for energy for the fungus, and also to build up fungal biomass within or on the surface of the wood structure. This initial stage of growth of decay and some stain fungi results in activation of biochemical "machinery" within the hyphae resulting in secretion of a diverse suite of extracellular metabolites and enzymes that can depolymerize and digest select polymeric components of wood. One of the most common ways to measure wood degradation in laboratory studies is to monitor mass loss as decay progresses and wood cell wall components are converted to CO_2 by the fungus. Although mass loss is not always an adequate indicator of structural strength loss, it is a straight-forward means for assessing some types of fungal degradation. Fungi have diverse suites of metabolites and enzymes that effect deterioration, and the discussion below provides information on how these fungi can be differentiated and how they may degrade wood in different ways.

Decay types: brown rot white rot, soft rot and their effects on wood chemistry: The fungi that degrade the cell wall polymers have long been segregated on the basis of the appearance of the damaged wood into brown, white and soft rot fungi. These separations are arbitrary and we now recognize that the decay types are more of a continuum [34], but the categories are helpful in generally classifying these fungal degradative agents, and for discussion.

7.3. Brown Rot Fungi

After the initial colonization stages, brown rot fungi primarily utilize the carbohydrate polymers of the wood cell wall, although they affect all three polymers. Brown rot fungi are considered to be more commercially important from a structural damage perspective because they tend to attack softwoods which represent the bulk of timber used in North American and European construction. However, they can also attack hardwoods and will readily do so in some environments [35]. The resulting decayed wood has a brownish, fractured appearance (Figure 1). The most important aspect of brown rot decay is the tendency for these fungi to produce degradation far in advance of the point of hyphal growth at the cellular level, and to depolymerize the carbohydrates much faster than they can be utilized. This results in very dramatic changes in many timber properties even when the macroscopic visual appearance of the wood is little changed. The result is very rapid losses in properties such as tension or bending at very low mass losses [26,36].

Figure 1. Classic advanced brown rot fungal decay in a softwood. Note the checks (cracks) which cut across the grain of the wood giving the wood a cubical appearance. In early stages of brown rot decay, the wood can appear unchanged from undecayed wood, yet significant mechanical property loss may be sustained even in early brown rot decay stages.

Brown rot fungi have a more limited suite of enzymes involved in the depolymerization and deconstruction of cellulose compared to white rot fungi, and they possess no peroxidase enzymes for lignin depolymerization. However, brown rot fungi rapidly attack and degrade cellulose while also using a non-enzymatic mechanism to depolymerize lignin to allow access to the cellulosic components. The lignin then rapidly repolymerizes, leaving a modified lignin that long led researchers to conclude that brown rot fungi had little effect on this polymer. This unique oxygen radical-based chemistry known as the "chelator-mediated Fenton" (CMF) system functions in advance of enzymatic action to deconstruct wood polymers [37]. CMF chemistry is more complex than conventional Fenton chemistry and allows brown rot fungi to generate powerful hydroxyl radicals within the wood cell wall rather than next to the fungal hyphae (which would kill the fungus). CMF chemistry causes lignin to depolymerize and then repolymerize as small, discrete irregular masses, separate from the cellulose [38,39], thereby opening the wood cell wall to further deconstruction. This type of chemistry is being studied as an efficient mechanism for lignin processing in future biorefineries. This later aspect may be of importance for modifying lignin for use in polymeric coatings and is reviewed further in the section on biotechnological applications.

7.4. White Rot Fungi

White rot fungi can depolymerize all three cell wall polymers using enzymes secreted by the hyphae, often leaving a whitened mass of fibers at the advanced stages of decay (Figure 2). White rot fungi metabolize all three cell wall polymers and can cause weight losses approaching 97% of the original wood. Most white rot fungi are known as "simultaneous" white rots and tend to digest and utilize the wood polymeric components as they are depolymerized. However, some "selective" white rot fungi can preferentially digest the hemicellulose and lignin components leaving much of the crystalline cellulose relatively undegraded. In either type of white rot, structural losses occur more slowly than in brown rots, and cell wall depolymerization is more directly linked to enzymatic erosion of the cell wall layers and also to the metabolic processes of the fungus. This coupling of degradation with utilization results in declines in properties such as tension or bending that parallel mass losses caused by decay. White rot fungi tend to attack hardwoods, but can also attack softwoods [35]. Carbohydrate active enzymes (CAZymes) and lignin degrading enzymes from white rot fungi have been previously explored for use in pulp and paper processing, and other industrial processes to free lignin from wood fiber with reduced use of harsher chemicals.

Figure 2. Advanced stage of white rot simultaneous decay. The wood in this advanced state of decay can be quite spongy and lightweight when dried. However, when decay is active, the wood will be moist and stands of wood can often be peeled from the surface along the grain. The flecks of white in the image are mycelial mats of the fungus.

Enzymes of specific importance in lignin degradation by white rot fungi are primarily oxidative and include: laccase, lignin peroxidase, manganese peroxidase, versatile peroxidase and the dye-decolorizing peroxidases [39]. Many of these enzymes are unique to white rot fungi with only a relatively few dye-decolorizing peroxidases, for example, found in select lignin-degrading bacteria. Relative to CAZymes, white rot fungi have a much more complete suite of glycoside hydrolases and also lytic polysaccharide monooxygenase enzymes that can attack cellulose both from the ends of the long-chain polymer (exo-glucanases) as well as centrally along the polymer chain (endo-glucanases). This dual action, combined with a suite of enzymes that can act on oligosaccharide depolymerization products, allows white rots to deconstruct a variety of holocellulose polymers.

7.5. Further Evolution of Brown rot Fungi from the Precursors of White Rot Species

Brown rot fungi evolved relatively recently from white rot fungi starting about 280 million years ago in the Permian period [31]. Only 20 years ago, brown rot fungi were looked upon by most mycologists as less evolutionarily advanced because many of the CAZymes and all of the lignin-degrading peroxidase enzymes in the precursor white rot fungi were lost as the brown rot fungi evolved. The brown rot fungi now dominate wood decay in the coniferous forests of the northern hemisphere, and there are chemical wood-processing efficiencies that can be learned from studying their evolutionary divergence from the white rots. Enzymes from both white and brown rot fungi are too large to penetrate the intact wood cell wall, and some form of chemical "pretreatment" is typically needed before enzymes can function efficiently in deconstructing the wood cell wall. The evolution of the bio-catalytic CMF mechanism by brown rot fungi is generally considered a key step that allowed these fungi to more efficiently deconstruct wood without the production of nutritionally costly enzymes. The CMF mechanism has considerable promise for use in the bioprocessing of lignocellulosic materials [39]. Enzymatic systems in industrial processes are widely used for a variety of applications to reduce energy inputs required for catalysis but enzymes can be fragile and often require highly specific reaction conditions including specific temperatures and buffering systems. Mimicking bio-catalytic processes without the use of enzymes may be a fruitful pathway for industrial processing for bio-based coatings.

7.6. Soft Rot Fungi

Although soft rot damage of wood was first observed in the 1860s, soft rot fungi were not classified as a decay type until the 1950s [40]. Most soft rot fungi are Ascomycota species. There are two types of soft rot attack. Type 1 soft rot involves formation of diamond-shaped cavities aligned with the cellulose microfibril angle within the S-2 cell wall layer, while Type 2 is a more generalized erosion of the S-2 cell wall layer from the lumen outward [20]. Type 2 attack is more prevalent, but some species can produce both types of damage depending on the timber as well as environmental conditions [41]. Soft rot fungi are often found in more extreme, and wetter conditions that are less suitable for traditional white and brown rot fungi [39]. Their damage tends to be confined to the external few mm of wood that is exposed to the environment, possibly because oxygen levels are too low in interior wood below ground to support more aggressive Basidiomycota fungal species. However, particularly in Scandinavian reports, soft rot has been observed to extend more deeply in some products such as utility poles. Soft rot damage presents an interesting mixture of white and brown rot characteristics in that these fungi utilize both cellulose and hemicellulose, but they can clearly degrade lignin as evidenced by the cavities and erosion they cause. Several soft rot fungi are known to produce laccase which is also involved in lignin degradation by white rot fungi [39].

Soft rot fungi tend to have very large, but localized effects on wood properties and these effects are magnified because the damage tends to be on the exterior of the timber where flexural properties for products such as utility poles become more important. In other products, such as boards that will be used for paneling, soft rot fungi can impart an appearance that some people consider as desirable for rustic interiors, and because only the surface wood is degraded, these fungi can sometimes be considered as enhancing the properties of certain wood products.

8. Specific Fungal Chemistries that Impact Polymeric Coatings and How These Can Be Harnessed for Biotechnological Applications

The characteristic ability of all three decay types to attack the three primary wood polymers highlights the effect of wood cell wall chemistry on convergence of fungal strategies for accessing these resources, but differences in processes creates potential opportunities for the using these fungi in industrial biomodification processes. The most heavily researched applications have been delignification for pulping and biodetoxification of xenobiotic pollutants, but fungi could also be used to modify various polymeric materials including those used for wood coatings.

8.1. White Rot Fungi

White rot fungi have been used in biotechnological and bio-processing applications for more than 40 years [42–44]. White rot fungi have been studied since the 1970s to free cellulose from lignin and release individual fibers in bio-based pulping systems. Wood composites have also developed by using isolated lignin from pulp liquors, or lignin residues that migrate to the surface of wood fibers during pulping. In both cases the lignin is modified to produce a "sticky" lignin radical by white rot fungal enzyme systems including peroxidase and laccase-mediator systems [45,46]. Similarly, laccases have also been used to create bioactive polymer coatings using soft plasma jet processing [47]. The ability of white rot fungi to depolymerize lignin has also been extensively explored for bioremediation of structurally similar pollutants and xenobiotics but the applications have been limited because many sites are anaerobic [48,49]

The emerging bio-economy, the development of biorefineries and the production of cellulose-derived sugars for fermentation and direct conversion to biofuels and platform chemicals have all created renewed interest in application of white rot fungi [50,51]. A number of white rot species, including *Pycnoporus cinnarbarinus*, *Phlebia subserialis*, *Dichomitus squalens* and *Ceriporiopsis subvermispora*, have been assessed for use in the bioprocessing of wood to make bio-based products and energy, but there is considerable opportunity to expand the suite of organisms to take advantage of the range of enzymatic capabilities [52]. Early researchers who explored bio-based deconstruction of wood for pulp fiber focused primarily on complete lignin removal, but subsequent studies showed that some white rot fungi had more subtle effects that resulted in reduced energy requirements in mechanical pulp production while improving other paper properties. Bio-bleaching of pulp using *Phanerochaete crassa*, *P. chrysosporium*, and *Pleurotus pulmonarius* previously has been studied for replacement of chlorine in conventional pulp bleaching processes [53,54], but bio-bleaching using these organisms has not been commercialized, illustrating the difficulty in scaling up laboratory results.

The need for oxygen and the filamentous nature of white rot fungi has largely limited most applications, but several of the CAZymes as well as the lignin degrading enzymes have been cloned into yeast and bacterial vectors for use in biorefinery applications. Many peroxidases have broad capabilities to oxidize phenolic substrates and could be used to activate phenolic-based coatings. Additional research in this area using enzyme cocktails, including the use of relatively newly discovered lytic polysaccharide monooxygenase (LPMO) enzymes, is needed to enhance platform chemical yield from biomass [55].

The desire for more sustainable processes for producing polymers and coatings using fungal peroxidases should encourage a re-evaluation of previous biomass conversion research, which was often discarded because the economics could not compete with less sustainable processes [56]. Fungal derived peroxidases could be used to provide platform chemicals for further synthesis. While lignin remains a puzzling and complex polymer that has largely defied technological advances geared to utilization, development of bio-processes that depolymerize lignin and utilize the broad array of monomeric lignin breakdown products created during industrial pulping could create an array of feedstocks similar to those derived from petroleum sources. Adapting microbial systems to utilize raw lignin, while producing platform chemicals useful for polymer development, is an important step in this process [57].

8.2. Brown Rot Fungi

Brown rot fungi have received less study for biotechnological applications, primarily because their ability to rapidly depolymerize the carbohydrate fraction of the wood was viewed as having little practical application. However, more recent studies indicate that some CAZymes, such as lytic polysaccharide monooxygenases (LMPOs), have the ability to work synergistically with peroxidases to promote depolymerization and solubilization of aromatic monomers from lignin [37,58]. These results suggest that cellulase enzymes from either white or brown rot fungi could play a greater role in lignin depolymerization for use as a chemical feedstock. CAZymes from brown rot fungi have not been explored to a great extent largely because they are less common compared to white rot fungi. LPMO enzymes from brown rot fungi have only been isolated within the last five years, and have not been fully explored for use in industrial applications. Preliminary reports suggest that an LMPO from the brown rot fungus *Gloeophyllum* spp. cloned into yeast has significant potential for biorefinery applications [37,59]. Studies have tended to look for one or a few mechanisms of action, but it may be necessary to change the paradigm by combining enzymatic and non-enzymatic mechanisms under controlled conditions to either modify fibers or create platform chemicals [38].

The non-enzymatic activity of brown rot fungi has previously been shown to activate lignin for production of laminated wood [60] and composite panels [61]. Brown rotted lignin modified with a sodium borohydride treatment was previously used to produce a formaldehyde-free adhesive resin with properties close to that of phenolic resins [62]. The CMF mechanism from fungal systems has been used to activate lignin on fiber surfaces for the producing fiber-based products at an experimental level and has been shown to have potential industrial applications [63–66]. Lignin has also been modified using a CMF system to produce water-soluble polymers that have potential applications as a high-value dispersant comparable to poly(acrylic) acid [67].

Brown rot fungi have also been explored for bioremediation, especially for removal of copper and other heavy metals. *Serpula* spp. has been shown to remove copper from preservative-treated wood [68]. CMF chemistry also has effectively degraded pollutants like 2,4-dichlorophenol [69], dichloro-diphenyl-trichloroethane (DDT) [70] and pentachlorophenol [71] as well as decolorizing recalcitrant dyes [72,73]. An improved understanding of brown rot mechanisms over the past 20 years has encouraged a re-examination of using these fungi in bioremediation.

Brown rot fungi have also successfully been used to pretreat bagasse, wood and other lignocellulose substrates to promote cellulose and hemicellulose depolymerization for biorefinery applications [74]. The Mycologix LTD company [75] successfully pretreated biomass in a commercial application, but dramatic decreases in the cost of competing fossil hydrocarbons led the company to declare corporate insolvency. The initial success of this process suggests that similar efforts will emerge as the economies of the process improve and hydrocarbon prices rise. While the need for biofuels may be somewhat mitigated by the emergence of other renewable energy sources, liquid fuels will continue to be required in many industries. The use of brown rot systems for biomass preparation in biorefineries will likely increase, and markets for the lignin residues produced in these processes will be needed, creating opportunities for developing lignin-based coatings as well as polymeric resins.

9. Relationship between Understanding Fungal Decay and Developing More Rational Wood Protection Methods

Humans have long labored to limit the risk of decay and prolong the useful life of timber. Initially, they did so without even knowing that fungi were the causal agents, but emerging knowledge led to the development of a range of biocides that could protect timber from decay. Interestingly, many of these advances, such as the development of creosote preceded that understanding. As a result, wood protection largely remains a combination of effective designs that limit moisture, coupled with the use of naturally durable timbers or impregnation with biocides where moisture control is not possible. However, the types of biocides being used, and their method of application, is changing as concerns emerge about the use of all chemicals in the environment. These concerns have created opportunities to develop more rationale methods for protecting timber based upon our understanding of wood science and fungal physiology.

Current efforts to protect wood without biocides have focused methods such as thermal modification and chemical wood modification with techniques such as acetylation. Both techniques were explored in the middle of the 20th Century but were dropped at that time because they were not considered economical, but their use is now being resurrected. Thermal modification and acetylation function to limit water uptake by wood, but there may be more subtle approaches to wood protection under in some environments. For example, our improved understanding of the roles of free radicals in the degradation process has also been exploited experimentally through the use of anti-oxidants to protect wood [76]. The use of many plant-based extracts follows a similar path but uses a less specific approach since these treatments are often mixtures of compounds. While still not commercially used, they illustrate the potential for using our knowledge of microbial decomposition mechanisms to identify more rationale prevention strategies.

10. Coatings for Bio-Based Materials and the Need for Better Protection against Fungi

While coatings have long been successfully used to impede and in some cases prevent fungal degradation while also limiting UV degradation, it is critical to understand both their function and limitations. The long-term effectiveness of any coating is dependent on its ability to prevent water penetration leading to coating failure and subsequent penetration by fungal hyphae or their enzymes and low-molecular weight metabolites through, around or behind the coating and into the wood. Coating efficacy is achieved via barrier technology and/or enhanced moisture exclusion. Loss of coating efficacy is often associated with increased moisture retention that can promote decay.

The general issues relating to coating failures for wood materials have been well summarized previously [77]. Decades of research at the U.S. Forest Products Laboratory showed that most paint coatings usually cracked at the wood joints or knots due to differential dimensional stability. Water enters the wood through these cracks and is trapped. Wood moisture content can then quickly reach the range suitable for fungal attack. Oil-based or alkyd-based paints contain oils that cure by reacting with oxygen to form cross-linked polymeric films. The more cross-linked the polymer, the more resistant it is to either liquid or vapor water. Latex paints form a film by coalescence of small spherical polymeric particles dispersed in water. These polymers are more flexible and not as highly cross-linked as cured oil-based paints, and generally remain more flexible with age. As a result, they provide less of a barrier to water ingress. Oil-based paint films generally provide superior initial water resistance; however, they become brittle over time.

As noted in the introduction to this chapter, moisture content changes with wood shrinkage and swelling are an issue which must be considered relative to the development of advanced coatings for wood products. Coatings must be designed to flex to accommodate changes in wood dimensions with wetting and drying. However, even under the best of circumstances, most coatings will fatigue and develop fine micro-fractures after thousands of shrinkage/swelling cycles. These fine micro-fractures are exploited by fungi that then penetrate into the wood beneath. This is particularly true in exterior environments where UV exposure weakens the coating thereby enhancing dimensional change stresses and promoting fractures (Figure 3). This can occur in seemingly impermeable barriers. For example, 1–2 mm thick polyurea coatings were penetrated by decay fungi over a four-year exposure in Hilo, Hawaii [78].

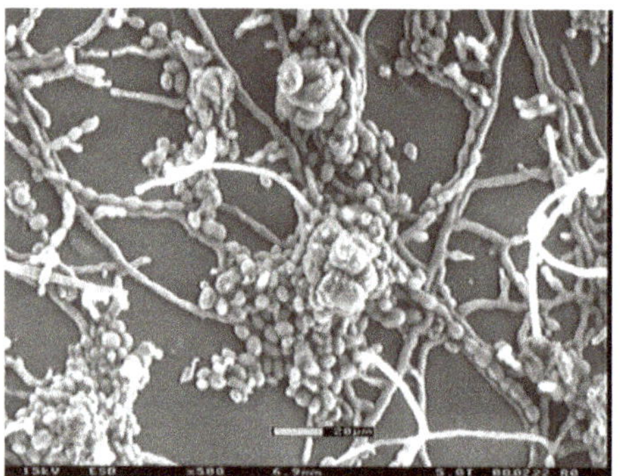

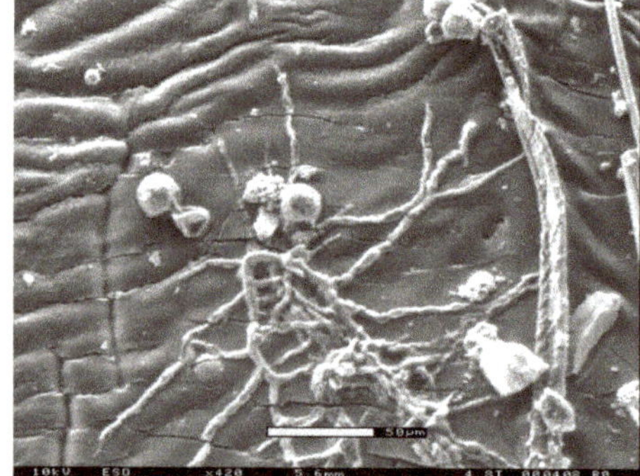

Figure 3. Scanning electron micrographs showing (**a**) fungal spores and hyphae growing on the surface of a new coating surface and (**b**) fungal spores and hyphae growing on an aged coating surface containing microcracks that can be exploited by fungal hyphae (Goodell images).

While we have previously discussed the decay process, hyphae of some decay and stain fungi also have the ability to directly penetrate wood cell walls using combinations of enzymatic/catalytic dissolution and mechanical/physical force. The mechanical forces generated by fungal hyphae can drive a fungal peg through a metal foil [79], even in the absence of enzymatic or catalytic action. This force is certainly great enough to allow fungi to penetrate many types and thicknesses of polymeric coatings, and the presence of surface micro-cracks undoubtedly aids in this process. Minimizing surface defects, slowing micro-crack development and even limiting hyphal growth on the coating surface (via the presence of biocides) can all contribute to limiting fungi compromising the coating barrier. A more difficult goal will be the development of clear coatings for exterior applications. Consumers generally prefer clear finishes so that they can see the wood; however, these coatings tend to be more susceptible to UV damage that leads to coating failures and, ultimately premature replacement of the timber [10,80].

Coated surfaces are bombarded with fungal spores along with dust, pollen, plant sap and a host of other potential nutrient sources for a fungus. Many paint film fungi utilize these nutrients to begin growth (Figure 3) and eventually exploit coating defects to penetrate into the wood beneath (Figure 4). Once inside, they preferentially colonize the layer between the coating and the wood, leading to further coating failure (Figure 5). The extent of this problem in both clear and opaque coatings highlights the need for coating formulations that contain long-lasting fungal biocides or other protection systems that can prevent fungi from causing this type of failure. It may also be useful to explore self-cleaning systems [81]. For example, developing superhydrophobic surfaces that mimic natural phenomena such as the self-cleaning capacity of lotus leaves would reduce accumulation of surface debris that supports fungal growth.

Figure 4. An example of fungal growth that has developed underneath a coating film (on the wood strip below hand) by penetrating the coating surface. This type of fungal attack occurs with opaque coatings as well, but it is readily observed when clear-polymer coatings are attacked. Clear coatings typically lack adequate UV resistance leading to development of micro-cracks in the coatings that fungi can exploit. (Goodell photo).

Figure 5. Example of a delaminated clear polymeric coating on wood being peeled from the surface after three years of exposure in a damp exterior environment. In this example, the finish was originally clear, similar to that seen in Figure 4. Fungi were involved in darkening of the finish and also, once they penetrated the coating film, in the delamination of the coating from the wood surface. In this example, the manufacturer's use of iron in the coating formulation also promoted the extreme darkening of the coating as, in this wet environment, wood extractives were driven to the surface, and these extractives reacted with the iron in the coating to produce the intense dark coloration (Goodell photo).

11. Future Needs and Opportunities

There is a critical need for an expanded understanding of how fungi attack and exploit current wood coatings to enable development of strategies to prevent premature damage to both the coating and the substrates they are intended to protect. Integrated technologies are needed to reduce fungal attack and enhance coating performance against physical and environmental factors. For example, enhanced UV protection coupled with better matching of long-term coating elasticity with expansion/contraction characteristics of the wood will limit the development of coating defects that fungi are known to exploit. Enhanced abrasion resistance and improved ability to shed environmental residues deposited on the coating will limit the ability of fungi to grow on and into the coating. Enhanced water repellency helps coatings shed water and reduces the time that a surface remains wet enough to permit fungal growth.

Biocides such as 3-iodo-2propynyl butyl carbamate and chlorothalonil have long been a component in many coating systems both for protecting the system against microbial attack prior to application as well as limiting fungal discoloration afterwards. While biocides ranging from tin and copper compounds to thiabendazoles are effective, they are not compatible with all polymeric systems. Novel approaches such as non-biocidal treatments that discourage attack by preventing fungal attachment to surfaces and preventing the development of biofilms have shown promise in the laboratory [82], but utilization of these coatings will likely remain limited to medical applications. While agricultural applications for this type of technology are proposed, it remains to be seen if cost-effective products can be developed that meet consumer needs for long-lasting surface protection under extreme UV exposures.

At present, a majority of coatings research remains proprietary, making it difficult to accurately assess the state of the science. However, market observations suggest that better coatings systems are still needed, and that fungal and UV resistance of coating systems has not dramatically improved in many consumer-available coatings in the past 20 years. The lack of long-lasting UV and moisture resistant coatings leads to premature replacement of functional structures, especially decking and this, in turn sharply reduces the normally positive life cycle attributes of timber. Developing a better understanding the mechanisms underlying fungal degradation of substrates, and the mechanisms involved in new biocidal and non-biocidal coating treatments will be essential for creating integrated coatings systems capable of limiting fungal attack while providing long term protection against environmental factors.

Funding: The first author was supported in part by the National Institute of Food and Agriculture, U.S. Department of Agriculture, the Center for Agriculture, Food and the Environment and the Microbiology department at University of Massachusetts Amherst, under project No. S1075-MAS00503. The contents are solely the responsibility of the authors and do not necessarily represent the official views of the USDA or NIFA.

Conflicts of Interest: The authors declare no conflict of interest with the content of this review.

References

1. Côté, W.A. Chemical Composition of Wood. In *Principles of Wood Science and Technology*; Solid Wood; Kollmann, F., Cote, W., Eds.; Springer: New York, NY, USA; Berlin/Heidelberg, Germany, 1968; Volume 1, pp. 55–78.
2. Mittal, N.; Ansari, F.; Gowda, V.K.; Brouzet, C.; Chen, P.; Larsson, P.T.; Söderberg, L.D. Multiscale Control of Nanocellulose Assembly: Transferring Remarkable Nanoscale Fibril Mechanics to Macroscale Fibers. *Acs Nano* **2018**, *12*, 6378–6388. [CrossRef] [PubMed]
3. Altgen, M.; Kyyrö, S.; Paajanen, O.; Rautkari, L. Resistance of thermally modified and pressurized hot water extracted Scots pine sapwood against decay by the brown–rot fungus Rhodonia placenta. *Eur. J. Wood Wood Prod.* **2020**, *78*, 161–171. [CrossRef]
4. Militz, H.; Altgen, M. Processes and properties of thermally modified wood manufactured in Europe. In *Deterioration and Protection of Sustainable Biomaterials*; Schultz, T.P., Goodell, B., Nicholas, D.D., Eds.; American Chemical Society: Washing, DC, USA, 2014; Volume 1158, pp. 269–286.
5. Winandy, J.E.; LeVan, S.L.; Ross, R.J.; Hoffman, S.P.; McIntyre, C.R. *Thermal Degradation of Fire-Retardant-treated Plywood: Development and Evaluation of a Test Protocol*; Research Paper FPL-RP-501; USDA, Forest Service: Madison, WI, USA, 1991.
6. Winandy, J.E. Relating wood chemistry and strength: Part II. Fundamental relationships between changes in wood chemistry and strength of wood. *Wood Fiber Sci.* **2017**, *47*, 2–11.
7. Winandy, J.E.; Morrell, J.J. Relationship between incipient decay, strength, and chemical composition of Douglas-fir heartwood. *Wood Fiber Sci.* **1993**, *25*.
8. Curling, S.; Clausen, C.A.; Winandy, J.E. Relationships between mechanical properties, weight loss, and chemical composition of wood during brown rot decay. *For. Prod. J.* **2002**, *52*, 34–39.
9. Kirker, G.; Winandy, J. Above Ground Deterioration of Wood and Wood–Based Materials. In *Deterioration and Protection of Sustainable Biomaterials*; American Chemical Society: Washington, DC, USA, 2014; Volume 1158, pp. 114–129.
10. Evans, P.D. Weathering and Photoprotection of Wood. In *Development of Commercial Wood Preservatives*; American Chemical Society: Washington, DC, USA, 2008; Volume 982, pp. 69–117.
11. Hon, D.N.S.; Feist, W.C. Weathering characteristics of hardwood surfaces. *Wood Sci. Technol.* **1986**, *20*, 169–183. [CrossRef]
12. Williams, R.S.; Winandy, J.E.; Feist, W.C. Adhesion of paint to weathered wood. *For. Prod. J.* **1987**, *37*, 29–31.
13. Williams, R.S.; Lacher, S.; Winandy, J.E.; Halpin, C.; Feist, W.C.; White, C. Comparison of traditional methods for testing paint service life with new methods for service life prediction. In Proceedings of the 3rd International Symposium on Surfacing and Finishing of Wood, Kyoto, Japan, 24–26 November 2004; pp. 167–190, Sponsored by IUFRO Division 5.04.12.
14. Lebow, P.K.; Winandy, J.E. Verification of the kinetics–based model for long–term effects of fire retardants on bending strength at elevated temperatures. *Wood Fiber Sci.* **1999**, *31*, 49–61.

15. LeVan, S.; Ross, R.J.; Winandy, J.E.; Laboratory, F.P. *Effects of Fire Retardant Chemicals on the Bending Properties of Wood at Elevated Temperatures*; U.S. Department of Agriculture, Forest Service, Forest Products Laboratory: Madison, WI, USA, 1990.
16. Sweet, M.S.; Winandy, J.E. Influence of Degree of Polymerization of Cellulose and Hemicellulose on Strength Loss in Fire–Retardant–Treated Southern Pine. *Holzforschung* **1964**, *53*, 311. [CrossRef]
17. Stamm, A.J. *Wood and Cellulose Science*; Ronald Press Co.: New York, NY, USA, 1964.
18. Cragg, S.M.; Beckham, G.T.; Bruce, N.C.; Bugg TD, H.; Distel, D.L.; Dupree, P.; Zimmer, M. Lignocellulose degradation mechanisms across the Tree of Life. *Curr. Opin. Chem. Biol.* **2015**, *29*, 108–119. [CrossRef]
19. Elam, J.; Björdal, C. A review and case studies of factors affecting the stability of wooden foundation piles in urban environments exposed to construction work. *Int. Biodeterior. Biodegrad.* **2020**, *148*, 104913. [CrossRef]
20. Zabel, R.A.; Morrell, J.J. *Wood Microbiology: Decay and Its Prevention*, 2nd ed.; Academic Press: San Diego, CA, USA, 2020; 556p, ISBN 978-0-12-819465-2.
21. Viitanen, H.; Toratti, T.; Makkonen, L.; Peuhkuri, R.; Ojanen, T.; Ruokolainen, L.; Räisänen, J. Towards modelling of decay risk of wooden materials. *Eur. J. Wood Wood Prod.* **2010**, *68*, 303–313. [CrossRef]
22. Jakes, J.E.; Zelinka, S.L.; Hunt, C.G.; Ciesielski, P.; Frihart, C.R.; Yelle, D.; Vogt, S. Measurement of moisture-dependent ion diffusion constants in wood cell wall layers using time-lapse micro X-ray fluorescence microscopy. *Sci. Rep.* **2020**, *10*, 9919. [CrossRef] [PubMed]
23. Miller, D.J.; Goodell, B. Blue staining in ponderosa pine sapwood at moderate and low temperatures. *For. Prod. J.* **1981**, *31*, 54–59.
24. Smith, R.S.; Ofosu-Asiedu, A. Distribution of thermophilic and thermotolerant fungi in a spruce-pine chip pile. *Can. J. For. Res.* **1972**, *2*, 16–26. [CrossRef]
25. Horvath, Z.; Marosvölgyi, B.; Idler, C.; Pecenka, R.; Lenz, H. Storage problems of poplar chips from short rotation plantations with special emphasis on fungal development. *Acta Silv. Lignaria Hung.* **2012**, *8*, 123–132. [CrossRef]
26. Wilcox, W.W. Review of literature on the effects of early stages of decay on wood strength. *Wood Fiber* **1978**, *9*, 252–257.
27. Gerke, J.; Köhler, A.M.; Meister, C.; Thieme, K.G.; Amoedo, H.; Braus, G.H. Coordination of fungal secondary metabolism and development. In *Genetics and Biotechnology. The Mycota (a Comprehensive Treatise on Fungi as Experimental Systems for Basic and Applied Research), vol. 2*; Benz, J.P., Schipper, K., Eds.; Springer: Cham, Switzerland; New York, NY, USA, 2020. [CrossRef]
28. Lücking, R.; Huhndorf, S.; Pfister, D.H.; Plata, E.R.; Lumbsch, H.T. Fungi evolved right on track. *Mycology* **2009**, *101*, 810–822. [CrossRef]
29. Prieto, M.; Wedin, M. Dating the Diversification of the Major Lineages of Ascomycota (Fungi). *PLoS ONE* **2013**, *8*, e65576. [CrossRef]
30. Bhatt, M.; Mistri, P.; Joshi, I.; Ram, H.; Raval, R.; Thoota, S.; Joshi, M. Molecular survey of basidiomycetes and divergence time estimation: An Indian perspective. *PLoS ONE* **2018**, *13*, e0197306. [CrossRef]
31. Floudas, D.; Binder, M.; Riley, R.; Barry, K.; Blanchette, R.A.; Henrissat, B.; Martínez, A.T.; Ortillar, R.; Spatafora, J.W.; Yadav, J.S.; et al. The paleozoic origin of white rot wood decay reconstructed using 31 fungal genomes. *Science* **2012**, *336*, 1715–1719. [CrossRef] [PubMed]
32. Eastwood, D.C. The evolution of fungal wood decay. In *Deterioration and Protection of Sustainable Biomaterials*; Nicholas, D.D., Goodell, B., Schultz, T.P., Eds.; American Chemical Society; Oxford University Press: Washington, DC, USA, 2014; Volume 1158, pp. 93–112.
33. Nelsen, M.P.; DiMichele, W.A.; Peters, S.E.; Boyce, C.K. Delayed fungal evolution did not cause the paleozoic peak in coal production. *Proc. Natl. Acad. Sci. USA* **2016**, *113*, 2442–2447. [CrossRef] [PubMed]
34. R Riley, R.; Salamov, A.A.; Brown, D.W.; Nagy, L.G.; Floudas, D.; Held, B.W.; Grigoriev, I.V. Extensive sampling of basidiomycete genomes demonstrates inadequacy of the white-rot/brown-rot paradigm for wood decay fungi. *Proc. Natl. Acad. Sci. USA* **2014**, *111*, 9923–9928. [CrossRef] [PubMed]
35. Krah, F.-S.; Bässler, C.; Heibl, C.; Soghigian, J.; Schaefer, H.; Hibbett, D.S. Evolutionary dynamics of host specialization in wood-decay fungi. *BMC Evol. Biol.* **2018**, *18*, 119. [CrossRef] [PubMed]
36. Curling, S.; Clausen, C.A.; Winandy, J.E. *The Effect of Hemicellulose Degradation on the Mechanical Properties of Wood during Brown Rot Decay*; International Research Group on Wood Preservation (IRG), Document No IRG/WP/01-20219; IRG: Stockholm, Sweden, 2001; pp. 1–10.

37. Zhu, Y.; Plaza, N.; Kojima, Y.; Yoshiida, M.; Zhang, J.; Jellison, J.; Pingali, S.V.; O'Neill, H.; Goodell, B. Nanostructural analysis of enzymatic and non-enzymatic brown-rot fungal deconstruction of the lignocellulose cell wall. *Front. Microbiol.* **2020**, *11*. [CrossRef]
38. Goodell, B.; Zhu, Y.; Kim, S.; Kafle, K.; Eastwood, D.; Daniel, G.; O'Neill, H. Modification of the nanostructure of lignocellulose cell walls via a non-enzymatic lignocellulose deconstruction system in brown rot wood-decay fungi. *Biotechnol. Biofuels* **2017**, *10*, 179. [CrossRef]
39. Goodell, B. Fungi involved in the biodeterioration and bioconversion of lignocellulose substrates. (chapter 15). In *The Mycota. Genetics and Biotechnology, (a Comprehensive Treatise on Fungi as Experimental Systems for Basic and Applied Research)*, 3rd ed.; Benz, J.P., Schipper, k., Eds.; Springer: Cham, Switzerland, 2020; Volume II, pp. 369–397. [CrossRef]
40. Savory, J.G. Damage to wood caused by microorganisms J. *Appl. Bacteriol.* **1954**, *17*, 213–218. [CrossRef]
41. Daniel, G.; Nilsson, T. Chapter 3. Development in the study of soft rot and bacterial decay. In *Forest Products Biotechnology*; Bruce, A., Palfreyman, J., Eds.; Taylor and Francis Publishers: London, UK, 1998; pp. 235–250.
42. Blanchette, R.A.; Burnes, T.A.; Leatham, G.F.; Effland, M.J. Selection of white-rot fungi for biopulping. *Biomass* **1988**, *15*, 93–101. [CrossRef]
43. Wolfaardt, F.; Taljaard, J.L.; Jacobs, A.; Male, J.R.; Rabie, C.J. Assessment of wood-inhabiting Basidiomycetes for biokraft pulping of softwood chips. *Bioresour. Technol.* **2004**, *95*, 25–30. [CrossRef]
44. Eriksson, K.E.L. Biopulping, biobleaching and treatment of kraft bleaching effluents with white–rot fungi. In *Comprehensive Biotechnology*; Moo, Y., Ed.; Pergamon Press: Toronto, ON, USA, 1985; Volume 4, pp. 271–294.
45. Widsten, P.; Kandelbauer, A. Adhesion improvement of lignocellulosic products by enzymatic pre–treatment. *Biotechnol. Adv.* **2008**, *26*, 379–386. [CrossRef]
46. Kharazipour, A.; Bergmann, K.; Nonninger, K.; Huttermann, A. Properties of fibre boards obtained by activation of the middle lamella lignin of wood fibres with peroxidase and H2O2 before conventional pressing. *J. Adhes. Sci. Technol.* **1998**, *12*, 1045–1053. [CrossRef]
47. Malinowski, S.; Herbert, P.A.F.; Rogalski, J.; Jaroszyńska-Wolińska, J. Laccase enzyme polymerization by soft plasma jet for durable bioactive coatings. *Polymers* **2018**, *10*, 532. [CrossRef] [PubMed]
48. Barr, D.P.; Aust, S.D. Mechanisms white rot fungi use to degrade pollutants. *Environ. Sci. Technol.* **1994**, *28*, 79A–87A. [CrossRef] [PubMed]
49. Magan, N.; Fragoeiro, S.; Bastos, C. Environmental factors and bioremediation of xenobiotics using white rot fungi. *Mycobiology* **2010**, *38*, 238–248. [CrossRef] [PubMed]
50. Dashtban, M.; Schraft, H.; Syed, T.A.; Qin, W. Fungal biodegradation and enzymatic modification of lignin. *Int. J. Biochem. Mol. Biol.* **2010**, *1*, 36–50.
51. Baba, Y.; Tanabe, T.; Shirai, N.; Watanabe, T.; Honda, Y.; Watanabe, T. Pretreatment of Japanese cedar wood by white rot fungi and ethanolysis for bioethanol production. *Biomass Bioenergy* **2011**, *35*, 320–324. [CrossRef]
52. Capolupo, L.; Faraco, V. Green methods of lignocellulose pretreatment for biorefinery development. *Appl. Microbiol. Biotechnol.* **2016**, *100*, 9451–9467. [CrossRef]
53. Singh, D.; Chen, S. The white-rot fungus Phanerochaete chrysosporium: Conditions for the production of lignin-degrading enzymes. *Appl. Microbiol. Biotechnol.* **2008**, *81*, 399–417. [CrossRef]
54. López, A.M.Q.; Silva, A.L.d.S.; Santos, E.C.L.d. The fungal ability for biobleaching/biopulping/bioremediation of lignin–like compounds of agro–industrial raw material. *Química Nova* **2017**, *40*, 916–931. [CrossRef]
55. Bissaro, B.; Várnai, A.; Røhr, Å.K.; Eijsink, V.G.H. Oxidoreductases and Reactive Oxygen Species in Conversion of Lignocellulosic Biomass. *Microbiol. Mol. Biol. Rev.* **2018**, *82*, e00029-18. [CrossRef]
56. Hwan Kim, Y.; Suk An, E.; Keun Song, B.; Shik Kim, D.; Chelikani, R. Polymerization of cardanol using soybean peroxidase and its potential application as anti-biofilm coating material. *Biotechnol. Lett.* **2003**, *25*, 1521–1524. [CrossRef] [PubMed]
57. Qian, Y.; Otsuka, Y.; Sonoki, T.; Mukhopadhyay, B.; Nakamura, M.; Jellison, J.; Goodell, B. Engineered Microbial Production of 2-Pyrone-4,6-Dicarboxylic Acid from Lignin Residues for Use as an Industrial Platform Chemical. *BioResources* **2016**, *11*, 6097–6109. [CrossRef]
58. Li, F.; Ma, F.; Zhao, H.; Zhang, S.; Wang, L.; Zhang, X.; Yu, H. A lytic polysaccharide monooxygenase from a white-rot fungus drives the degradation of lignin by a versatile peroxidase. *Appl. Environ. Microbiol.* **2019**, *85*, e02803–e02818. [CrossRef] [PubMed]

59. Kojima, Y.; Varnai, A.; Ishida, T.; Sunagawa, N.; Petrovic, D.M.; Igarashi, K.; Jellison, J.; Goodell, B.; Alfredsen, G.; Westereng, B.; et al. A lytic polysaccharide monooxygenase with broad xyloglucan specificity from the brown–rot fungus gloeophyllum trabeum and its action on cellulose–xyloglucan complexes. *Appl. Environ. Microbiol.* **2016**, *82*, 6557–6572. [CrossRef] [PubMed]
60. Jin, L.; Nicholas, D.D.; Schultz, T.P. Wood laminates glued by enzymatic oxidation of brown–rotted lignin. *Holzforschung* **1991**, *45*, 467–468.
61. Jin, L.; Sellers, T., Jr.; Schultz, T.P.; Nicholas, D.D. Utilization of Lignin Modified by Brown-Rot Fungi. I. Properties of Flakeboard Produced with a Brown-rotted Lignin Modified Phenolic Adhesive. *Holzforschung* **1990**, *44*, 207–210. [CrossRef]
62. Li, K.; Geng, X. Formaldehyde-Free Wood Adhesives from Decayed Wood. *Macromol. Rapid Commun.* **2005**, *26*, 529–532. [CrossRef]
63. Qian, Y.; Goodell, B.; Genco, J.M. The effect of a chelator mediated fenton system on the fiber and paper properties of hardwood kraft pulp. *J. Wood Chem. Technol.* **2002**, *22*, 267–284. [CrossRef]
64. Qian, Y.; Goodell, B.; Jellison, J.; Felix, C.C. The Effect of Hydroxyl Radical Generation on Free-Radical Activation of TMP Fibers. *J. Polym. Environ.* **2004**, *12*, 147–155. [CrossRef]
65. Goodell, B.; Jellison, J. Oxidation Using a Non-Enzymatic Free Radical System Mediated by Redox Cycling Chelators. U.S. Patent No. 7396974B2. Available online: https://patents.google.com/patent/US7396974B2/en:USPTO (accessed on 13 September 2019).
66. Yelle, D.; Goodell, B.; Gardner, D.; Amirbahman, A.; Winistofer, P.; Shaler, S. Bonding of wood fiber composites using a synthetic chelator-lignin activation system. *For. Prod. J.* **2004**, *54*, 73–78.
67. Kent, M.S.; Zeng, J.; Rader, N.; Avina, I.C.; Simoes, C.T.; Brenden, C.K.; Sale, K.L. Efficient conversion of lignin into a water-soluble polymer by a chelator-mediated Fenton reaction: Optimization of H_2O_2 use and performance as a dispersant. *Green Chem.* **2018**, *20*, 3024–3037. [CrossRef]
68. Kartal, S.N.; Terzi, E.; Yılmaz, H.; Goodell, B. Bioremediation and decay of wood treated with ACQ, micronized ACQ, nano-CuO and CCA wood preservatives. *Int. Biodeterior. Biodegrad.* **2015**, *99*, 95–101. [CrossRef]
69. Ren, H.; Jin, X.; Li, C.; Li, T.; Liu, Y.; Zhou, R. Rosmarinic acid enhanced Fe(III)-mediated Fenton oxidation removal of organic pollutants at near neutral pH. *Sci. Total Environ.* **2020**, *736*, 139528. [CrossRef] [PubMed]
70. Purnomo, A.S.; Kamei, I.; Kondo, R. Degradation of 1,1,1-trichloro-2,2-bis (4-chlorophenyl) ethane (DDT) by brown-rot fungi. *J. Biosci. Bioeng.* **2008**, *105*, 614–621. [CrossRef] [PubMed]
71. Goodell, B.; Jellison, J.; Liu, J.; Daniel, G.; Paszczynski, A.; Fekete, F.; Xu, G. Low Molecular Weight Chelators And Phenolic Compounds Isolated from Wood Decay Fungi and their Role in the Fungal Biodegradation of Wood. *J. Biotechnol.* **1997**, *53*, 133–162. [CrossRef]
72. Goodell, B.; Qian, Y.; Jellison, J.; Richard, M.; Qi, W. Lignocellulose oxidation by low molecular weight metal–binding compounds isolated from wood degrading fungi: A comparison of brown rot and white rot systems and the potential application of chelator–mediated fenton reactions. In *Progress in Biotechnology 21. Biotechnology in the Pulp and Paper Industry*; Lantto, L.V.a.R., Ed.; Elsevier Press: Amsterdam, The Netherlands, 2002; pp. 37–49.
73. Goodell, B.; Qian, Y.; Jellison, J.; Richard, M. Decolorization and degradation of dyes with mediated Fenton reaction. *Water Environ. Res.* **2004**, *76*, 2703–2707. [CrossRef]
74. Ray, M.J.; Leak, D.J.; Spanu, P.D.; Murphy, R.J. Brown rot fungal early stage decay mechanism as a biological pretreatment for softwood biomass in biofuel production. *Biomass Bioenergy* **2010**, *34*, 1257–1262. [CrossRef]
75. CompaniesHouse.Gov.UK; Mycologix Ltd. CompaniesHouse.Gov.UK. 2013. Available online: https://beta.companieshouse.gov.uk/company/07002759 (accessed on 13 September 2019).
76. Schultz, T.P.; Nicholas, D.D. Improving the Performance of Organic Biocides by Using Economical and Benign Additives. In *Development of Commercial Wood Preservatives*; American Chemical Society: Washington, DC, USA, 2008; Volume 982, pp. 272–284.
77. Williams, R.S.; Knaebe, M.T.; Feist, W.C. *Finishes for Exterior Wood: Selection, Application & Maintenance*; USDA. Forest Products Society: Madison, WI, USA, 1996; p. 127.
78. Konkler, M.J.; Cappellazzi, J.; Morrell, J.J. Performance of polyurea-coated Douglas-fir timbers exposed in Hilo Hawaii. *Int. Wood Prod. J.* **2019**, *10*, 31–36. [CrossRef]
79. Liese, W. Ultrastructural Aspects of Woody Tissue Disintegration. *Annu. Rev. Phytopathol.* **1970**, *8*, 231–245. [CrossRef]

80. Evans, P.D.; Thay, P.D.; Schmalzl, K.J. Degradation of wood surfaces during natural weathering. Effects on lignin and cellulose and on the adhesion of acrylic latex primers. *Wood Sci. Technol.* **1996**, *30*, 411–422. [CrossRef]
81. Yao, Y.; Gellerich, A.; Zauner, M.; Wang, X.; Zhang, K. Differential anti-fungal effects from hydrophobic and superhydrophobic wood based on cellulose and glycerol stearoyl esters. *Cellulose* **2018**, *25*, 1329–1338. [CrossRef]
82. Vallieres, C.; Hook, A.L.; He, Y.; Crucitti, V.C.; Figueredo, G.; Davies, C.R.; Avery, S.V. Discovery of (meth)acrylate polymers that resist colonization by fungi associated with pathogenesis and biodeterioration. *Sci. Adv.* **2020**, *6*, eaba6574. [CrossRef] [PubMed]

Publisher's Note: MDPI stays neutral with regard to jurisdictional claims in published maps and institutional affiliations.

Article

A Method for Accelerated Natural Weathering of Wood Subsurface and Its Multilevel Characterization

Anna Sandak [1,2], Jakub Sandak [1,3,*], Marion Noël [4] and Athanasios Dimitriou [5]

1 Wood Modification, InnoRenew CoE, Livade 6, 6310 Izola, Slovenia; anna.sandak@innorenew.eu
2 Faculty of Mathematics, Natural Sciences and Information Technologies, University of Primorska, Glagoljaška 8, 6000 Koper, Slovenia
3 Andrej Marušič Institute, University of Primorska, Titov trg 4, 6000 Koper, Slovenia
4 Institute Materials and Wood Technology, Bern University of Applied Sciences, Solothurnstrasse 102, Postfach 6096, CH-2500 Biel 6, Switzerland; marion@lekolabs.com
5 The BioComposites Centre, Bangor University, Bangor, Gwynedd LL57 2UW, UK; a.dimitriou@bangor.ac.uk
* Correspondence: jakub.sandak@innorenew.eu; Tel.: +386-4028-2959

Abstract: The function of altering weathering factors and degradation mechanisms are essential for understanding the weathering process of materials. The goal of this work was to develop a method for the acceleration of natural weathering and to investigate the molecular, microstructure and macrostructure degradation of wood caused by the process. Tests were performed in the whole month of July, which, according to previous research, is considered as the most severe for weathering of wood micro-sections. Sample appearance was evaluated by colour measurement. Scanning electron microscopy was used for evaluation of the structural integrity and changes in the microstructure of wood morphological components. Changes on the molecular level were assessed by means of FT-IR spectroscopy. Observation of the effects of weathering allowed a better understanding of the degradation process. Typical structural damage, such as cracks on bordered pits and cross-field pits, and, as a consequence, their erosion, revealed the sequence of the degradation process. It was confirmed that earlywood was more susceptible to damage than latewood. Even if the weathering test was conducted for a relatively short time (28 days) the ultra-thin wood samples changed noticeably. The progress of alteration was similar as usually noticed for wood surfaces, but occurred at shorter exposure times. The estimated acceleration factor was ×3, compare to the natural weathering kinetics of wood. The research methodology presented can be used for the determination of the weather dose-response models essential to estimate the future service life performance of timber elements.

Keywords: photodegradation; wood weathering; SEM imaging; FT-IR spectroscopy; thin samples; accelerated natural weathering

Citation: Sandak, A.; Sandak, J.; Noël, M.; Dimitriou, A. A Method for Accelerated Natural Weathering of Wood Subsurface and Its Multilevel Characterization. *Coatings* **2021**, *11*, 126. https://doi.org/10.3390/coatings11020126

Received: 4 January 2021
Accepted: 19 January 2021
Published: 23 January 2021

1. Introduction

Wood as a building material has traditionally been used for different types of load-bearing structures, decking, façades cladding, doors and windows. Wood, due to its versatile character, is often used in both rustic and modern buildings, as well as frequently applied as a material for retrofitting of existing structures. The trend for raising sustainable buildings, in addition to increasing environmental awareness observed nowadays, leads to the reactivation of bio-architecture as an alternative to other construction techniques [1,2]. However, the major limitation for choosing natural wood for external use is the lack of confidence that many architects, designers and customers have towards this material, mostly due to aesthetic preferences. The speed and pattern of the colour change are difficult to predict and are rarely taken into consideration at the early phases of building design [3].

Today's bio-based building materials, even if well characterized from the technical point of view, are often lacking reliable models describing their performance during service life and the operational durability is still a limiting factor in many applications and environments. Wooden elements often suffer due to mechanical, environmental or biological alterations during their use phase. The most susceptible parts are unprotected surfaces since they are primarily exposed to ageing, weathering or decay.

Weathering is the general term used to define the slow degradation of materials exposed to weather condition [4]. The intensity of weathering depends on timber species, design solution, function of product and finishing technology applied for wood protection and on the specific local conditions. The corrosion rate of wood is low: approximately 6mm per century in the case of European softwood species [4]. In general, the process of wood weathering leads to a slow breaking down of surface fibers, their removal and, in consequence, a roughening of the surface and reduction of the glossiness. Formation of discontinuities on the surface simplify moisture penetration and allow access of wood-decaying biological agents into the material structure. As a result, mechanical performance of the load-bearing members might decrease.

A critical limitation of the state-of-the-art natural weathering procedures is a fact that the test is very long lasting. Diverse approaches for its acceleration were proposed, including amplification of the absorbed radiation by implementing additional solar light focusing optics, scheduled water spraying, temperature control of the sample, or automatic systems for exposing samples that continuously follow the Sun's position. An example of that sophisticated instruments is Equatorial Mount with Mirrors for Acceleration with Water (EMMAQUA) that is commercially available for testing diverse materials [5]. Even though, the amplification of the weathering process is limited, not to mention elevated costs for such complicated equipment and demanding procedure for the test itself. An alternative approach for accelerating the natural weathering of bio-based materials is reported here. The solution is based on the fact that the most affected part of the sample during natural weathering is the sample surface (subsurface) of few hundreds of micrometre thickness. A pilot experiment was designed to prepare wood micro-sections (ultra-thin samples) and to confront the mechanism of changes to the wood due to the action of atmospheric agents occurring in short term. Similar approach had already been used successfully for investigating the resistance of wood against biological corrosion [6], chemical corrosion [7] as well as atmospheric corrosion [8,9]. Use of extra thin samples permits direct investigation of the surface layer and its disintegration due to photochemical or mechanical degradation.

An artificial weathering is an established alternative for testing a material's resistance to deterioration due to climate related factors. The European standard EN 927-6 [10] describes such a method that is frequently used to assess wood coatings. It uses a specially designed device equipped with temperature control, light radiation by means of fluorescent ultraviolet range A (UVA) lamps, water spray and water condensation units. Another, equally common methodology relays on testing weathering resistance by simulating the Sunlight irradiance by means of xenon lamps [11]. Podgorski et al. reported relatively good agreement for the development of cracking when compared to results from natural weathering of coated wood, with the estimated acceleration factor of $\times 5$ [12]. Even a higher acceleration factor of $\times 10$ was identified there for a loss of surface glossiness. Development of cracking was faster in artificial weathering tests due to higher doses of degrading factors provided in shorter cycles but with precisely controlled exposure conditions. Most artificial weathering methods lack biological impacts as well as the influence of dirt that are present in typical exterior exposure environments [13]. Özgenç noticed significant differences in the colour alteration as well as in compression strength along fibres for samples subjected to the natural and artificial weathering conditions [14]. It was demonstrated that the reason for such differences laid in the lack of several minor factors that are not included in the artificial weathering protocols. These factors include air pollution, biological pests, and snow, as well as difference in the spectral composition of light irradiation, nature of the rain-driven wetting, and humidity variations occurring under natural conditions. Similar observations were reported by Ruther and Jelle, where solar radiation and wind-driven rain were identified as the main factor inducing colour changes [15]. However, the surface greying typically observed for the natural wood exposed to exterior conditions is caused by mould growth and is hardly replicated in up-to-date artificial weathering tests. Outdoor exposed samples become usually darker at the first stage of weathering, to turn in to lighter tones afterward. This is supported by both visual assessment of inspectors as well as by objective spectra/colour measurements. On the contrary, samples exposed to laboratory weathering cycles turn darker to varying extend and do not exhibit initial increase in International Commission on Illumination (CIE) a^* (green–red) and CIE b^* (blue–yellow) values, typical to samples exposed outdoors. In addition, it was found that samples exposed over a longer period to artificial weathering show a form of unnatural bleaching, not present in wood samples exposed to natural weathering [15]. The overall conclusion is, therefore, that even if several advantages of artificial weathering make this an interesting testing methodology, the results from laboratory weathering can hardly be compared to the real outdoor weathering. The reason lays in different mechanisms and kinetics of the material deterioration, which are especially diverse in complex biological composites such as wood.

Several attempts to understand and model wood weathering have been conducted by various researchers. A review published by Brischke and Thelandersson [16] presents an overview on modelling methodologies used for outdoor performance prediction of wood products. According to the authors, dose-response models seem to be most frequently applied, where the changes in any material property are related to the dosage of the relevant environmental causes. In general, solar radiation and wind-driven rain are the two main driving factors influencing wood surface appearance [17]. They entail further internal stresses of bulk material imposed by cyclic wetting and temperature changes. Methodology for determining the weather dose of the building façade was also reported by Thiis et al. [18]. The authors stated that the surface temperature can be very different than the ambient air temperature, especially when the building elevation is exposed to direct sunlight. Therefore, variation of the exterior microclimate is also important for deterioration extent. Access to an extensive data set generated by different types of sensors is indispensable for the development of high-quality numerical models. The attempt for defining an original algorithm for calculation of "the weathering indicator", by merging the multi-sensor data and linking these to the surface performance indicators, was also presented by the authors [19,20].

New knowledge regarding the role of matrix polymers and their interaction on a molecular level permit discovering of complex biological structures [21]. In particular, microscopic and spectroscopic techniques allow better understanding of the ultrastructure of wood, which is a composite of diverse cell types and chemical components acting together to serve the needs of a plant [22]. Tracheids are the major component of softwood. Their cell wall is composed of several layers: middle lamella, primary wall and secondary walls (S1–S3) [23]. The thickness, chemical composition and structure (microfibril orientations) of those layers vary due to their role. Both primary and secondary walls contain pectin, cellulose, and hemicellulose, although in different proportions. The function of the middle lamella is adhesion of adjacent cells. The primary wall consists of a rigid skeleton of cellulose microfibrils. A layered secondary cell wall, composed of lignin, cellulose, and hemicellulose, provides strong mechanical strength [24].

Microstructure analyses of wood investigated by scanning electron microscopy (SEM) were conducted by several researchers. Hamed et al. [25] and Jingran et al. [26] used SEM for monitoring changes in archaeological wood. Nicole et al. [27] and Popescu et al. [28] investigated decayed wood. Singh et al. [29] used light and scanning electron microscopy for visualization of wood impregnation and imaging of the wood–matrix polymer interface in bio-composites [30]. Microscopic structural analyses of weathered wood were conducted by several researchers [31–33]. Ganne-Chédeville et al. [34] used SEM images for the evaluation of wood façade elements after cleaning treatments. SEM is an essential tool allowing observation of the structural changes in degraded wood tissues at the level of cell wall [25]. Even if the obtained information refers mostly morphological and structural changes of the wood, it might also indicate changes in its chemical structure. Spectroscopy allows even more detailed investigation of chemical composition by revealing molecular structure of wood and functional groups of its constituents [35]. Specific utilization of spectroscopy for the assessment and monitoring of timber members has been proposed by several researchers and an overview was previously presented by the authors [36]. Two-dimensional correlation spectroscopy (2D-COS) used for the analysis of spectral data allowed the identification of small and weak peaks that are masked in one-dimensional spectrum by highlighting spectral changes caused by external disturbances (exposure time). The method was previously used for the evaluation of the degradation stage of wood [37], evaluation of heat-treated wood [38] or wood species discrimination [39].

The goal of this research was to investigate the feasibility of the novel approach for accelerated natural weathering of wood. An extent of the microstructure degradation and other alterations on the molecular, microscopic, and macroscopic levels recorded in ultra-thin wood samples were confronted with thick references. As a result, an acceleration factor relating weathering kinetics of ultra-thin samples with references was estimated. An additional objective was to develop a novel testing protocol resulting in at least comparable or even more detailed characterization with simplified sample preparation, but still assuring significant shortening of the natural weathering procedure.

2. Materials and Methods

2.1. Experimental Samples

Experimental samples were prepared from Norway spruce wood (*Picea abies* L. Karst.) originated from Italian Dolomite mountains (Trentino Region, Italy). A single log 0.5 m long, extracted from the bottom part of the trunk was spitted to assure radial surface opening. A board 30 mm × 50 mm × 500 mm (radial × tangential × longitudinal directions respectively) was cut out from the sapwood assuring defect free and uniform yearly ring structure. The average yearly ring width was 1.9 mm and latewood width 0.8 mm. No drying process was used to minimize wood properties alteration due to elevated temperature. Wet board was finally processed on the slicing planner (Super MECA-S, Marunaka Tekkonsho Inc., Shizuoka, Japan) to prepare ultra-thin experimental samples. This method allowed the preparation of a high number of defect-free specimens from a limited source of material, minimizing natural heterogeneity of biological material. Moreover, samples prepared in this way might be immediately characterized after weathering (e.g., direct SEM observation or spectroscopic measurement in transmission mode) without supplementary interference (e.g., microtome sectioning), which might affect their original state [40]. The thickness of samples was ~100 µm and the efficient surface exposed to weathering was 30 mm × 30 mm (width × length).

An additional set of reference (thick) samples was prepared from the same log as ultra-thin samples to assure minimal variance of the tested wood properties. Twelve wooden blocks of dimensions 30 mm × 10 mm × 50 mm were extracted and used for natural weathering.

2.2. Weathering Test

Sets of both ultra-thin and thick samples (Figure 1b) were placed for natural exposure at 45° to the horizon, facing four cardinal directions: north, west, east and south in San Michele, Italy (46°11′15″ N, 11°08′00″ E, 206 m above sea level). The sample stand (Figure 1a) was installed on the roof of the building assuring no shadowing of surrounding structures and vegetation. The test was carried out from July 2014 to June 2015. Ultra-thin samples were collected before exposition (0) and after 1, 2, 4, 7, 9, 11, 14, 17, 21, 24 and 28 days of natural weathering. Thick samples were collected once per month for a duration of one year. The weather data observed during the test are summarized in Table 1. In addition, the early climate data for exposure location are provided in Table 2. Experimental samples together with non-weathered references were conditioned after collection in the dark climatic chamber (20 °C, 60% Relative Humidity (RH)) to the equilibrium moisture content of ~12%.

(**a**)

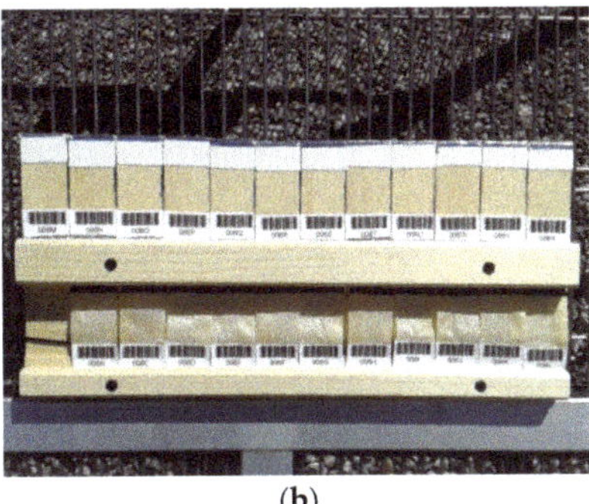

(**b**)

Figure 1. Experimental stand (**a**) and set of ultra-thin and thick wood samples (**b**) exposed to natural weathering in San Michele All'Adige (Italy).

Table 1. Weather data acquired during weathering progress of ultra-thin samples.

Day of Exposure	0	1	2	4	7	9	11	14	17	21	24	28
Mean temp (°C)	–	18	20	22	17	18	20	22	24	20	21	19
∑ radiation (MJ/m^2)	–	30	60	105	162	202	251	324	407	469	527	581
∑ insolation (h)	–	14	26	46	70	90	108	139	173	202	229	255
total rain (mm)	–	11	0	2	43	0	2	0	0	0	0	13
RH (%)	–	84	59	73	93	68	68	68	60	80	70	89
mean wind speed (m/s)	–	0.8	1.9	0.5	0.5	0.9	1.2	0.9	1.0	0.1	0.8	0.5

Table 2. Weather data acquired during weathering progress of reference (thick) samples.

Month	July	August	September	October	November	December	January	February	March	April	May	June
Mean temp (°C)	19	17	10	6	4	1	−4	−2	0	6	8	13
RH (%)	68	69	77	83	71	76	84	84	79	76	76	78
Wind speed (m/s)	0.6	0.5	0.4	0.3	1.0	1.0	0.9	0.4	0.4	0.7	0.3	0.6
Wind direction (°)	199	165	159	62	333	338	317	337	1	187	168	185
∑ rainfall (kg/m^2)	99	176	147	192	4	2	26	61	21	115	237	312
∑ snowfall (kg/m^2)	0	1	8	24	11	1	25	102	81	29	16	8
∑ irradiation (kWh/m^2)	195	175	126	87	68	54	55	70	134	157	179	171

2.3. Macroscopic Observation: Colour Measurement

Sample colour was measured by using ultraviolet-visible-near infrared (UV–VIS–NIR) spectrometer Maya2000pro, produced by Ocean Optics (Largo, FL, USA). Equipment calibration was carried out using white and black standards provided by the manufacturer. Changes in colour were also assessed by a colorimeter following the CIE $L^*a^*b^*$ system where colour is expressed as three coordinates: L^* (lightness), a^* (red-green tone) and b^* (yellow-blue tone). CIE $L^*a^*b^*$ colours were measured using MicroFlash 200D spectrophotometer (DataColor Int, Lawrenceville, NJ, USA). The selected illuminant was D65 and viewer angle was 10°. All specimens were measured on three different spots over the weathered surface. Digital images of sample surfaces were also captured on an G2710 (Hewlett-Packard, Palo Alto, CA, USA) office scanner, in parallel to the colour measurements.

2.4. Microscopic Observation: Scanning Electron Microscopy

Small pieces of investigated samples were cut out and glued with carbon tape sticker to the sample holder. The samples were placed in SC7620 (Quorum Technologies, Laughton, UK) Mini Sputter Coater/Glow Discharge System device and then plasma coated for 90 s with 10 nm gold/palladium (Au/Pd) layer. Prepared samples were investigated by using TM 3030 (Hitachi, Tokyo, Japan) Scanning Electron Microscope (SEM). An acceleration voltage of 15 kV was used for imaging of samples. Images were acquired with a dedicated software provided by the equipment producer.

2.5. Molecular Observation: Fourier Transform–Infra Red Measurement

All experimental samples were measured with an Alpha Fourier transform infrared (FT-IR) spectrometer produced by Bruker Optics GmbH (Ettlingen, Germany) equipped with universal sampling module allowing measurement in transmission. The spectral range measured was between 4000 and 600 cm^{-1}. The spectral resolution of the spectrophotometer was 4 cm^{-1}; each spectrum has been computed as an average of 32 successive measurements in order to minimize the measurement error. Three measurements were performed on each sample. Opus 6.5 (Bruker, Ettlingen, Germany), PLS_Toolbox (Eigenvector Inc, Manson, IA, USA), available as an extension of the Matlab package (Mathworks Inc, Natick, MA, USA) and LabView 2019 (National Instruments Inc, Austin, TX, USA) software packages were used for spectral pre-processing and data mining. IR spectra before interpretation were corrected by extended multiplicative scatter correction (EMSC). EMSC was applied to minimize an effect of nonuniform optical density for measured wood specimens and related differences in the measured amplitude and baseline deviation [41]. Three replicate measurements on each sample were averaged and the resulting spectrum was baseline-corrected. A lot of research efforts have been directed toward systematizing the infrared peaks interpretations that resulted in several scientific publications [42–47]. These papers were an inspiration to review the available know-how and adopt it for the specific needs of this research (wood weathering). The complete list of peaks observed in investigated experimental specimens is summarized in Table 3.

Table 3. Band assignment for FT-IR transmittance spectra of Norway spruce [42–47].

Number	Peak Position	Band Assignment
1	985	CO valence vibrations
2	1024	CO stretching in cellulose and non-cellulose polysaccharides
3	1060	C–O + C–C stretch of cellulose
4	1110	Aromatic skeletal; C–C stretch
5	1160	CO stretching in ester groups in lignin
6	1201	untreated cellulose (ρ OH; δ CH)
7	1226	syringil ring; C–O stretching lignin and xylan
8	1265	guaiacyl ring vibrations and CO stretching lignin
9	1315	C–H vibration in cellulose; C–O in syringil derivatives
10	1334	vibration in CH and stretching in CO related to syringil ring
11	1370	CH bending in cellulose and non-cellulosic polysaccharides
12	1425	vibration of aromatic structures in lignin
13	1452	aromatic skeletal vibrations of lignin and CH_2 vibration in cellulose
14	1508	aromatic skeletal vibrations of lignin and C=O stretching
15	1598	aromatic skeletal vibrations of lignin and C=O stretching
16	1652	stretching vibrations of conjugated C=O
17	1732	C=O–OH stretching in glucuronic acid in xylan
18	2902	symmetric CH_2 valence vibrations
19	2940	asymmetric stretching vibrations of CH related to methyl and methylene in lignin, cellulose and hemicellulose
20	3149	–OH stretching in cellulose
21	3280	C–H stretching in methyl and methylene groups
22	3375	–O(3)H O(5) intramolecular hydrogen bonds in cellulose
23	3496	moderately H-bonded water
24	3565	valence vibration of H bonded OH groups
25	3602	weakly H-bonded water

Standard spectra interpretation was aided by two-dimensional spectra correlation (2DCOS), which permits direct deconvolution and determination of correlations between bands in the spectra; 2D-IR spectroscopy allows new information to be obtained, which cannot be acquired by using conventional IR and its derivative spectra. The algorithm proposed by Noda [48] was implemented for the needs of this research. All spectra were interpolated before analysis in order to compensate uneven exposure time, considered here as the 2DCOS disturbance factor.

3. Results

3.1. Macroscopic Level

The appearance of investigated samples after natural weathering in diverse exposure times is presented in Figure 2. The changes in colour are observed in parallel with samples disintegration evident from day 17. Samples become yellowish with the progress of degradation. It is noticed that weathering caused removal of wood fibers, presence of cracks and increase in brittleness of samples.

Figure 2. Appearance changes of ultra-thin (**a**) and reference thick (**b**) wood samples exposed to natural weathering in San Michele All'Adige (Italy). Note: arrow indicate weathering-induced damages to wood cells.

Visible range reflectance spectra of the test samples in this study shows that spruce tends to darken with weathering time, as shown by a decrease of the reflectance value (Figure 3). The reflectance value dropped after one day of exposure. Further changes, as observed after day 7 until the end of the experiment, were more consistent and resulted in a nearly constant VIS spectrum after three weeks of weathering. Analogous results were obtained by measurement of colour with colorimeter. The CIE $L^*a^*b^*$ colour coordinates system is widely used to represent colours in three-dimensional space. According to the CIE Lab definition, L^* represents lightness (varying from 100-white, to 0-black). Similarly, a^* and b^* are chromaticity coordinates, where $+a^*/-a^*$ corresponds to red/green, and $+b^*/-b^*$ to yellow/blue tonality. A decrease of L^* was observed for all exposure directions (Figure 4). The value of a^* increased in the first days of exposure and from day 4 seems to be stable. The noticeable increase in b^*

value, visible as yellowing of wood, is also observed for all exposure directions; however, it is followed by b^* values drop after one week of exposure.

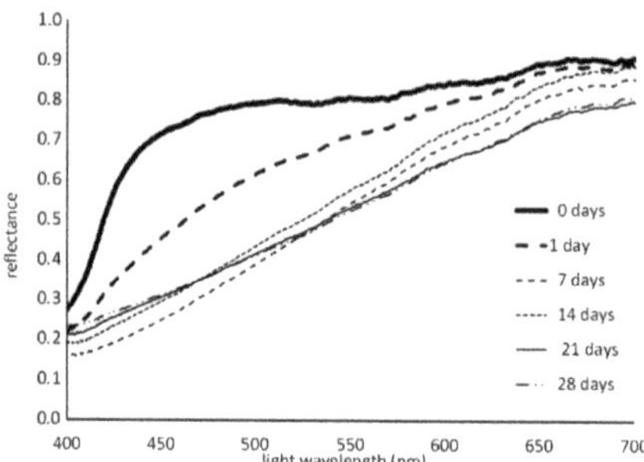

Figure 3. Reflectance VIS spectra of the wood exposed to natural weathering.

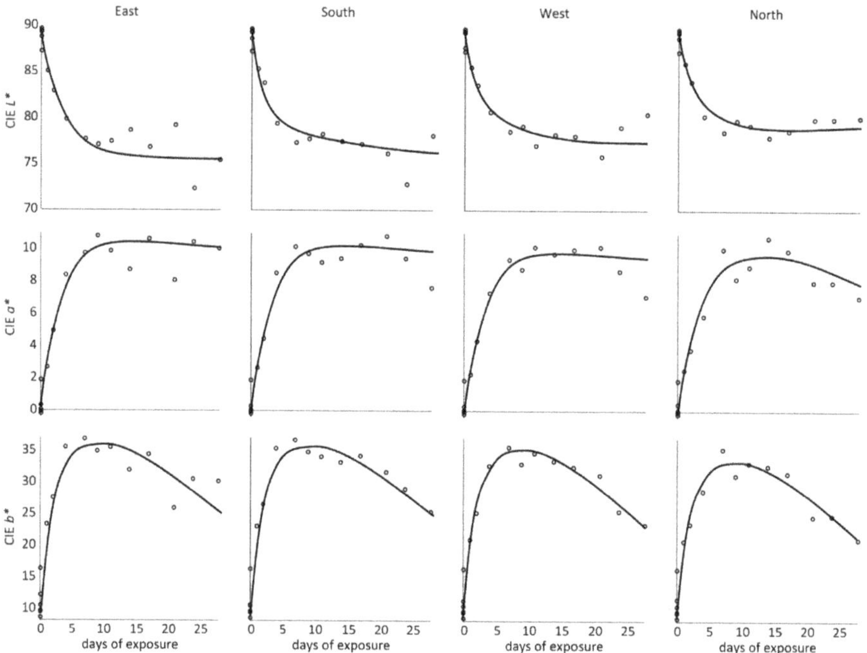

Figure 4. Change of CIE $L^*a^*b^*$ colour coordinates in ultra-thin wood samples exposed to natural weathering.

3.2. Cellular Level

The first sign of deterioration visible on the SEM images was openings of bordered pits membranes (torus) in radial walls of earlywood tracheids (Figure 5a). In the successive step, the membrane covering the piths was broken (Figure 5b) and presence of small diagonally oriented micro-checks was observed (Figure 5c). With the progress of degradation, an enlargement of the pith crack was noticed as a result of contraction of the cell wall caused by moisture variation (Figure 5d).

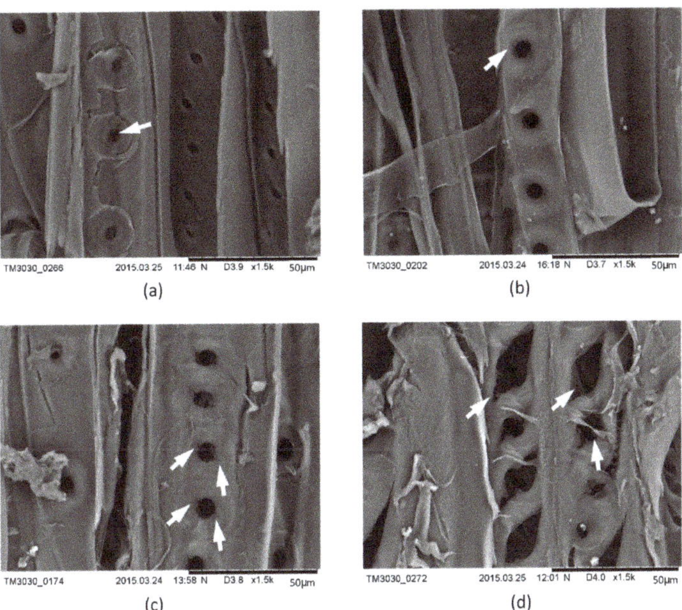

Figure 5. SEM images of earlywood exposed to the south for 2 days (**a**), 7 days (**b**), 9 days (**c**) and 21 days (**d**). Note: arrow indicate weathering-induced damages to wood cells.

Evidence of fungal infection (not visible by naked eye) was found after 17 days of natural exposure, indicating that the growth of microorganisms begins by deposit of their spores into the micro-cracks. Evaluation of the earlywood zones of samples exposed to four directions after 9 days proved that the degradation kinetics vary according to exposure direction. The experimental samples images at the final stage (day 28) are shown in Figure 6, which presents SEM images of latewood zone. Cracks propagate through the cell wall along the progress of decomposition. Pits are completely eroded, and the degradation products are continuously removed with rain. However, the northern samples seem to be relatively less affected by the weathering process (Figure 6c). The time-related progress of morphological changes of the investigated samples is systematically presented in Figure 7.

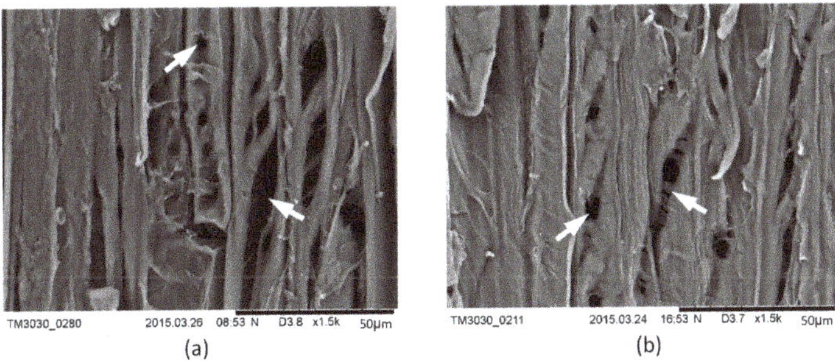

Figure 6. *Cont.*

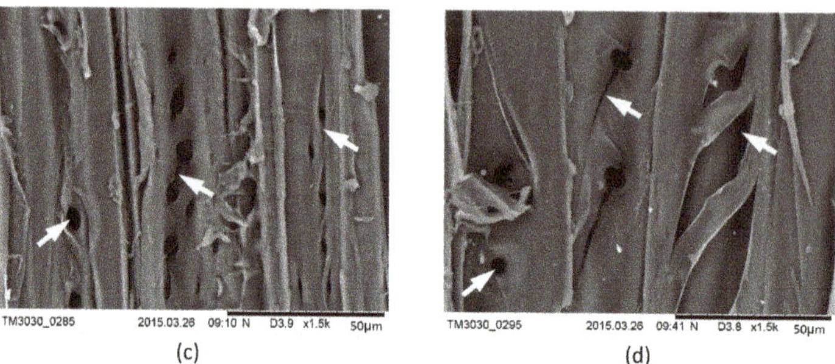

Figure 6. SEM images of latewood exposed to different exposure sites: south (**a**), west (**b**), north (**c**) and east (**d**). Note: arrows indicate weathering-induced damage to wood cells.

Day of exposure	0	1	2	4	7	9	11	14	17	21	24	28	
spores deposition	×	×	×	×	×	×	×	×	·	·	·	●	× - no · - seldom ● - intense
degraded surface	×	×	×	×	·	·	·	·	·	●	●	●	× - no · - little ● - very
big cracks in cell wall	×	×	×	×	×	×	×	·	·	●	●	●	× - no · - some ● - many
diagonally oriented micro-checks in cell wall	×	×	×	×	●	●	●	●	●	●	●	●	× - no ● - yes
dust deposition	×	×	·	·	·	·	·	●	●	●	●	●	× - no · - some ● - intense
cell wall delamination	×	×	×	·	·	·	·	·	●	●	●	●	× - no · - some ● - intense
border piths crack	×	×	×	·	·	·	·	●	●	●	●	●	× - no · - seldom ● - all
border piths broken membrane	×	×	×	×	●	●	●	●	●	●	●	●	× - no ● - yes
border piths opening membrane	×	×	●	●	●	●	●	●	●	●	●	●	× - no ● - yes
cross field pits crack	×	×	×	×	×	×	×	×	·	·	●	●	× - no · - present ● - large

Figure 7. Microscopic observation for samples exposed to natural weathering for the south direction.

3.3. *Molecular Level*

FT-IR spectra measured in transmission mode on samples at various stages of degradation clearly show the progress of changes related to functional groups of wood polymers (Figure 8). The band assignment (Table 1) allowed the identification of 25 peaks that might be observed and interpreted. The most evident are changes in spectra intensity of bands #7, #8, #14 assigned to aromatic skeletal vibrations of lignin. In all the above-mentioned peaks, the intensity diminishes with weathering progress, consequently, the peaks disappear, indicating decomposition of corresponding chemical component (or functional group). The region 3900–2700 cm^{-1} reflects essentially the amounts and structure of hydroxyl groups in various types of hydrogen bonding of wood components [44]. Even if bands assigned to hydroxyl groups in IR spectrum are better resolved after deconvolution, changes noticed in #24 and #25 reflect reduction in hygroscopic properties of wood exposed to the weathering process [43].

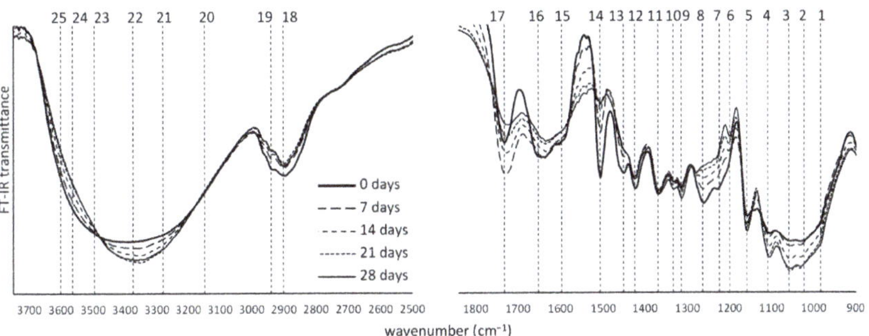

Figure 8. Changes to FT-IR transmittance spectra of Norway spruce wood due to southern exposure to natural weathering.

Peak #14, assigned to aromatic skeletal vibrations of lignin and C=O stretching, was selected to illustrate changes of its intensity due to weathering progress in different cardinal directions (Figure 9). In general, similar progress of intensity changes can be observed; however, the deviations are slightly less intense in the case of samples exposure to north compared with other directions. It is especially evident in the last week of exposure, where intensity of degradation is apparently lower than in the other three cardinal directions. For southern exposure, the highest changes are observed at the initial stage (up to first two weeks), followed by less intense variation in the second part of the experiment. Eastern exposure seems to follow the southern; however, after 14 days, the intensity is less pronounced. Western exposure is less affected up to three weeks. After that period, both (western and eastern) exposures reach the same degradation level as south.

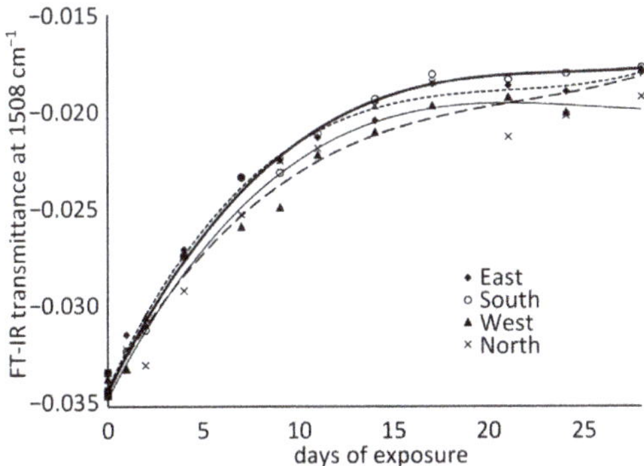

Figure 9. Changes to FT-IR lignin peak (1508 cm^{-1}) along the natural weathering test duration in samples exposed to diverse cardinal directions.

Two-dimensional correlation spectroscopy (2D-COS) was developed by Noda and involved a series of spectra being recorded from one or more samples during or following perturbation [48]. The resulting spectral data matrix allowed identification of compositional differences within and between samples. Changes in data are visualized in 2D spectral maps. Correlations between measured variables is included in the synchronous matrix. The asynchronous matrix represents sequential variations occurring in the data. The peaks might be located on the diagonal or off the diagonal space. Peaks located on the diagonal (autopeaks) show regions of the spectrum that are changing with respect to the average spectrum of the series. Peaks off the diagonal (cross peaks) show the correlation intensity (positive or negative) of different bands [48].

Figure 10 presents homospectral correlation, where two identical assemblies of data (spectra) are used. The synchronous 2D IR correlation spectrum shows strong positive autopeak at #14. The magnitude of autopeak corresponds to the overall extent of spectral intensity variation observed for specific wavenumber. Presence of other regions assigned to functional groups in lignin (#12, #13 and #15) indicated changes in intensity during weathering progress, meaning that they are susceptible to external perturbation. In fact, the changes in intensity of peak #14 as observed on Figure 8 are very evident. Positive peaks are highlighted by red colour, in contrast to negative peaks marked in violet. The positive peak on the synchronous spectrum indicates that spectral bands corresponding to coordinates of that peak change continuously in the way that direction of changes (increase or decrease) is identical for both wavelengths. Conversely, the negative peak appears when one wavelength increases, while the second wavelength decreases. The sign of asynchronous spectrum allows identification the sequence of changes occurring at both confronted wavelengths [49]. Crosspeak at coordinates 17–14 represent negative intensity. It indicates that changes of the intensity of both bands are substantially different (aromatic skeletal vibrations of lignin and C=O stretching versus C=O–OH stretching in glucuronic acid in xylan).

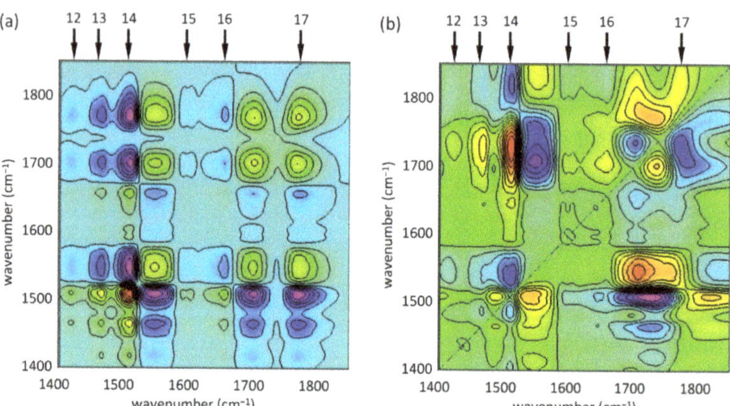

Figure 10. 2D spectral correlation of the FT-IR spectra of Norway spruce wood exposed to south direction in San Michele All'Adige (Italy): (**a**) synchronous spectrum; (**b**) asynchronous spectrum

4. Discussion

According to Raczkowski [9], changes in wood structure, caused by natural weathering and expressed by the reduction in tensile strength, are more intensive during the summer months in comparison with autumn and winter; therefore, our tests were carried out in July. During the weathering experiment, the mean temperature oscillated around 20 °C, while relative humidity was slightly above 70%. Table 1 contains meteorological data captured during sample exposure. Three major rain events and an additional two days with small (2 mm) precipitation were observed. The total hours of insolation were 255, accounting for 581 MJ/m^2. The test duration was limited to 28 days; however, according to Evans [50], the weathering conditions in 45° south exposure are accelerated relative to vertically installed elements. The configuration of experimental samples (~100 μm thickness of strips) allowed their non-destructive and direct analysis with several methods, such as IR spectroscopy in transmission mode or SEM observation without microtome sectioning. Changes in surface texture, colour and chemical composition being an overall result of exposure to various environmental conditions such as moisture, heat and solar radiation were assessed at different levels providing a wide range information regarding degradation process.

The relatively short exposure time affected appearance of investigated wood. Yellowing of the wood is caused by the formation of chromophoric groups of secondary origin [51,52]. When the chromophoric groups leach from the wood, the wood colour changes further, eventually converting to grey shades. Due to limitation of experiment to one month, this phenomenon was not observed within this research. Analogous results regarding colour changes of wood due to weathering was reported previously by authors while performing artificial weathering test according to EN 927-6 [53]. The changes in colour were already evident after 28 h of artificial weathering. Specific shape of curves representing colour coordinates (CIE b^* measured for all cardinal direction and CIE a^* measured for north) disqualify their use for modelling general trends in the appearance change (Figure 4). In fact, wood might have similar colour values after receiving different weathering doses (e.g., CIE b^* of samples exposed for one and four weeks). The pattern of initial increase of CIE b^*, followed by its decrease, was also reported by other authors [51,54,55]. It was explained by lignin degradation and photooxidation of $-CH_2$ and $-CH(OH)$ groups, with the formation of secondary chromophoric structures (quinones).

Microscopic methods provide detailed information about surface morphology. Scanning electron microscopy is considered a valuable tool for the examination of the morphological characteristics of degraded wood at the level of cell wall, evaluating their damage and detecting eventual presence of decay [25]. As stated by Turkulin [56], observation of the effects caused by photo degradation may provide information about structural integrity of wood surface and help understanding the weathering process. The process of thin wood section destruction due to natural weathering was previously investigated by Raczkowski [9]. He stated that the degradation penetrates relatively deep and observed that the decrease of mechanical strength properties is very significant. According to Turkulin et al. [31], a weathered wood surface shows damage to several anatomical elements: development of cracks of bordered pits, presence of cracks and loss of membranes of the ray pits, and expansion of cracks that can be seen on the lumen surfaces, originating at single pits (Figure 5). Thinning of the cell walls is due to breakdown of lignin in the S2 layer of the cell wall [56]. Progress of weathering (UV degradation, leaching effect of water and mechanical damages caused by wind blow) leads to extensive destruction of the pit membrane and further crack propagation (Figure 5d). Pit openings are then enlarged and, in consequence, the whole pit erodes away. According to Sandberg [57], the orientation of these cracks suggests that these follow the fibril orientation in the S2-layer of the cell wall. Pandley and Pitman [58] reported that specific ridges on the S3 wall layer, wall checking, pit degradation and middle lamella breakdown are typical microscopic features characteristic for natural wood exposed to weathering.

Fungal infestation of investigated samples was noticed after the third week of exposure. Turkulin [56] observed first spores after only 4 days of exposure; however, more intense fungal growth was noticed by him after 41 days. In general, fungal development is constrained on the surfaces exposed to high sun radiation; therefore, intense UV absorbance slows down speed of colonization by fungi. However, summer 2014 was rainy and relatively cooler than usual. Consequently, it promoted growth of the fungi, even on samples with southern exposure. Owen et al. [59] observed that degradation caused by the combined weathering stimuli, including both UV light and precipitation, is considerably more intensive and rapid than exposure to each parameter separately. UV radiation leads to lignin decomposition (photolysis) and, in consequence, to delamination of surface fibres. In particular, the chromophoric groups of lignin are strong UV absorbers and responsible for yellowing of the surface [51]. The other apparent result of the intensive delignification process is brittleness of the samples. Degradation of lignin was also observed while analyzing IR spectra. The most pronounced reduction of intensity was evident for band #14 and is related to degradation of lignin caused by UV radiation. Similar statements were derived by other researchers [51,60], who observed a decrease of peaks assigned to the lignin along the irradiation time in laboratory experiments. Lignin and extractives are considered as absorbers of UV radiation; therefore, these components suffer major degradation. When the sample system is under an external perturbation, various chemical constituents of the system are selectively excited, and the degrees and orders of the excitations are different [61]. In our case, weathering progress being external perturbation in the system affects molecular structure of investigated material. 2DCOS reveals characteristic behaviours of individual molecular constituents, in particular lignin susceptible to weathering process. Since the intensity of peaks on the autocorrelation spectrum is directly proportional to the relative importance of the intensity change in the original spectra, it can be stated that aromatic skeletal vibrations of lignin are the most vulnerable functional groups.

The observation of weathering process on three different scales, macroscopic, microscopic and molecular, allowed comparison of mechanisms occurring on various levels. It was assumed that the southern location might cause slightly more degradation of wooden surfaces, even in a relatively short time. It was related to longer exposure of UV radiation and, in consequence, more rapid changes of surface moisture. Northern location seems to be less susceptible to degradation, especially at the beginning of weathering. It corresponds to changes of integrated cumulative solar radiation (SR) calculated according to the method proposed by Grossman [62]. It can be clearly seen that the value of cumulative SR of southern exposure is double than northern (80 and 40 $kW \cdot m^{-2}$, respectively) [63]. This phenomenon might explain variation in weathering intensity between exposure sites.

The use of ultra-thin wood samples for investigation of material resistance to natural weathering as reported here reveals that mechanism of degradation as well as sequence of material deterioration follows directly state-of-the-art understanding of wood weathering processes. Investigation of the effect of natural weathering on the microstructure of radiata pine by means of SEM shows analogous deterioration sequence. Degradation of the pit membranes was followed by the gradual enlargement of the pit aperture to the approximate limit of the pit chamber, development of microchecks in the pit borders and to ultimate destruction of pit borders in earlywood tracheids in about 6 months [64].

The particular sample geometry (ultra-thin thickness) determined elevated kinetics of such changes and shortened response time of the material for the weather dose. Analogous approach was proposed by Wang et al. for other types of materials [65]. Their study was related to the thickness-dependent accelerated ageing method that leaded to the more severe ageing degree of the thinner specimens. The phenomenon was explained by the ratio of aged area to the total area. It was reported that in the thinner specimens the ratio was larger than in the thicker specimens. It is always a challenge to determine a specific quantifier of the weathering acceleration factor (AF). Jelle proposed calculation of AF by comparing natural and artificial weathering [66]. He assumed that the higher UV intensity (W/m^2) and total energy (kWh/m^2) in the weathering apparatus is proportional to the aging process. However, the natural weathering process is more complex, as includes several other than UV irradiation factors influencing the degradation rate [67]. This might be a reason to the wide array of reported AF values, ranging between 5 to 250, while comparing the laboratory results with outdoor tests in natural climate.

For the needs of this research a direct comparison approach was implemented by confronting above presented results with corresponding characterizations for thick wood samples (10 mm thickness) made of the same wood piece. A set of such thick samples was exposed during the natural weathering test in vicinity of thin samples. In that case, weathered samples were collected from the stand for duration of one year, one sample each lunar month. "Thick" samples were assessed with same analytical techniques as ultra-thin strips, by applying necessary amendments (such as Attenuated Total Reflectance (ATR) instead of transmission mode of infrared spectrometer, or reflectance instead of transflectance when measuring colour spectrum). The trends of material changes can be then directly compared between thin and thick wood samples to understand the weathering process acceleration. Assuming that the sole difference between both sample types is kinetics of deterioration, it become possible to identify an acceleration factor. This factor was determined empirically by adjusting the time scale for diverse deterioration kinetics curves measured on ultra-thin and thick samples to assure an overlap of both trends. An example of such analysis is presented in Figure 11 where acceleration factor of ×3 was applied for the CIE *L* and CIE *a* colour curves in combination with the lignin deterioration trends at peak #14 (1507 cm^{-1}). The numerical factor (determined through visual observations) can be therefore used to quantify the time relation between both confronted natural weathering configurations. It is clear that such an approach introduces certain over-simplification as nature of scanned parameters is different, as well as the weather dose distribution is very different during periods of one month (thin samples) and one year (thick samples). In any case, it provides a scientific basis for a statement that the use of ultra-thin samples for natural weathering may lead to shorten the test duration still preserving the complex physical and chemical mechanism of the material, particularly surface, deterioration when exposed to natural weather conditions.

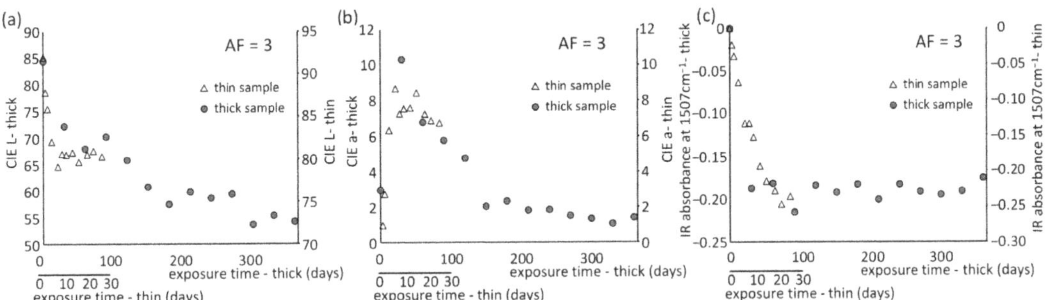

Figure 11. Confrontation of (**a**) CIE *L*, (**b**) CIE *a* and (**c**) lignin peak #14 of IR spectrum deterioration kinetics curves for experimental data acquired from thin and thick wood samples exposed to natural weathering, assuming acceleration factor AF = 3.

An important limitation of using ultra-thin samples for natural weathering testing is a fact of losing a complex interaction of different layers within material bulk. It implicates a dramatic change to the diffusion of moisture and heat within bulk material. However, in the context of the wood weathering it is clear that the most relevant changes occur in the subsurface of the material. The majority of extensive chemical–physical reactions (photolysis, hydrolysis, etc.), deposition of biological and abiotic agents, as well as moisture induced deformations take place on the surface of sample and progress into bulk interior only after erosion of the external layers. The electromagnetic radiation in the range of photodegrading UV light penetrates wood to the extent of 50–100 μm, that corresponds to the thickness of samples used in this research. Other important missing aspect in the proposed "accelerated natural weathering" approach is a fact that an ultra-thin sample is missing the mechanical support of the bulk. The support constrains moisture-induced deformations and serves as a physical barrier (or buffer) slowing down the moisture and heat transfer. It affects also the number and extend of microcracks formation. Despite the above limitations, using ultra-thin wood samples is a correct approach for assuring replication of the real mechanisms of the weathering occurring in "thick samples". That cannot be assured in alternative up-to-date artificial weathering methods.

5. Conclusions

Understanding the mechanisms of weathering and the role of altering factors is fundamental to assess the actual conditions of timber structures. It is also essential for future performance, and, possibly, to ensure long-term preservation and maintenance. Even if it is impossible to fully integrate results obtained by assessment of thin wood samples with standard wooden samples, the presented approach allowed rapid and reliable assessment of the complex chemical and structural changes occurring within the surface layers.

With the progress of degradation, samples become yellowish. Weathering also caused removal of wood fibers, presence of cracks and, an increase in brittleness of samples. Spectroscopic analysis provided information regarding susceptibility of functional groups to natural weathering. Gradual changes in absorbance intensity of CH deformation of aromatic skeletal vibration of lignin confirmed the vulnerability of this component to photodegradation. Scanning electron microscopy (SEM) was used here for examination of the morphological characteristics of naturally weathered thin wooden samples. The structural degradation was initiated from openings of bordered pits membranes in radial walls of earlywood tracheids. In the following steps, the membrane covering the piths was broken and diagonally oriented micro-checks were observed. The progress of weathering led to entire destruction of the pit membrane and whole pit erosion.

It was confirmed that earlywood was more susceptible to damage than latewood, which was explained by the fact that cells in earlywood have thinner and weaker walls and, in consequence, lower density. It was also observed that western and northern exposure sites are slightly less affected by weathering. First signs of fungal infestation, not visible by naked eye, were observed in the third week of natural weathering.

Even if the weathering test was conducted for a relatively short time (28 days), the progress of ultra-thin wood sample alteration was similar as usually noticed for thick wood surfaces. However, such an extent of changes occurred at shorter exposure times. The estimated acceleration factor was $\times 3$. The research methodology presented here can be directly used for determining the weather dose-response models essential to predict the future service life performance of timber elements. It may also serve as a screening technique for a rapid estimate of the material's resistance to deterioration. In that case, all the natural weathering factors are fully considered, in contrast to artificial weathering procedures limiting the real exposure conditions.

Author Contributions: Conceptualization: A.S., J.S., M.N., and A.D.; methodology: A.S., J.S., M.N., and A.D.; software: J.S.; validation: A.S., J.S., and M.N.; formal analysis: A.S.; investigation: A.S., J.S., M.N., and A.D.; resources: A.S. and J.S.; data curation: J.S. and A.S.; visualization: J.S.; supervision: J.S. and A.S.; project administration: A.S.; funding acquisition: A.S., J.S., M.N., and A.D. All authors have read and agreed to the published version of the manuscript.

Funding: The authors gratefully acknowledge the European Commission for funding the InnoRenew project (Grant agreement #739574 under the Horizon2020 Widespread-Teaming program) and the Republic of Slovenia (investment funding of the Republic of Slovenia and the European Union European Regional Development Fund) and infrastructural ARRS program IO-0035. Part of this work was conducted during project Archi-BIO (BI/US-20-054) funded by ARRS and CLICK DESIGN "Delivering fingertip knowledge to enable service life performance specification of wood"–(No. 773324) supported under the umbrella of ERA-NET Cofund ForestValue by the Ministry of Education, Science and Sport of the Republic of Slovenia. ForestValue has received funding from the European Union's Horizon 2020 research and innovation programme.

Institutional Review Board Statement: Not applicable.

Informed Consent Statement: Not applicable.

Data Availability Statement: The data presented in this study are available on request from the corresponding author.

Acknowledgments: The experimental samples were prepared and weathered during Round Robin test of COST FP1006. The support of COST FP1303 for funding Short Term Scientific Mission of Anna

Sandak is highly acknowledged. Authors would also like to thank Thomas Volkmer for inspiring discussion.

Conflicts of Interest: The authors declare no conflict of interest.

References

1. Ivanović-Šekularac, J.; Čikić-Tovarović, J.; Šekularac, N. Application of wood as an element of façade cladding in construction and reconstruction of architectural objects to improve their energy efficiency. *Energy Build.* **2016**, *115*, 85–93. [CrossRef]
2. De Vetter, L.; Cnudde, V.; Masschaele, B.; Jacobs, P.J.S.; Van Acker, J. Detection and distribution analysis of organosilicon compounds in wood by means of SEM-EDX and micro-CT. *Mater. Charact.* **2006**, *56*, 39–48. [CrossRef]
3. Petrillo, M.; Sandak, J.; Grossi, P.; Sandak, A. Chemical and appearance changes of wood due to artificial weathering—Dose-response model. *J. Near Infrared Spectrosc.* **2019**, *27*, 26–37. [CrossRef]
4. Williams, R.R. Weathering of wood. In *Handbook of Wood Chemistry and Wood Composites*, 1st ed.; Rowell, R.M., Ed.; CRC Press: Boca Raton, FL, USA, 2005; pp. 139–185.
5. Low-Temperature EMMA®/EMMAQUA®. Available online: https://www.atlas-mts.com/products/accelerated-weathering-test-services/natural-weathering/accelerated-weathering-test/emmaqua-weathering-test (accessed on 3 January 2021).
6. Kennedy, R.W.; Ifju, G. Application of microtensile testing to thin wood sections. *Tappi* **1962**, *45*, 725–733.
7. Manwiller, F.G.; Godfrey, P.R. Microtensile strength of spruce pine after exposure to acids and bases. *Wood Sci.* **1973**, *5*, 295–297.
8. Raczkowski, J. Natural atmospheric corrosion of microtome wood sections. In Proceedings of the 10th Symp. on Wood Protection, Rogów, Poland, 17–19 October 1978. (In Polish).
9. Raczkowski, J. Seasonal effects on the atmospheric corrosion of spruce micro-sections. *Holz Roh Werkst.* **1980**, *38*, 231–234. [CrossRef]
10. *BS EN 927-6 Paints and Varnishes. Coating Materials and Coating Systems for Exterior Wood*; European Committee for Standardization: Brussels, Belgium, 2018.
11. Arnold, M.; Sell, J.; Feist, W.C. Wood weathering in fluorescent ultraviolet and xenon arc chambers. *For. Prod. J.* **1991**, *41*, 40–44.
12. Podgorski, L.; Arnold, M.; Hora, G. A reliable artificial weathering test for wood coatings. *Coat. World* **2003**, *2*, 39–48.
13. Grüll, G.; Tscherne, F.; Spitaler, I.; Forsthuber, B. Comparison of wood coating durability in natural weathering and artificial weathering using fluorescent UV-lamps and water. *Eur. J. Wood Prod.* **2014**, *72*, 367–376. [CrossRef]
14. Ozgenc, O. Comparing durability of wood material in natural and artificial weathering conditions. *ProLigno* **2016**, *12*, 3–11.
15. Rüther, P.; Jelle, B.P. Color changes of wood and wood-based materials due to natural and artificial weathering. *Wood Mater. Sci. Eng.* **2013**, *8*, 13–25.
16. Brischke, C.; Thelandersson, S. Modelling the outdoor performance of wood products—A review on existing approaches. *Constr. Build. Mater.* **2014**, *66*, 384–397. [CrossRef]
17. Rüther, P.; Time, B. External wood claddings—Performance criteria, driving rain and large-scale water penetration methods. *Wood Mater. Sci. Eng.* **2016**, *10*, 287–299. [CrossRef]
18. Thiis, T.K.; Burud, I.; Kranitis, D.; Gobakken, L.R. Simulation of surface climate and mould growth on wooden facades. In *The International Research Group on Wood Protection IRG/WP 16-20585*; IRG Secretariat: Stockholm, Sweden, 2016.
19. Sandak, J.; Sandak, A.; Riggio, M. Characterization and monitoring of surface weathering on exposed timber structures with multi-sensor approach. *Int. J. Archit. Herit.* **2015**, *9*, 674–688. [CrossRef]
20. Sandak, J.; Sandak, A.; Burud, I. Modelling of weathering. In *Performance of Bio-Based Building Materials*, 1st ed.; Jones, D., Brischke, C., Eds.; Woodhead Publishing: Cambridge, UK, 2017; pp. 502–510.
21. Salmén, L. Wood morphology and properties from molecular perspectives. *Ann. For. Sci.* **2015**, *72*, 679–684. [CrossRef]
22. Wiedenhoeft, A. Structure and function of wood. In *Wood Handbook: Wood as an Engineering Material*; Rowell, R.M., Ed.; Forest Products Laboratory: Madison, WI, USA, 2010; pp. 3.1–3.18.
23. Fengel, D.; Wegener, G. *Wood—Chemistry, Ultrastructure, Reactions*; Walter de Gruyter: New York, NY, USA, 1989.
24. Chinga-Carrasco, G.; Johnsen, P.O.; Øyaas, K. Structural quantification of wood fibre surfaces—Morphological effects of pulping and enzymatic treatment. *Micron* **2010**, *41*, 648–659. [CrossRef] [PubMed]
25. Hamed, S.A.M.; Ali, M.F.; El Hadidi, N.M.N. Using SEM in monitoring changes in archaeological wood: A review. In *Current Microscopy Contributions to Advances in Science and Technology*; Méndez-Vilas, A., Ed.; Formatex Research Center: Badajoz, Spain, 2012; pp. 1077–1084.
26. Jingran, G.; Jian, L.; Jian, Q.; Menglin, G. Degradation assessment of waterlogged wood at Haimenkou site. *Frat. Integrità Strutt.* **2014**, *30*, 495–501. [CrossRef]
27. Nicole, M.; Chamberland, H.; Rioux, D.; Xixuan, X.; Blanchette, R.A.; Geiger, J.P.; Ouellette, G.B. Wood degradation by Phellinus noxius: ultrastructure and cytochemistry. *Can. J. Microbiol.* **1995**, *41*, 253–265. [CrossRef]
28. Popescu, C.M.; Tibirna, C.M.; Nanoliu, A.; Gradinariu, P.; Vasile, C. Microscopic study of lime wood degraded by Chaetomium globosum. *Cellul. Chem. Technol.* **2011**, *45*, 565–569.
29. Singh, A.P.; Singh, T.; Rickard, C.L. Visualising impregnated chitosan in Pinus radiata early wood cells using light and scanning electron microscopy. *Micron* **2010**, *41*, 263–267. [CrossRef]
30. Singh, A.P.; Anderson, R.; Parka, B.D.; Nuryawana, A. A novel approach for FE-SEM imaging of wood-matrix polymer interface in a biocomposite. *Micron* **2013**, *54–55*, 87–90. [CrossRef] [PubMed]

31. Turkulin, H.; Arnold, M.; Derbyshire, H.; Sell, J. Structural and fractographic SEM analysis of exterior coated wood. *Surf. Coat. Int. Part B Coat. Trans.* **2001**, *84*, 67–75. [CrossRef]
32. Pfeffer, A.; Mai, C.; Militz, H. Weathering characteristics of wood treated with water glass, siloxane or DMDHEU. *Eur. J. Wood Prod.* **2012**, *70*, 165–176. [CrossRef]
33. Kocaefe, D.; Huang, X.; Kocaefe, Y. Comparison of weathering behaviours of heat-treated jack pine during different artificial weathering conditions. In Proceedings of the 4th International Conference on Fluid Mechanics and Heat & Mass Transfer (FLUIDSHEAT'13), Dubrovnik, Croatia, 25–27 June 2013; pp. 74–79.
34. Ganne-Chedeville, C.; Volkmer, T.; Letsch, B.; Lehmann, M. Measures for the maintenance of untreated wood facades. In Proceedings of the WCTE, Auckland, New Zealand, 15–19 July 2012; p. 6.
35. Fackler, K.; Schwanninger, M. How spectroscopy and microspectroscopy of degraded wood contribute to understand fungal wood decay. *Appl. Microbiol. Biotechnol.* **2012**, *96*, 587–599. [CrossRef]
36. Sandak, A.; Sandak, J.; Riggio, M. Assessment of wood structural members degradation by means of infrared spectroscopy: An overview. *Struct. Control Health* **2016**, *23*, 396–408. [CrossRef]
37. Popescu, C.M.; Popescu, M.C.; Vasile, C. Structural changes in biodegraded lime wood. *Carbohydr. Polym.* **2010**, *79*, 362–372. [CrossRef]
38. Popescu, C.M.; Froidevaux, J.; Navi, P.; Popescu, M.C. Structural modifications of Tilia cordata wood during heat treatment investigated by FT-IR and 2D IR correlation spectroscopy. *J. Mol. Struct.* **2013**, *1033*, 176–186. [CrossRef]
39. Huang, A.; Zhou, Q.; Liu, J.; Fei, B.; Sun, S. Distinction of three wood species by Fourier transform infrared spectroscopy and two-dimensional correlation IR spectroscopy. *J. Mol. Struct.* **2008**, *883*, 160–166. [CrossRef]
40. Sandak, A.; Sandak, J.; Burud, I.; Gobakken, L.R. Weathering kinetics of thin wood veneers assessed with near infrared spectroscopy. *J. Near Infrared Spectrosc.* **2016**, *24*, 549–553. [CrossRef]
41. Gallagher, N.B.; Blake, T.A.; Gassman, P.L. Application of extended inverse scatter correction to mid-infrared reflectance spectra of soil. *J. Chemometr.* **2005**, *19*, 271–281. [CrossRef]
42. Faix, O. Classification of lignins from different botanical origins by FT-IR spectroscopy. *Holzforschung* **1991**, *45*, 21–27. [CrossRef]
43. Schwanninger, M.; Rodrigues, J.; Pereira, H.; Hinterstoisser, B. Effects of short-time vibratory ball milling on the shape of FT-IR spectra of wood and cellulose. *Vib. Spectrosc.* **2004**, *36*, 23–40. [CrossRef]
44. Popescu, C.M.; Vasile, C.; Popescu, M.C. Degradation of lime wood painting supports II. Spectral characterisation. *Cellul. Chem. Technol.* **2006**, *40*, 649–658.
45. McLean, J.P.; Jin, G.; Brennan, M.; Nieuwoudt, M.K.; Harris, P.J. Using NIR and ATR-FTIR spectroscopy to rapidly detect compression wood in Pinus radiata. *Can. J. For. Res.* **2014**, *44*, 820–830. [CrossRef]
46. Guo, X.; Wu, Y.; Yan, N. In situ micro-FTIR observation of molecular association of adsorbed water with heat-treated wood. *Wood Sci. Technol.* **2018**, *52*, 971–985. [CrossRef]
47. Nejad, S.M.M.; Madhoushi, M.; Vakili, M.; Rasouli, D. Evaluation of degradation in chemical compounds of wood in historical buildings using FT-IR and FT-Raman vibrational spectroscopy. *Maderas Cienc. Tecnol.* **2019**, *21*, 381–392.
48. Noda, I.; Ozaki, Y. *Two-Dimensional Correlation Spectroscopy*; John Willey & Sons Ltd.: Chichester, UK, 2005.
49. Noda, I. Two-dimensional infrared spectroscopy. *J. Am. Chem. Soc.* **1989**, *111*, 8116–8118. [CrossRef]
50. Evans, P.D. Effect of angle of exposure on the weathering of wood surfaces. *Polym. Degrad. Stabil.* **1989**, *24*, 81–87. [CrossRef]
51. Cogulet, A.; Blanchet, P.; Landry, V. Wood degradation under UV irradiation: A lignin characterization. *J. Photochem. Photobiol. B* **2016**, *158*, 184–191. [CrossRef]
52. Kishino, M.; Nakano, T. Artificial weathering of tropical woods. Part1: Changes in wettability. *Holzforschung* **2004**, *58*, 552–557. [CrossRef]
53. *EN 927-3 Paints and Varnishes—Coating Materials and Coating Systems for Exterior Wood—Part 3: Natural Weathering Test*; European Committee for Standardization: Brussels, Belgium, 2000.
54. Müller, U.; Rätzsch, M.; Schwanninger, M.; Steiner, M.; Zöbl, H. Yellowing and IR-changes of spruce wood as result of UV-irradiation. *J. Photochem. Photobiol. B* **2003**, *69*, 97–105. [CrossRef]
55. Tolvaj, L.; Faix, O. Artificial ageing of wood monitored by DRIFT spectroscopy and CIE $L^*a^*b^*$ colour measurements. *Holzforschung* **1995**, *45*, 397–404. [CrossRef]
56. Turkulin, H. *SEM Methods in Surface Research on Wood*; COST E 18 Final Seminar; CTBA: Paris, France, 2004.
57. Sandberg, D. Weathering of radial and tangential wood surfaces of pine and spruce. *Holzforschung* **1999**, *53*, 355–364. [CrossRef]
58. Pandley, K.K.; Pitman, A.J. Weathering characteristics of modified rubberwood (*Hevea brasiliensis*). *J. Appl. Polym. Sci.* **2002**, *85*, 622–631. [CrossRef]
59. Owen, J.A.; Owen, N.L.; Feist, W.C. Scanning electron microscope and infrared studies of weathering in southern pine. *J. Mol. Struct.* **1993**, *300*, 105–114. [CrossRef]
60. Timar, M.C.; Varodi, A.M.; Gurău, L. Comparative study of photodegradation of six wood species after short-time UV exposure. *Wood Sci. Technol.* **2016**, *50*, 135–163.
61. Noda, I. Advances in two-dimensional correlation spectroscopy (2DCOS). In *Frontiers and Advances in Molecular Spectroscopy*, 1st ed.; Laane, J., Ed.; Elsevier Science Publishing Co Inc.: Cambridge, MA, USA, 2017; pp. 47–75.
62. Grossman, D.M. Errors caused by using Joules to time laboratory and outdoor exposure tests. In *Accelerated and Outdoor Durability Testing of Organic Materials*; Grossman, D., Ketola, W., Eds.; ASTM International: West Conshohocken, PA, USA, 1994; pp. 68–87.

63. Burud, I.; Sandak, A.; Sandak, J.; Flö, A.; Thiis, T.; Gobakken, R.L.; Smeland, K.A.; Kvaal, K. Weather degradation of thin wood samples assessed with NIR hypersepctral imaging in transmission mode. In Proceedings of the 17th International Conference on Near Infrared Spectroscopy, Foz do Iguassu, Brazil, 18–23 October 2015; pp. 23–25.
64. Groves, K.W.; Banana, A.Y. Weathering characteristics of Australian grown radiata pine. *J. Inst. Wood Sci.* **1986**, *10*, 210–213.
65. Wang, Y.; Zhu, W.; Zhang, X.; Cai, G.; Wan, B. Influence of thickness on water absorption and tensile strength of BFRP Laminates in water or alkaline solution and a thickness-dependent accelerated ageing method for BFRP laminates. *Appl. Sci.* **2020**, *10*, 3618. [CrossRef]
66. Jelle, B.P. Accelerated climate ageing of building materials, components and structures in the laboratory. *J. Mater. Sci.* **2012**, *47*, 6475–6496.
67. Jankowska, A.; Rybak, K.; Nowacka, M.; Boruszewski, P. Insight of weathering processes based on monitoring surface characteristic of tropical wood species. *Coatings* **2020**, *10*, 877. [CrossRef]

Article

Characterization of Arctic Driftwood as Naturally Modified Material. Part 1: Machinability

Daniel Chuchala [1,*], Anna Sandak [2,3], Kazimierz A. Orlowski [1], Jakub Sandak [2,4], Olafur Eggertsson [5] and Michal Landowski [1]

[1] Faculty of Mechanical Engineering and Ship Technology, Gdańsk University of Technology, 80-233 Gdańsk, Poland; kazimierz.orlowski@pg.edu.pl (K.A.O.); michal.landowski@pg.edu.pl (M.L.)
[2] InnoRenew CoE, 6310 Izola, Slovenia; anna.sandak@innorenew.eu (A.S.); jakub.sandak@innorenew.eu (J.S.)
[3] Faculty of Mathematics, Natural Sciences and Information Technologies, University of Primorska, 6000 Koper, Slovenia
[4] Andrej Marušič Institute, University of Primorska, 6000 Koper, Slovenia
[5] Icelandic Forest Research, Mógilsá, 162 Reykjavik, Iceland; olie@skogur.is
[*] Correspondence: daniel.chuchala@pg.edu.pl; Tel.: +48-58-3471450

Abstract: Arctic driftwood has reached the coast of Iceland for centuries. This material was used by the inhabitants of the island as a building material for houses, boats, churches and pasture fences. Nowadays, the driftwood is used in the furniture industry, for the finishing of internal and external walls of buildings and also by artists. The properties of driftwood differ to that of original resource due the long-term effects of exposure to Arctic Sea water and ice. This process can be considered as a natural modification, even if its effect on various wood properties and the potential use of driftwood are not yet fully understand. This research is focused on the comparison of cutting forces measured for Siberian larch (*Larix sibirica* L.) from Siberia provenance and driftwood found on the coast of Iceland. The cutting forces were determined directly from the cutting power signal that was recorded during the frame sawing process. A new procedure for compensation of the late/early wood ratio variation within annual rings is proposed to homogenize mechanical properties of wood. It allows a direct comparison of machinability for both types of larch wood investigated (driftwood and natural). Noticeable differences of normalized cutting force values were noticed for both wood types, which were statistically significant for two set values of feed per tooth. These results provide a new understanding of the effect of the drifting process in the Arctic Sea (natural modification) on mechanical and physical properties of wood. Such a natural modification may influence transformation processes of driftwood as well as performance of the coating systems applied on its surface.

Keywords: Arctic driftwood; natural modification; cutting forces; larch wood; sawing process

Citation: Chuchala, D.; Sandak, A.; Orlowski, K.A.; Sandak, J.; Eggertsson, O.; Landowski, M. Characterization of Arctic Driftwood as Naturally Modified Material. Part 1: Machinability. *Coatings* **2021**, *11*, 278. https://doi.org/10.3390/coatings11030278

Academic Editor: Mariaenrica Frigione

Received: 2 February 2021
Accepted: 22 February 2021
Published: 26 February 2021

1. Introduction

Arctic driftwood has reached the coast of Iceland for centuries. In the early Middle Ages, 40% of the island area was covered by birch forests [1]. However, the low durability of birch combined with weak mechanical properties and small sizes of trees limited its use in construction. For that reason, driftwood was the main resource used by locals to build houses, boats, churches or bridges [1–4]. The majority of wood types reaching the coast of Iceland are softwood species, especially pine (*Pinus* sp.), spruce (*Picea* sp.) and larch (*Larix* sp.) [5]. These species possess weaker mechanical properties than birch wood from inland Europe [6,7] but are still widely used for construction elsewhere. Historical sources reveal that driftwood was an appreciated building material due to its high suitability for use as structural members as well as superior durability [1,3]. Eggertsson [5] discovered that driftwood arrives to the coasts of Iceland with sea ice and surface currents from the north. By applying the method of dendrochronology on the driftwood, the main origin was revealed to be mostly from the boreal forests of northern Europe and from the Russian part of Asia (Figure 1a). Alternatively, some logs originate from Alaska and Northern Canada [8]. Hellmann et al. [8–11] reported that in the majority of cases, pine, spruce and larch logs were identified on the northern coast of Iceland. The same studies revealed that larch wood found on the northern beaches in Iceland originates from Central and Eastern Siberia. The majority of logs harvested in Siberian forests are transported using rivers to ports located on the coast of the Arctic Sea [10]. Some of these logs are not captured at the final location and cruise farther to the open seas. In addition, whole trees felled due to natural forest processes are taken by rivers each year, especially during intensive snow melting periods and ice breakup on rivers in spring. Similar events occur in North American forests, supplying logs identified as a second major source of Arctic driftwood [12].

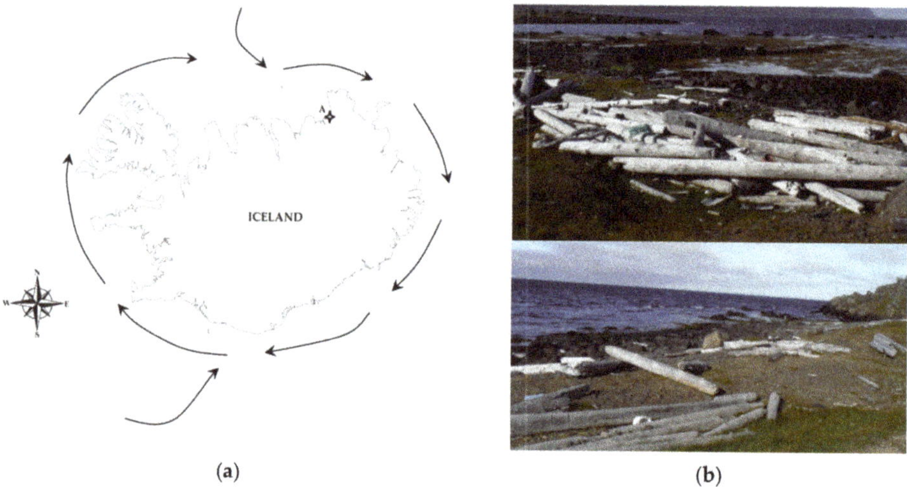

Figure 1. The arctic driftwood on the Icelandic coasts: (**a**) general view of surface water currents around Iceland [5] and sampling location (A) of the investigated driftwood log collection; (**b**) the Arctic driftwood on the North Icelandic coast (photo by Chuchala).

Driftwood is defined here as remains of trees that flow into the ocean as a consequence of river bank erosion and flooding, storms, winds or other natural occurrences as well as a side effect of logging. These trees are carried by rivers and sea currents for distances of hundreds or thousands of kilometers. Drifting logs can repeatedly be covered by Arctic ice along the drift, which may result in a substantial extension of the exposure time to sea water. Observations of the drifted wood have been used to analyze the dynamics of Arctic Sea ice cover [13]. Dalaiden et al. [14] have created a model that predicts trails of runoff wood from different locations. Arctic driftwood is exposed to sea water and Arctic ice for a few months to several years before it reaches the coasts of Iceland, Greenland or other drylands. It is estimated that the average distance the driftwood travels following sea currents corresponds to 400–1000 km (250–620 miles) per year [15]. Therefore, it is estimated that it takes at least 4 to 5 years for a log from Siberia to reach the coast of Iceland. Such harsh conditions have a major impact on the properties of the wood. A study carried out by Komorowicz et al. [16] showed differences in the physical and mechanical properties of pine driftwood compared to the reference material. A reduced compressive strength along the grain, lower calorific value and increased equilibrium moisture content were observed for driftwood. Reported lower compressive strength along the fibers of Arctic driftwood may be caused by the presence of fungi on the drifting logs, which develop during the wood's journey in the sea [17–20].

Differences in thermal degradation of driftwood during combustion, gasification and pyrolysis processes were also reported [21,22]. Endeavors [21–24] proposed using driftwood as a low-cost thermal energy source and demonstrated that the calorific values are slightly lower than those of natural wood. Furthermore, the same studies have shown that driftwood, when compared to the reference wood species from land, contains much larger amounts of chlorine, sodium, calcium and magnesium. More slags are generated during the burning of driftwood, compared to reference wood directly delivered from the forest [21–24].

Arctic driftwood is relatively well researched when considering silviculture, dendrology and dendrochronology [5,8–11]. However, this material was not systematically studied regarding its mechanical and physical properties. These are highly relevant when considering its use in the modern sustainable construction sector and/or furniture industry. For these particular sectors, the quality of the wood surface obtained after machining is very important, as these surfaces are often covered with protective and decorative coatings. Wood surface quality is affected by several wood–machine–tool interaction factors. Especially, wood species, its density and moisture content, as well as cutting parameters including feed speed, cutting speed, cutting depth, processing direction and cutting forces, influence the resulting surface quality [25]. The roughness of the wood surface affects the adhesion strength of coatings. The contact area for the mechanical interlocking between coating and wood substrate increases following the roughness of the surface, which results in an increase in adhesion [25,26]. The early/late wood ratio is another important factor affecting the adhesion of coating. In the majority of wood species, early wood is more porous than late wood. Consequently, the coating penetration depth within the early wood zone is higher due to simplified impregnation of the finishing product into capillaries of the wood [27]. Arctic driftwood is gaining popularity for use in architecture, furniture design and art [28]. Archeological investigations in the territory of Iceland reveal that driftwood was a very common and important construction resource several centuries ago [1–4]. It is not proved, however, if its popularity was related to the superior performance of drifted (naturally modified) wood or simply easy access to such resources.

The aim of this study was to investigate an effect of the long-term travel of larch wood in the Arctic Sea on its machinability properties. For that reason, an experimental testing was performed to measure cutting forces while sawing wood on a frame sawing machine and then the measured values of the forces were subjected to double normalization (by the wood density and by the ratio of late to early wood). The double normalization was carried out in order to eliminate the effect of morphological differences of the tested wood on the values of the cutting forces. The new knowledge regarding mechanical properties may lead to optimization of the surface treatment methodologies adopted for driftwood as well as proper selection of cutting process conditions used for generating quality products.

2. Materials and Methods

2.1. Material

Arctic driftwood was used for the preparation of experimental samples. The driftwood was collected in Kópasker, the north coast of Iceland (latitude: 66.167418° N, longitude: −16.647679° W) by a local farmer, who was the owner of that land (Figure 1a). Collected logs were cut into boards, which were stored in outdoor conditions for a period of two years (Figure 2a). A single, randomly selected board was used for the preparation of six rectangular blocks (Figure 2b) of dimensions 50 × 50 × 600 mm^3 (width (W) × height (H) × length (L), respectively). This operation was performed in the carpentry workshop Trésmiðja H Ben ehf in Akureyri (Iceland). All experimental samples after transportation to Poland were conditioned in laboratory conditions for 12 months, assuring a constant air temperature (T_a) of 20 °C and relative humidity (RH) of 65%.

(**a**)

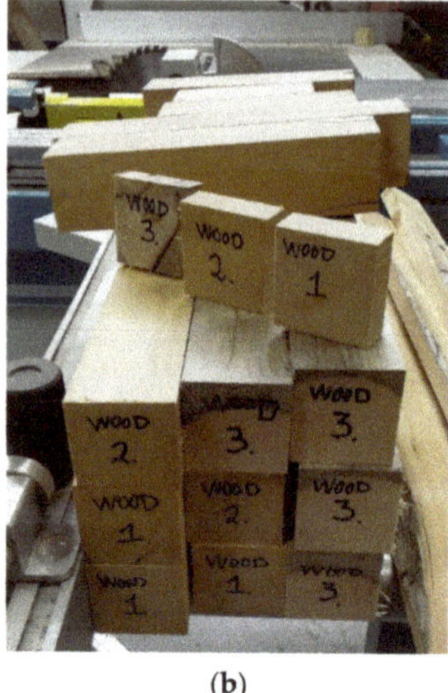

(**b**)

Figure 2. The investigated driftwood: (**a**) driftwood boards stored in outdoor conditions on North Icelandic farm; (**b**) derived experimental samples (only "WOOD 1" samples were used for determination of the driftwood fracture properties).

The biological species of experimental samples was verified as Siberian larch (*Larix sibirica* L.) by microscopic observation on Leica DM2500 light microscope (Leica Microsystems, Wetzlar, Germany) with magnification of 1000×. Ultrathin (10–20 μm) samples of transverse, tangential and radial sections were cut-out by the microtome and mounted on the microscope slide. The origin of drifted logs was estimated by combining microscopic observations and literature references as Central or Eastern Siberia [5,8–11]. More detailed identification of the origin based on dendrological analysis was not possible due to lack of access to the full radial section of log. Non-treated reference wood samples were prepared from a log of Siberian larch imported by the sawmill Sylva Sp. z o. o. from Wiele (Poland) from Eastern Siberia. A material matching (as much as possible) the driftwood samples dimensions, annual ring orientation as well as late/early wood ratio was prepared by the sawmill for comparative analysis. The reference wood samples were not exposed to any treatment.

The wood density ρ, defined here as a ratio of the wood mass to its volume at the air-dry state (moisture content MC = 12%), was measured separately on each block of both driftwood and reference sample groups. All results are summarized in Table 1 together with the average width of annual ring (WAR), the average width of late (LW) and early (EW) wood in annual rings and ratio of late and early wood (L/E).

Table 1. Physical properties of examined wood samples (average values with standard deviations).

Sample Code	Sample Name	ρ	WAR	LW	EW	*L*/*E*
		kg m^{-3}	mm	mm	mm	–
DL	Driftwood Larch	554.7 ± 15.5	0.72 ± 0.23	0.15 ± 0.04	0.57 ± 0.20	0.28 ± 0.09
L	Larch	694.9 ± 31.5	0.52 ± 0.09	0.15 ± 0.04	0.37 ± 0.07	0.41 ± 0.10

Legend: ρ—density of air-dry wood at MC = 12%; WAR—width of annual rings; LW—width of the late wood; EW—width of the early wood; ratio E/L—ratio of the late to early wood.

2.2. Machinability Tests Methodology

Sawing tests were performed on the PRW15M sash-gang saw with a hybrid dynamically balanced driving system and elliptical teeth trajectory movement [29]. The concept of the machine was developed at the Gdańsk University of Technology (Gdańsk, Poland) [29] and a prototype manufactured by REMA S.A. (Reszel, Poland). Detailed technical parameters of the sash-gang saw and saw blades used in experimental cutting tests are presented in Table 2.

Table 2. Technical parameters of sash-gang saw and its saw blade used in the experimental cuttings.

Parameter		Symbol	Value	Unit
Machine Parameters				
number of strokes of saw frame per min		n_F	685	spm
saw frame stroke		H_F	162	mm
number of saws in the gang		m	5	–
average cutting speed		v_c	3.69	$m\cdot s^{-1}$
feed speed	slow	v_{f1}	0.99	$m\cdot min^{-1}$
	fast	v_{f2}	1.45	$m\cdot min^{-1}$
feed per tooth	slow	f_{z1}	0.116	mm
	fast	f_{z2}	0.171	mm
Tool Parameters				
the sharp saw blades with stellite tipped teeth		–	–	–
overall set (kerf width)		S_t	2	mm
saw blade thickness		s	0.9	mm
free length of the saw blade		L_0	318	mm
blade width		b	30	mm
tooth pitch		t_p	13	mm
tool side rake angle		γ_f	9	°
tool side clearance		α_f	14	°
tension stresses of saws in the gang		σ_N	300	MPa

The use of electric power (active and passive) during idling and working cycles was continuously monitored with the power converter PP54 (LUMEL S.A., Zielona Góra, Poland). The data was collected with the acquisition converter μDAQ USB30 A/D (Eagle Technology, Zonnebloem, Cape Town, South Africa) and further processed to determine energetic effects of cutting. The mean value of feed per tooth f_z (mm) for a sash-gang saw was calculated as in Equations (1) and (2) [30–32]:

$$f_z = \frac{1000 \times v_f \times t_p}{n_F \times H_F} \quad (1)$$

$$v_f = \frac{L}{t_c} \quad (2)$$

where: v_f—feed speed ($m\cdot min^{-1}$), t_p—tooth pitch (mm), L—length of the sample (m), H_F—saw frame stroke (mm), n_F—number of strokes of saw frame per min (spm) and t_c—cutting time (min) necessary to process sample of the length L.

The average cutting power Pc (W) was calculated as the difference of the mean total power PT and the average idle power Pi [28], as expressed in Equation (3):

$$P_c = P_T - P_i \quad (3)$$

The average idle power P_i (W) of the frame saw PRW15-M was determined each time before beginning the proper cutting cycle. It allowed the minimization of the effect of the varying temperature of the machine components (such as hydraulic oil, gear boxes, etc.) on the energetic effects corresponding directly to the cutting process. The average cutting power in a working stroke P_{cw} (W) was calculated as in Equation (4), following the proceeding works of authors [31,32]:

$$P_{cw} = 2 \times P_c \quad (4)$$

The corresponding cutting forces F_c (N) (as related to one tooth of the saw blade) were calculated as a ratio of the obtained average cutting power in a working stroke P_{cw} at the average cutting speed v_c, and the average number of teeth in contact with the kerf z_a (Equation (5)) [33].

$$z_a = \frac{H}{t_p} \quad (5)$$

2.3. Normalization and Statistical Analysis

The obtained cutting force values were subjected to a single normalization (by density), followed by a novel two-level normalization (by density and by the late to early wood ratio). The normalization processes consisted of dividing values of derived cutting forces by the density of machined wood, that was followed by addition division by the corresponding late to early wood ratio assessed on the machined wood samples. The two-level normalization was intended to compensate for the effect of the tested wood morphology on the cutting forces.

The differences between the morphological parameters of the tested wood, as well as differences between the obtained values of cutting forces, were statistically analyzed using one-way and multi-factor analysis of variance (ANOVA).

3. Results and Discussion

The values of cutting forces per cutting tooth obtained from the experimental series of sawing samples of larch wood, both drifting in the Arctic Sea and not subjected to any modification and thermal treatment processes, are summarized in Figure 3. Two test point groups for each type of sawn wood are presented in the chart. These groups correspond to values of cutting forces obtained while sawing wood at two levels of feed speed v_f. The variation of feed speed is represented as a basic geometrical parameter of the cutting process, i.e., uncut chip thickness h. Two values of uncut chip thicknesses h presented in Figure 3 are h_1 = 0.116 mm and h_2 = 0.171 mm, which correspond to v_{f1} = 0.99 m·min^{-1} and v_{f2} = 1.45 m·min^{-1}. Linear regressions were created for each group of points representing both types of wood. Relatively high values of determination coefficients r^2 were noticed as a result of regression with r^2 = 0.98 for natural (reference) larch wood and r^2 = 0.95 for driftwood larch.

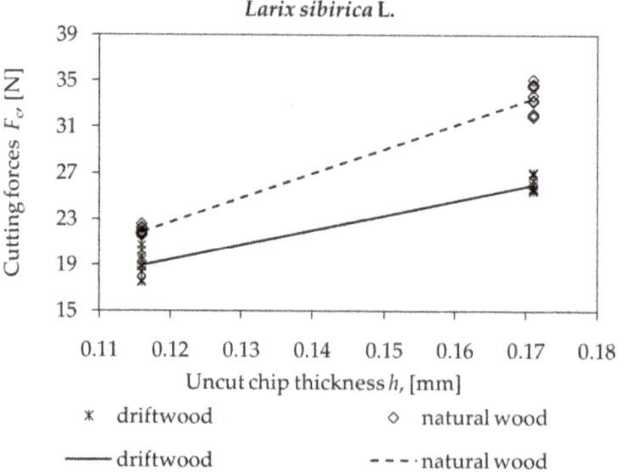

Figure 3. Relation between cutting forces per cutting tooth and uncut chip thickness when sawing drifted larch wood and natural larch wood.

Analyses of the results presented in Figure 3 allow the formulation of a statement that differences in cutting force values of drift and natural woods are noticeable. However, even if all the efforts were directed to assure homogenous and comparable experimental materials representing drift and natural woods, the differences between the average density (Table 1) may affect measured cutting forces [33]. Therefore, following protocols proposed in previous studies [30–32,34], the cutting forces were normalized to eliminate the effect of differences in the density of the tested wood samples on the values of these forces. Results obtained after data normalization are presented in Figure 4. It becomes evident that differences noted in the original cutting force values (Figure 3) are associated with the density differences within tested wood samples. Cutting forces normalized by the density are not noticeably different between driftwood and reference larch samples, even if slopes of the regression curves vary. This indicates different fracture properties of both types of wood, particularly shear yield strength [35,36]. It should be noticed, however, that the density variation is not the lone source of variation within experimental samples. The annual ring morphology, including its average width as well as the ratio of late and early woods, is also varied (Table 1).

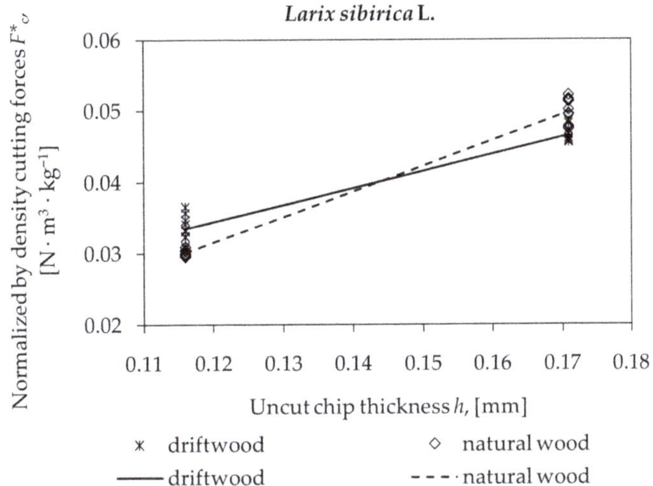

Figure 4. Relation between values normalized by density cutting forces per cutting tooth and uncut chip thickness when sawing drifted larch wood and natural larch wood.

The wood morphological features, such as width of annual rings, have a significant effect on the cutting resistance during the drilling process [37]. These observations were evidenced for pine, beech and oak wood species, other than the Siberian larch investigated here. Nevertheless, Koizumi et al. [38], Zhu et al. [39] and Luostarinen and Heräjärvi [40,41] reported that the width of annual rings of the larch wood strongly affects its mechanical properties. Likewise, the share of late wood within an annual ring has a substantial effect on the density value among other wood properties. Mikkola and Korhonen [42] reported a similar observation noticed for pine wood trees growing in cold weather climate zones. Several studies [43–47] revealed that properties of Siberian larch may vary significantly within diverse regions of Siberia. This was confirmed in the case of the experimental samples researched in this study. Even if the late wood width was similar in both investigated Siberian larch samples, the average ring width varied noticeably. Consequently, the late/early wood ratio L/E was different in the case of the studied drift and natural woods (Table 1). The results of the ANOVA summarized in Table 3 confirmed the statistical significance of differences within yearly ring anatomical structures.

Table 3. Significance of differences between annual ring characteristics noticed for drifting and natural larch samples (ANOVA, α = 0.05), where: DL—driftwood larch, L—natural larch.

Larch Wood (*Larix siberica* L.)								
Sample Code		**Source**	**DF**	**Adj SS**	**Adj MS**	**F—Value**	**P—Value**	**F—Critical**
Density, ρ (kg m^{-3})								
DL		between groups	1	59,009	59,009	99.09	1.66×10^{-6}	4.96
L		within groups	10	5955	595	-	-	-
	total		11	64,964	-	-	-	-
Width annual rings, WAR (mm)								
DL		between groups	1	0.566	0.566	19.75	4.21×10^{-5}	4.01
L		within groups	56	1.604	0.029	-	-	-
	total		57	2.170	-	-	-	-
Width late wood in annual rings, LW (mm)								
DL		between groups	1	9.22×10^{-5}	9.22×10^{-5}	0.058	0.81	4.01
L		within groups	56	0.088	0.002	-	-	-
	total		57	0.088	-	-	-	-
Width early wood in annual rings, EW (mm)								
DL		between groups	1	0.580	0.580	26.98	2.98×10^{-6}	4.01
L		within groups	56	1.205	0.021	-	-	-
	total		57	1.785	-	-	-	-
Ratio late wood to early wood in annual rings, Ratio L/E (−)								
DL		between groups	1	0.272	0.272	32.72	4.32×10^{-7}	4.01
L		within groups	56	0.465	0.008	-	-	-
	total		57	0.737	-	-	-	-

Koizumi et al. [38], Zhu et al. [39] and Luostarinen and Heräjärvi [40,41] evidenced a significant effect of the late wood ratio on diverse mechanical properties of wood. A second level of cutting force normalization was, therefore, proposed as a novelty in this research. The alternative approach aims to account for the effect of differences in the morphology of the tested wood on the cutting process energy requirements. The physical interpretation of such normalized values corresponds to the double-normalized cutting force that is required to cut wood of a unit density $\rho = 1$ kg·m^{-3} and a ratio of late to early wood $L/E = 1$. The values of such double-normalized specific cutting forces are presented for experimental samples in Figure 5. It becomes evident that the mutual position of both trends has reversed and become almost parallel to each other. An increase in the double-normalized cutting forces in relation to the uncut chip thickness is apparent. This corresponds to literature references [30–32,34,48], and confirms the versatility of the fracture toughness theory of the wood cutting process as proposed by Atkinks [35] and Orlowski et al. [36]. This theory, based on a linear regression describing the phenomenon of an increase in the cutting forces with an increase in the uncut chip thickness, allows determination of the values of fracture toughness and shear yield stress in the shear plane [35,36].

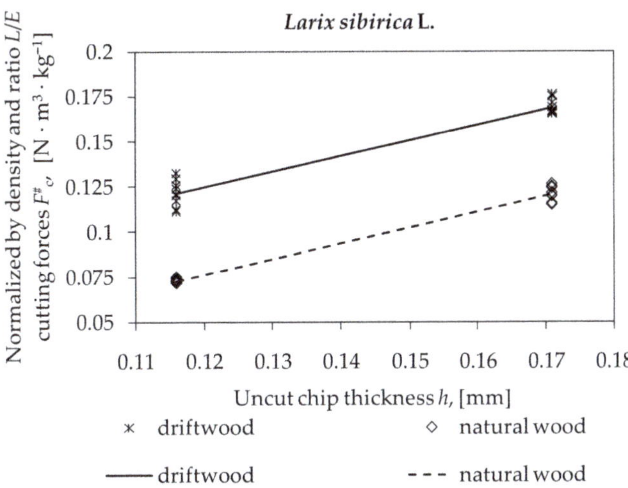

Figure 5. Relation between double normalized (by density and by ratio L/E) cutting forces per one cutting tooth and uncut chip thickness when sawing drifting and natural larch woods.

The level of difference between cutting force values obtained for driftwood and natural wood is evident and amounts to approximately 0.05 $N{\cdot}m^3{\cdot}kg^{-1}$ for both smaller and larger uncut chip thicknesses. These differences are statistically significant, as tested by the multi-factor ANOVA (Table 4). It can be stated, therefore, that the double normalization removes the effect of density and of late to early wood ratio from the cutting force quantifiers.

Table 4. Significance of differences between values of double normalized cutting forces obtained for two levels of uncut chip thickness while sawing process of drifting and natural larch samples (ANOVA) (α = 0.05). DL—driftwood larch, L—natural larch.

Larch Wood (*Larix siberica* L.)							
Sample Code	**Source**	**DF**	**Adj SS**	**Adj MS**	**F—Value**	**P—Value**	**F—Critical**
Double normalized cutting forces, $F^{\#}_c(h_1)$ ($N{\cdot}m^3{\cdot}kg^{-1}$)							
DL	between groups	1	0.012	0.012	487.4	1.74×10^{-14}	4.414
L	within groups	18	0.0004	2.4×10^{-5}	-	-	-
total		19	0.012	-	-	-	-
Double normalized cutting forces, $F^{\#}_c(h_2)$ ($N{\cdot}m^3{\cdot}kg^{-1}$)							
DL	between groups	1	0.011	0.011	670.1	1.08×10^{-15}	4.414
L	within groups	18	0.0003	1.72×10^{-5}	-	-	-
total		19	0.012	-	-	-	-

Such significant differences in the double-normalized cutting force values as presented in Figure 5 indicate that the long period of wood contact with the Arctic Sea water, in combination with the effect of low temperatures of the freezing Arctic ice, significantly alter the machinability of larch wood. It is evident that the double-normalized cutting forces are higher in the case of driftwood. Based on previous research, this might be related to the higher mineral (or ash) content deposited in driftwood [23]. It was reported by Lhate et al. [49] that a high mineral content may be attributed to poor machining properties by increasing of the friction between the tool and the processed wood. This results in increased values of cutting forces, elevated heat release and intensive cutting edge dullness. The other explanation for higher values of double-normalized cutting forces noticed in the case of sawing driftwood may be analogous to those observed by Orłowski and Sandak [50] and Orlowski et al. [51] during cutting of frozen pine wood. An increase in cutting forces was associated with the effect of frozen water crystals deposited inside of wood fibers that provided additional strength and material stiffness. Likewise, the presence of sodium chloride, among other minerals, inside driftwood fibers was reported by Cotana et al. [22] and Bartocci et al. [21]. Such crystals may increase the strength of the cell wall structure, which results in higher cutting forces.

The statistically significant increase in double-normalized cutting forces while processing driftwood is a confirmation of the modification of the native wood properties induced during its journey in the Arctic sea. It remains an open question, however, to what extent such natural modification has an effect on other properties of the drifted wood. Further studies are also necessary to confirm if such modification remains beneficial for the use of drifted wood in diverse applications.

4. Conclusions

The presented research allows the derivation of the following conclusions:

- The long period of larch wood logs' exposure to the Arctic Sea water, combined with the effect of cyclic freezing conditions of Arctic ice, resulted in the natural modification of driftwood and the alteration of several material properties. Exploring the exact causes of this phenomenon requires more detailed research and a broad reference sample set covering the whole variance of altered material properties. The new knowledge reported here may result, however, in a better management of driftwood. The appropriate selection and optimization of manufacturing processes may lead to the superior quality of protective or decorative coatings on products made of drifted wood.
- The wood modification observed in driftwood affects the overall machinability properties of this material, revealed as changes of double-normalized cutting forces. It is known that elevated cutting forces result in a deterioration of the machined surface smoothness. As a consequence, the adhesion of coatings on these machined surfaces is altered.
- The ratio of late and early woods within the annual ring is an important factor affecting the values of cutting forces. A higher content of late wood increases the cutting power required for wood sawing. Special attention should be directed toward controlling the L/E ratio while researching the effect of wood modification on machinability.
- The double normalization of cutting forces by density and followed by *L/E* ratio allows the direct comparison of machinability properties not possible for wood samples with different properties of annual growth.

The obtained results showed that the drifting of larch wood across the Arctic Sea significantly increases the cutting force values during the frame sawing process. However, the details of the effect of this natural modification on other wood properties (chemical, mechanical and physical) are still unknown. Further research is therefore being conducted by authors to reveal the mechanisms of material changes due to the drifting of wood in the Arctic Sea.

Author Contributions: Conceptualization, D.C., K.A.O., A.S. and J.S.; methodology, D.C., K.A.O. and O.E.; formal analysis, D.C., O.E. and M.L.; investigation, D.C. and K.A.O.; data curation, D.C.; writing—original draft preparation, D.C., A.S., K.A.O. and J.S.; writing—review and editing, D.C., A.S., K.A.O. and J.S.; supervision, K.A.O., J.S. and A.S.; project administration, D.C.; funding acquisition, D.C. and K.A.O. All authors have read and agreed to the published version of the manuscript.

Funding: This research was funded by the Ministry of Science and Higher Education, Poland, for funding the maintenance of scientific and research equipment—PRW-15M frame saw, Grant No. 21/E-359/SPUB/SP/2019.

Institutional Review Board Statement: Not applicable.

Informed Consent Statement: Not applicable.

Data Availability Statement: The data presented in this study are available on request from the corresponding author. The data are not publicly available due to still ongoing research activities.

Acknowledgments: The authors would like to acknowledge Rögnvaldur Harðarson and Harðar Benediktsson from Trésmiðja H Ben ehf (Akureyri, Iceland) and Piotr Taube from Sylva Sp. z o. o. (Wiele, Poland) for their support and wood samples used in the experiment. The authors gratefully acknowledge the Gdańsk University of Technology for funding the project "Golden Membership in Living Lab InnoRenew" in the program Hydrogenium Supporting Membership In International Networks supported under the umbrella of Excellence Initiative—Research University by the Ministry of Science and Higher Education of Poland. The authors gratefully acknowledge the European Commission for funding the InnoRenew project (Grant Agreement # 739574) under the Horizon2020 Widespread-Teaming program, the Republic of Slovenia (investment funding from the Republic of Slovenia and the European Union European Regional Development Fund) and infrastructural ARRS program IO-0035. "Delivering fingertip knowledge to enable service life performance specification of wood"— CLICK DESIGN (N°773324) was supported under the umbrella of ERA-NET Cofund ForestValue by the Ministry of Education, Science and Sport of the Republic of Slovenia. ForestValue has received funding from the European Union's Horizon 2020 research and innovation program.

Conflicts of Interest: The authors declare no conflict of interest.

References

1. Mooney, D.E. Examining possible driftwood use in Viking Age Icelandic boats. *Nor. Archaeol. Rev.* **2016**, *49*, 156–176. [CrossRef]
2. The First Churches Were Made of Driftwood and Icelandic Birch. Available online: https://english.hi.is/the_first_churches_were_made_of_driftwood_and_icelandic_birch (accessed on 14 December 2020).
3. Mooney, D.E. A 'North Atlantic island signature' of timber exploitation: Evidence from wooden artefact assemblages from Viking Age and Medieval Iceland. *J. Archaeol. Sci. Rep.* **2016**, *7*, 280–289. [CrossRef]
4. Mooney, D.E. Does the 'Marine Signature' of driftwood persist in the archaeological record? An experimental case study from Iceland. *Environ. Archaeol.* **2018**, *23*, 217–227. [CrossRef]
5. Eggertsson, O. Origin of the driftwood on the coasts of Iceland: A dendrochronological study. *Jökull* **1993**, *43*, 15–32.
6. Peltola, H.; Kellomäki, S.; Hassinen, A.; Granander, M. Mechanical stability of Scots pine, Norway spruce and birch: An analysis of tree-pulling experiments in Finland. *For. Ecol. Manag.* **2000**, *135*, 143–153. [CrossRef]
7. Danielewicz, D.; Surma-Ślusarska, B. Properties and fibre characterisation of bleached hemp, birch and pine pulps: A comparison. *Cellulose* **2017**, *24*, 5173–5186. [CrossRef]
8. Hellmann, L.; Tegel, W.; Geyer, J.; Kirdyanov, A.V.; Nikolaev, A.N.; Eggertsson, Ó.; Altman, J.; Reinig, F.; Morganti, S.; Wacker, L.; et al. Dendro-provenancing of Arctic driftwood. *Quat. Sci. Rev.* **2017**, *162*, 1–11. [CrossRef]
9. Hellmann, L.; Tegel, W.; Eggertsson, Ó.; Schweingruber, F.H.; Blanchette, R.; Kirdyanov, A.; Gärtner, H.; Büntgen, U. Tracing the origin of Arctic driftwood. *J. Geophys. Res. Biogeosci.* **2013**, *118*, 68–76. [CrossRef]
10. Hellmann, L.; Tegel, W.; Kirdyanov, A.V.; Eggertsson, Ó.; Esper, J.; Agafonov, L.; Nikolaev, N.A.; Knorre, A.A.; Myglan, V.S.; Churakova, O.; et al. Timber logging in Central Siberia is the main source for recent Arctic driftwood. *Arct. Antarct. Alp. Res.* **2015**, *47*, 449–460. [CrossRef]
11. Hellmann, L.; Kirdyanov, A.V.; Büntgen, U. Effects of Boreal timber rafting on the composition of Arctic driftwood. *Forests* **2016**, *7*, 257. [CrossRef]
12. Steelandt, S.; Marguerie, D.; Bhiry, N.; Delwaide, A. A study of the composition, characteristics, and origin of modern driftwood on the western coast of Nunavik (Quebec, Canada). *J. Geophys. Res. Biogeosci.* **2015**, *120*, 480–501. [CrossRef]
13. Hole, G.M.; Macias-Fauria, M. Out of the woods: Driftwood insights into Holocene pan-Arctic sea ice dynamics. *J. Geophys. Res. Oceans* **2017**, *122*, 7612–7629. [CrossRef]
14. Dalaiden, Q.; Goosse, H.; Lecomte, O.; Docquier, D. A model to interpret driftwood transport in the Arctic. *Quat. Sci. Rev.* **2018**, *191*, 89–100. [CrossRef]
15. An International Team of Scientists Studying Driftwood along Icelandic Shores. Available online: https://icelandmag.is/article/international-team-scientists-studying-driftwood-along-icelandic-shores (accessed on 10 January 2021).
16. Komorowicz, M.; Wróblewska, H.; Fojutowski, A.; Kropacz, A.; Noskowiak, A.; Gajek, G.; Franczak, Ł.; Łęczyński, L. Properties of driftwood from Bellsund coast (Svalbard): Preliminary results. New perspectives in polar research. In *Proceedings of the 35th Polar Symposium Diversity and State of Polar Ecosystems, Wrocław, Poland, 4–7 June 2014*; Migała, K., Owczarek, P., Kasprzak, M., Mateusz, M.C., Eds.; Institute of Geography and Regional Development, University of Wrocław: Wroclav, Poland, 2014.
17. Blanchette, R.A.; Held, B.W.; Hellmann, L.; Millman, L.; Büntgen, U. Arctic driftwood reveals unexpectedly rich fungal diversity. *Fungal Ecol.* **2016**, *23*, 58–65. [CrossRef]

18. Rämä, T.; Davey, M.L.; Nordén, J.; Halvorsen, R.; Blaalid, R.; Mathiassen, G.H.; Alsos, I.G.; Kauserud, H. Fungi sailing the Arctic Ocean: Speciose communities in North Atlantic driftwood as revealed by high-throughput amplicon sequencing. *Microb. Ecol.* **2016**, *72*, 295–304. [CrossRef]
19. Rämä, T.; Hassett, B.T.; Bubnova, E. Arctic marine fungi: From filaments and flagella to operational taxonomic units and beyond. *Bot. Mar.* **2017**, *60*, 433–452. [CrossRef]
20. Kunttu, P.; Pasanen, H.; Rämä, T.; Kulju, M.; Kunttu, S.-M.; Kotiranta, H. Diversity and ecology of aphyllophoroid fungi on driftwood logs on the shores of the Baltic Sea. *Nord. J. Bot.* **2020**, e02735. [CrossRef]
21. Bartocci, P.; Barbanera, M.; D'Amico, M.; Laranci, P.; Cavalaglio, G.; Gelosia, M.; Ingles, D.; Bidini, G.; Buratti, C.; Cotana, F.; et al. Thermal degradation of driftwood: Determination of the concentration of sodium, calcium, magnesium, chlorine and sulfur containing compounds. *Waste Manag.* **2016**, *60*, 151–157. [CrossRef]
22. Cotana, F.; Buratti, C.; Barbanera, M.; Cavalaglio, G.; Foschini, D.; Nicolini, A.; Pisello, A.L. Driftwood biomass in Italy: Estimation and characterization. *Sustainability* **2016**, *8*, 725. [CrossRef]
23. Tsai, W.T.; Tsai, Y.-L.; Liu, S.-C. Utilization of driftwood as an energy source and its environmental and economic benefit analysis in Taiwan. *BioResources* **2011**, *6*, 4781–4789.
24. Shaw, J.D. Economies of driftwood: Fuel harvesting strategies in the Kodiak Archipelago. *Études/Inuit/Studies* **2012**, *36*, 63–88. [CrossRef]
25. Salca, E.A.; Krystofiak, T.; Lis, B. Evaluation of selected properties of alder wood as functions of sanding and coating. *Coatings* **2017**, *7*, 176. [CrossRef]
26. Vitosyte, J.; Ukvalbergiene, K.; Keturakis, G. The effects of surface roughness on adhesion strength of coated ash (*Fraxinus excelsior* L.) and birch (*Betula* L.) wood. *Mater. Sci.* **2012**, *18*, 347–351. [CrossRef]
27. Meijer, M.; Thurich, K.; Militz, N. Comparative study on penetration characteristics of modern wood coatings. *Wood Sci. Technol.* **1998**, *32*, 347–365. [CrossRef]
28. Hidden Wood—Driftwood in Design. Available online: https://hadesignmag.is/2015/08/10/hidden-wood-driftwood-in-design/?lang=en (accessed on 14 December 2020).
29. Wasielewski, R.; Orlowski, K. Hybrid dynamically balanced saw frame drive. *Holz als Roh- und Werkst.* **2002**, *60*, 202–206. [CrossRef]
30. Chuchala, D.; Ochrymiuk, T.; Orlowski, K.A.; Lackowski, M.; Taube, P. Predicting cutting power for band sawing process of pine and beech wood dried with the use of four different methods. *BioResources* **2020**, *15*, 1844–1860. [CrossRef]
31. Sinn, G.; Chuchała, D.; Orlowski, K.A.; Taube, P. Cutting model parameters from frame sawing of natural and impregnated Scots pine (*Pinus sylvestris* L.). *Eur. J. Wood Wood Prod.* **2020**, *78*, 777–784. [CrossRef]
32. Chuchala, D.; Sandak, J.; Orlowski, K.A.; Muzinski, T.; Lackowski, M.; Ochrymiuk, T. Effect of the drying method of pine and beech wood on fracture toughness and shear yield stress. *Materials* **2020**, *13*, 4692. [CrossRef]
33. Chuchala, D.; Orlowski, K.A.; Sandak, A.; Sandak, J.; Pauliny, D.; Barański, J. The effect of wood provenance and density on cutting forces while sawing Scots pine (*Pinus sylvestris* L.). *BioResources* **2014**, *9*, 5349–5361. [CrossRef]
34. Licow, R.; Chuchala, D.; Deja, M.; Orlowski, K.A.; Taube, P. Effect of pine impregnation and feed speed on sound level and cutting power in wood sawing. *J. Clean. Prod.* **2020**, *272*, 122833. [CrossRef]
35. Atkins, A. Toughness and cutting: A new way of simultaneously determining ductile fracture toughness and strength. *Eng. Fract. Mech.* **2005**, *72*, 849–860. [CrossRef]
36. Orlowski, K.A.; Ochrymiuk, T.; Sandak, J.; Sandak, J. Estimation of fracture toughness and shear yield stress of orthotropic materials in cutting with rotating tools. *Eng. Fract. Mech.* **2017**, *178*, 433–444. [CrossRef]
37. Sharapov, E.; Brischke, C.; Militz, H. Effect of grain direction on drilling resistance measurements in wood. *Int. J. Archit. Herit.* **2020**. [CrossRef]
38. Koizumi, A.; Kitagawa, M.; Hirai, T. Effects of growth ring parameters on mechanical properties of Japanese larch (*Larix kaempferi*) from various provenances. *Eurasian J. For. Res.* **2005**, *8*, 85–90.
39. Zhu, J.; Nakano, T.; Hirakawa, Y. Effect of growth on wood properties for Japanese larch (*Larix kaempferi*): Differences of annual ring structure between corewood and outerwood. *J. Wood Sci.* **1998**, *44*, 392–396. [CrossRef]
40. Luostarinen, K.; Heräjärvi, H. Dependence of shear strength on wood properties in cultivated *Larix sibirica*. *Wood Mater. Sci. Eng.* **2011**, *6*, 177–184. [CrossRef]
41. Luostarinen, K.; Heräjärvi, H. Relation of arabinogalactans to density, growth rate and shear strength in wood of cultivated Siberian larch. *Eur. J. Wood Wood Prod.* **2013**, *71*, 29–36. [CrossRef]
42. Mikkola, M.T.; Korhonen, R.K. Effect of latewood proportion on mechanical properties of Finnish pine wood modified with compression drying. *Wood Fiber Sci.* **2013**, *45*, 335–342.
43. Ishiguri, F.; Tumenjargal, B.; Baasan, B.; Jigjjav, A.; Pertiwi, Y.A.B.; Aiso-Sanada, H.; Takashima, Y.; Iki, T.; Ohshima, J.; Iizuka, K.; et al. Wood properties of *Larix sibirica* naturally grown in Tosontsengel, Mongolia. *Int. Wood Prod. J.* **2018**, *9*, 127–133. [CrossRef]
44. Neverov, N.A.; Belyaev, V.V.; Chistova, Z.B.; Kutinov, Y.G.; Staritsyn, V.V.; Polyakova, E.V.; Mineev, A.L. Effects of geo-ecological conditions on larch wood variations in the North European part of Russia (Arkhangelsk region). *J. For. Sci.* **2017**, *63*, 192–197. [CrossRef]
45. Tumenjargal, B.; Ishiguri, F.; Aiso-Sanada, H.; Takahashi, Y.; Baasan, B.; Chultem, G.; Ohshima, J.; Yokota, S. Geographic variations of wood properties of *Larix sibirica* naturally grown in Mongolia. *Silva Fenn.* **2018**, *52*, 10002. [CrossRef]

46. Horacek, M.; Jakusch, M.; Krehan, H. Control of origin of larch wood: Discrimination between European (Austrian) and Siberian origin by stable isotope analysis. *Rapid Commun. Mass Spectrom.* **2009**, *23*, 3688–3692. [CrossRef] [PubMed]
47. Sykacek, E.; Gierlinger, N.; Wimmer, R.; Schwanninger, M. Prediction of natural durability of commercial available European and Siberian larch by near-infrared spectroscopy. *Holzforschung* **2006**, *60*, 643–647. [CrossRef]
48. Hlásková, L.; Kopecký, Z.; Novák, V. Influence of wood modification on cutting force, specific cutting resistance and fracture parameters during the sawing process using circular sawing machine. *Eur. J. Wood Wood Prod.* **2020**, *78*, 1173–1182. [CrossRef]
49. Lhate, I.; Cuvilas, C.; Terziev, N.; Jirjis, R. Chemical composition of traditionally and lesser used wood species from Mozambique. *Wood Mater. Sci. Eng.* **2010**, *5*, 143–150. [CrossRef]
50. Orlowski, K.A.; Sandak, J. Analysis of specific cutting resistance while cutting frozen pine blocks with narrow-kerf stellite tipped saws on frame sawing machines. In Proceedings of the COST E35 Workshop on Processing of Frozen Wood, Lappeenranta University of Technology, Lappeenranta, Finland, 16–17 June 2005.
51. Orlowski, K.; Sandak, J.; Negri, M.; Dzurenda, L. Sawing frozen wood with narrow kerf saws: Energy and quality effects. *For. Prod. J.* **2009**, *59*, 79–83.

Article

Thermal Transmittance, Dimensional Stability, and Mechanical Properties of a Three-Layer Laminated Wood Made from Fir and Meranti and Its Potential Application for Wood-Frame Windows

Dimitrios Koutsianitis [1], Konstantinos Ninikas [1], Andromachi Mitani [1], George Ntalos [1], Nikolakakos Miltiadis [1], Argyris Vasilios [1], Hamid R. Taghiyari [2] and Antonios N. Papadopoulos [3,*]

1 Department of Forestry, Wood Science and Design, University of Thessaly, GR-43100 Karditsa, Greece; dkoutsianitis@uth.gr (D.K.); kninikas@uth.gr (K.N.); amitani@uth.gr (A.M.); gntalos@uth.gr (G.N.); miltosnikolak@gmail.com (N.M.); argbill@sch.gr (A.V.)
2 Wood Science and Technology Department, Faculty of Materials Engineering & New Technologies, Shahid Rajaee Teacher Training University, Tehran 16788-15811, Iran; htaghiyari@sru.ac.ir
3 Laboratory of Wood Chemistry and Technology, Department of Forestry and Natural Environment, International Hellenic University, GR-66100 Drama, Greece
* Correspondence: antpap@for.ihu.gr

Citation: Koutsianitis, D.; Ninikas, K.; Mitani, A.; Ntalos, G.; Miltiadis, N.; Vasilios, A.; Taghiyari, H.R.; Papadopoulos, A.N. Thermal Transmittance, Dimensional Stability, and Mechanical Properties of a Three-Layer Laminated Wood Made from Fir and Meranti and Its Potential Application for Wood-Frame Windows. *Coatings* **2021**, *11*, 304. https://doi.org/10.3390/coatings11030304

Academic Editor: Anna Sandak

Received: 8 February 2021
Accepted: 3 March 2021
Published: 7 March 2021

Abstract: The aim of this paper was to investigate the physical (thermal transmittance and dimensional stability) and mechanical properties of two types of three layer laminated wood made from fir and meranti; fir in surface layers and meranti in core (FMF) and vice versa (MFM) and to examine its potential application for wood-frame windows. An additional objective was to compare the properties of the laminated wood with those of solid wood, namely meranti and fir. Both types of laminated wood had by far substantial lower bending properties than solid wood. MFM laminated wood performed better than the FMF as far as the physical and mechanical properties are concerned. Water absorption and thickness swelling of MFM laminated wood were substantially lower than those of the FMF type, and all the differences were statistically significant. Longitudinal width swelling, and bending properties of MFM laminated wood were higher than those of FMF but these differences were not statistically significant. The thermal transmittance (rate of the heat transferred) of the FMF window is 13.3% better (less) compared to the MFM window. The main reason for this is believed to be the lower overall density of the FMF window, which also makes it more competitive as a result of the reduced manufacturing cost since fir is less expensive compared tomeranti. It was concluded that wood-frame windows can be successfully made from these types of laminated wood, employing therefore easily renewable materials, with low environmental impact, recyclable and manageable in the medium term.

Keywords: laminated wood; meranti; fir; physical properties; mechanical properties; wood-frame windows; thermal transmittance

1. Introduction

An important distinction between timber and other structural materials is that timber is quite difficult to meet performance requirements, whereas man-made products like steel and concrete can be easily modified through the manufacturing process for a specified use. In addition, the currently growing demand for engineered wood products is largely attributable to their outstanding ecological performance. The substitution of materials having a larger ecological footprint, such as steel and concrete, by structural engineered wood products like laminated wood has proven to be effective in minimizing the environmental impacts of the building sector [1]. Furthermore, laminated wood is a material of a great aesthetic value, with various applications in the field of building components, such as doors and windows, and in the field of timber structures of big and small spans [2,3].

Nearly any species or mixed-species combination can be used to produce laminated wood provided its physical and mechanical properties are suitable and the timbers can be glued together. According to Moody and Hernandez [4], species and mixed-species combination commonly used for glued-laminated timber in the United States include southern pine (*Pinus* spp.), Douglas fir (*Pseudotsugamenziesii*), larch (*Larixoccidentalis*), hemlock (*Tsugaheterophylla*), and spruce (*Picea* spp.). Red maple (*Acer* spp.) species was also used by Janowiak et al. [5] to study the performance of glulam beam made from two distinct timber resources, namely sawn logs and lower-grade, smaller-dimension timber. Komariah et al. [6] developed glulam made from sengon, manii, and mangium, with either the same wood species being used for all layers or mangium being used for the face and back layers, with a core layer of manii or sengon. The physical and mechanical properties of the glulam did not show any significant difference from that of solid wood of the same species. Such samples successfully fulfilled the JAS 234 (Japanese Agricultural Standard 2003) standard [7]. Similar results were reported by Hayashi and Miyatake [8] for structural laminated timber made from Japanese cedar.

The tremendous diversity of tropical hardwoods available for structural applications significantly compounds the complexity of matching a particular species of timber with specific performance requirements. The number of merchantable species of timber available has been reported at 650 for Malaysia [9] and 2500 in the Amazon [10]. There is a tendency in most regions to use clear material from a few species for which considerable experience is available on long-term structural performance. Corigliano et al. [11] performed theoretical and experimental analyses of Iroko wood laminates. The experimental results demonstrated that the presence of scarf joints only affected the strength of the glued laminate while the stiffness properties remained the same. Various bamboo species have also been used for the manufacture of laminated wood [12].

Meranti is a common name applied commercially to four groups of species from the genus *Shorea*. The four groups of meranti are separated on the basis of heartwood color and weight. About 70 species of *Shorea* belong to the light and dark red meranti groups, 22 species to the white meranti group, and 33 species to the yellow meranti group. *Shoreapauciflora*, a dark red meranti, is very commonly used in southeast Asia, and there is an abundance of variety between the difference species.Few studies have been published regarding the application of meranti as a raw material for laminated wood manufacture. Hartono and Sucipto [13] made three layer laminated wood from oil palm trunk in the core, a low quality raw material, and applied high density woods in the surface layers, namely sengon and meranti. They reported strong relationship between density and bending properties and they concluded that the best combination was that of oil palm trunk in the core layer and meranti in the surface layers. Modulus of rupture (MOR) and modulus of elasticity (MOE) values for laminated wood made purely from oil palm trunk was 168.79 and 30,115 (Kg/cm^2), whereas the ones made from oil palm trunk in the core layer and meranti in the surface layers was 438.29 and 100,454 respectively. Puluhulawa et al. [14] manufactured three-layer laminated wood using a light wood in the core (mahang) and meranti in the surface layers. They reported that the laminated wood had by far substantial lower bending properties than meranti solid wood. Ong [15] investigated the performance of glue-laminated beams from Malaysian dark red meranti timber. Phenol resorcinol formaldehyde, commonly used in structural glulam production, was used in the fabrication of finger joints and laminations of the glulam beams. Overall, it was found that dark red merantifinger joints exhibited better bending strength than spruce finger joints which represented softwood used in European glulam. Wood density and end pressure were shown to affect the strength properties of the finger joints.

It was mentioned earlier that physical and mechanical properties of the low-quality timber can be improved by processing it into glued laminated wood or to be combined with tropical woods to obtain value added wood products. Consequently, the aim of this paper was two-fold:

(1) To investigate the physical (thermal transmittance and dimensional stability) and mechanical properties of two types of three layer laminated wood made from fir and meranti; fir in surface layers and meranti in core and vice versaand to examine its potential application for wood-frame windows.

(2) An additional objective was to compare the properties of the laminated wood with those of solid wood, namely meranti and fir.

2. Materials and Methods

2.1. Specimen Preparation

Fir (*Abies alba* Mill) and meranti (*Shoreapauciflora*) boards were supplied by a local plant (Halucom, Karditsa, Greece); they were kept in the wood workshop for five months before being cut to size, 360 mm in length, 25 mm in width (plain sawn), and 20 mm in thickness. Specimens were free from any fungal or insect attack, checks or cracks, and knots. The density of fir and meranti specimens was measured to be 0.43 and 0.47 $Kg{\cdot}m^{-3}$, respectively, at moisture contents of 11.56% and 10.13% respectively. Physical and mechanical properties—namely density, water absorption after 24, 48 and 72 h immersion in water, and swelling (both longitudinal, width and thickness) after 24, 48 and 72 h, modulus of rupture (MOR), and modulus of elasticity (MOE)—were determined in accordance with ASTM D0143-94 [16]. Tests were carried out at room conditions (25 ± 3 °C; 35% ± 3% relative humidity).

2.2. Manufacture of Laminated Wood

Fir and meranti specimens, 20 × 25 × 360 mm^3 (thickness × width × length), were conditioned at 20 °C temperature and 65% relative humidity until the weight of the wood was stabilized. Two types of three-layer laminated wood produced from fir and meranti, namely fir in surface layers and meranti in core (FMF), and vice versa (MFM), as depicted in Figure 1. Laminated wood made from meranti in both core and surface layers was not manufactured due to the high cost involved. On the other hand, laminated wood made from fir in both core and surface layers was not manufactured, since it was expected to show very low dimensional stability. A two-component polyvinyl acetate wood adhesive for outdoor use was applied, with the following properties. viscosity: 11,000−14,000 mPas (Brookfield RVT, 20 rpm, at 23 °C), pH value: approx. 4−5.5, at 20 °C, density: ca. 1.13 g/cm^3 at 20 °C, mixing rate adhesive/hardener (by weight) 15:1, mixture pH value: approximately 3.2.Polyvinyl acetate (PVAc) based glue was applied by brush on one side of the joints. PVAc was selected due its low cost and its easy applicability. Once glued, they were kept together for 60 s, using a manually operated press. The wood samples were subjected to a constant hydraulic pressure of 5 bar (per surface) and an ambient temperature of 20 °C for 12 h. The laminated wood specimens were conditioned and cut to the final dimensions of 20 × 75 × 360 mm^3 (thickness × width × length). Bending strength (modulus of rupture; MOR and modulus of elasticity; MOE) of the specimens was determined according to the procedures detailed in the EN 385:2001 [17]. Also, physical properties, namely density, water absorption after 24, 48, and 72 h and swelling (both longitudinal, width, and thickness) after 24, 48, and 72 h immersion in water, were determinedaccording to the procedures detailed in the EN 385:2001.

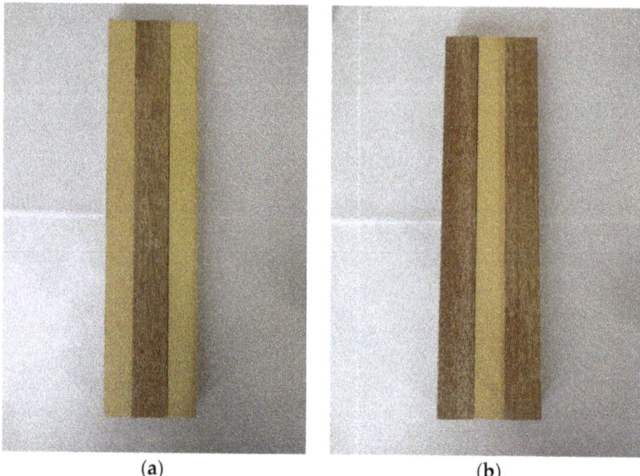

(**a**) (**b**)

Figure 1. Types of laminated wood; fir in surface layers and meranti in core (FMF—**a**) and meranti in surface layers and fir in core (MFM—**b**).

2.3. Thermal Transmittance Determination

Commercial wood frame windows manufactured with the two types of the investigated laminated wood, namely fir in surface layers and meranti in core (FMF) and vice versa (MFM), were supplied from a local plant and their thermal transmittance is determined. The thermal transmittance of solid wood, fir and meranti, is also determined.

The thermal transmittance, also referred as U-value, is the heat flow rate divided by the area and temperature difference in the surroundings of both sides of a system at a steady state [18] and is a concept employed to describe the insulation properties for the building materials, in our case for windows. The inside pane of a double-glazed window captivates heat from the inner side and transmits it through conduction and convection to the cooler pane (outside of the window) [19]. Heat conduction in windows is ruled by the material of the frame. Wood has a rather high thermal insulation properties (average thermal conductivity λw = 0.13 W/mK) [20–24]. This, in a wood window frame, results in less heat loss compared to the window frames constructed from other materials which have greater thermal conductivity values.

A heat flux measuring device (Testo North America, West Chester, USA) (HFM-Testo 635-2/ accuracy: ±0.2 °C/operating temperature: −20 to +50 °C with a wireless humidity/temperature probe) has been used to calculate the window's thermal transmittance. This apparatus measures the specific heat flow (q) and the air temperatures for the two sides (T_w'warm' and T_c'cold') of the window with the assistance of a wireless temperature probe [25]. Having reading from these three parameters, the U-value is being calculated automatically using the Equation (1)

$$U = \frac{q}{(T_w - T_c)} \left(\mathrm{W/m^2 \cdot K}\right) \tag{1}$$

The external dimensions of the window frame for the two sample windows were 1000 × 1000 mm^2 with a frame height of 100 mm and frame thickness of 68 mm. The double glazed window had the following setup: "4-12-4". This is translated into 4 mm glass thickness and 12 mm gap between the two glass panes. Double glazed windows are principally the combination of double glass window panes separated by an air or other inert gas filled space in order to reduce heat loss through a part of the building envelope.

The thermal transmittance (U-value) readings for the two windows were undertaken according to the ISO 9869-1:2014 [18] and to the BS EN ISO 10077-2 [26]. The required parameters to calculate the thermal transmittance of a window (U_w) are the following: The value of the window's frame (U_f), the value of the double-glazed pane (U_g) and also the length (area) of the window frame and the glass.

At both windows, the same thermal insulating glass was used with a thermal transmittance value given by the manufacturer: U_g = 1.8 W/m^2·K.

The calculation method of the window's thermal transmittance is based on the following Equation (2) explained with details in Figure 2:

$$U_w = \frac{\Sigma A_g U_g + \Sigma A_f U_f + \Sigma l_g \Psi_g}{\Sigma A_g + \Sigma A_f} \left(\mathrm{W/m^2{\cdot}K}\right) \quad (2)$$

where:

ΣA_g: summation of the window's glass pane area (m^2)
ΣA_f: summation of the window's frame area (m^2)
U_g: thermal transmittance of the glass pane (W/m^{2}K)
U_f: thermal transmittance of the window's frame (W/m^{2}K)
Σl_g: summation of the window's glass pane length—perimeter (m)
Ψ_g: linear heat transfer coefficient of the glass pane (W/mK)
Incorporating the above data on to the Equation (2),we can derive:

$$U_w = \frac{\Sigma A_g U_g + \Sigma A_f U_f + \Sigma l_g \Psi_g}{\Sigma A_g + \Sigma A_f} = \frac{0.64 \times 1.8 + 0.36 \times U_f + 3.2 \times 0.08}{0.64 + 0.36} \left(\mathrm{W/m^2{\cdot}K}\right) \quad (3)$$

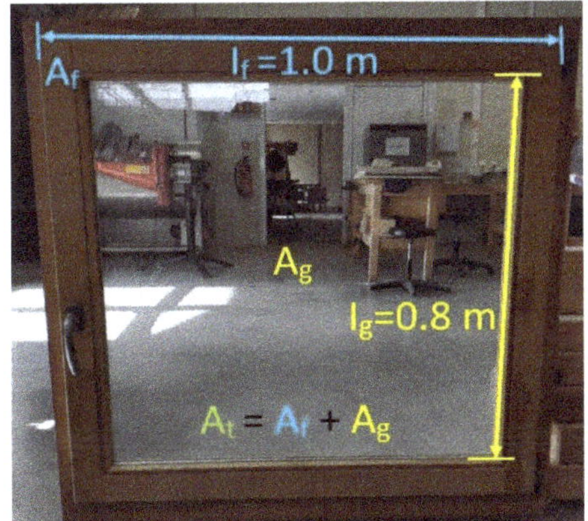

U_g = 1.8 W/(m^2·K) (manufacturer)
U_f = window frame thermal transmittance
A_g = 0.8 × 0.8m= 0.64 m^2
A_f = A_t − A_g = 1^2 − 0.64 = 0.36 m^2(A_t: window's total area)
Ψ_g = 0.08 (for a timber frame with a double glazed window with air between the two panes)

Figure 2. The tested wood-frame window.

The calculation of the window frame's thermal transmittance (U_f) was based on the following setup. This was achieved with the use of window testing apparatus where the 'cold' side temperature is being adjusted electronically by a wireless probe. The view of the tested windows mounted into the 'window tester' is given in Figure 3. The temperature of the 'warm' side, is also adjustable electronically by a digital thermostat.

Before starting the experiments for the window frame's thermal transmittance (U_f), the 'cold' and 'warm' side temperatures were adjusted at 6 °C (±1 °C) and 15 °C (±1 °C) respectively and remained within these boundaries for 76 h. The three HFM's thermocouples were positioned in a 0.10 m triangle setup to capture the heat flow connected to the HFM (Figure 3). During this period, the U-values for each window were recorded by the HFM device.

FMF MFM

Figure 3. The HFM device with the attached thermocouples (green-cable edges at the yellow circle) taking reading for a 72 h period.

2.4. Statistical Analysis

SAS software program was used to carry out statistical analysis in the present study (version 9.2; 2010). To discern significant difference among different treatments and produced panels, one-way analysis of variance was performed at 95% level of confidence. Then, Duncan's multiple range test (DMRT) was done for grouping among treatments for each property. In order to find degrees of similarities among different treatments based on all properties studied here, Hierarchical cluster analysis from SPSS/18 (2010) software (version 9.2) was used. In a cluster analysis, there is a scale bar on top; those treatments that are connected at lower scale levels (that is, lower numbers on the scale bar) by a vertical line are considered to be more similar, in comparison to those that are connected at higher scale levels. The maximum scale (number 25) indicates that treatments have practically very little in common and therefore, they are to be considered as two completely different ones [27]. In DMRT, different treatments are grouped based on only one property. However, cluster analysis can do groupings among different treatments based on more than one property. This capability provides useful information for managers who are to make decisions based on all aspects and perspectives, and to choose the most appropriate option for their production program. This type of statistical analysis can be used for all kinds of materials, including wood species, wood-based composites, papers, and even cloths [28,29]. For graphical statistics (fitted-line, contour, and surface plots), Minitab software was utilized (version 16.2.2; 2010).

3. Results

The moisture content of the FMF and MFM laminated wood was determined to be 10.36% and 10.41% respectively and these values are nearly the same with those of solid wood, as mentioned in session 2.1. The density of the FMF and MFM laminated wood was 0.44 and 0.49 Kg·m^{-3} respectively, values close to the density of fir and meranti solid wood (0.43 and 0.47 Kg·m^{-3} respectively), a fact that indicates that the density of the laminated wood was not affected by the density of the solid wood.

The physical properties of both fir and meranti wood are presented in Table 1. Water absorption and thickness swelling of meranti specimens after 24, 48, and 72 h immersion in water were substantially lower than those of the fir wood and this difference is statistically significant. Table 2 shows the bending properties of the two wood species. It was revealed that both MOR and MOE values of fir wood were higher than those of meranti wood and the difference was statistically significant.

Table 1. Physical properties of fir and meranti wood.

Wood Species	Physical Properties [a]		
	Water Absorption (%)		
	24 h	**48 h**	**72 h**
fir	39.44 (±7.75) [b] A [c]	51.02 (±9.81) A	58.84 (±10.9) A
meranti	18.05 (±0.70) B	23.08 (±0.70) B	25.84 (±0.68) B
Wood Species	Longitudinal Swelling (Axial) (%)		
	24 h	48 h	72 h
fir	0.06 (±0.01) A	0.10 (±0.09) A	0.12 (±0.09) A
meranti	−0.16 (±0.04) B	−0.11 (±0.04) B	−0.09 (±0.04) B
Wood Species	Width Swelling (Tangential) (%)		
	24 h	48 h	72 h
fir	2.03 (±0.97) A	3.08 (±0.95) A	3.3 (±1.27) A
meranti	0.31 (±0.02) B	1.41 (±0.42) B	1.87 (±0.34) B
Wood Species	Thickness Swelling (Radial) (%)		
	24 h	48 h	72 h
fir	1.79 (±0.86) A	2.46 (±0.73) A	2.8 (±0.73) A
meranti	0.67 (±0.031) B	1.67 (±0.29) B	2.26 (±0.36) B

[a] Each value is the mean of six replicates; [b] Standard deviation; [c] Letters on each column represent Duncan groupings at 95% level of confidence.

Table 2. Bending properties of fir and meranti wood.

Wood Species	Bending Properties [a]	
	MOR (N/mm²)	**MOE (N/mm²)**
fir	68.66 (±6.54) [b] A [c]	8796.73 (±1254.32) A
meranti	48.81 (±3.92) B	6771.6 (±780.16) B

[a] Each value is the mean of four replicates (determined in the radial direction of the solid wood samples); [b] Standard deviation; [c] Letters on each column represent Duncan groupings at 95% level of confidence.

Table 3 depicts the physical properties of the two types of laminated wood investigated in this study. It was revealed that those laminated specimens with fir wood on the surface layers had higher water absorption and thickness swelling values than those with meranti on the core layer. The bending properties of the two types of laminated wood are presented in Table 4. MFM laminated wood presented better bending properties than FMF. Going back to the Equation (2), the overall window thermal transmittance is as follows:

For the FMF window, as shown in Equation (4),

$$U_{w-FMF} = \frac{\Sigma A_g\ U_g\ +\Sigma A_f\ U_f\ +\Sigma l_g\ \Psi_g}{\Sigma A_g+\Sigma A_f} = \frac{0.64\times 1.8+\ 0.36\times\ U_f\ +3.2\times 0.08}{0.64+0.36} = \frac{0.64\times 1.8+\ 0.36\times 0.775+3.2\times 0.08}{0.64+0.36} = 1.687\ \mathrm{W/m^2{\cdot}K} \quad (4)$$

and for MFM window, as shown in Equation (5),

$$U_{w-MFM} = \frac{\Sigma A_g U_g+\ \Sigma A_f U_f+\Sigma l_g \Psi_g}{\Sigma A_g+\Sigma A_f} = \frac{0.64\times 1.8+\ 0.36\times U_f+3.2\times 0.08}{0.64+0.36} = \frac{0.64\times 1.8+\ 0.36\times 1.494+3.2\times 0.08}{0.64+0.36} = 1.946\ \mathrm{W/m^2{\cdot}K} \quad (5)$$

Table 3. Physical properties of the two types of laminated wood.

Laminated Wood	Physical Properties [a]		
	Water Absorption (%)		
	24 h	**48 h**	**72 h**
MFM	20.51 (±1.48) [b] A [c]	25.4 (±1.53) A	27.60 (±1.34) A
FMF	26.21 (±0.61) B	32.34 (±0.64) B	34.93 (±0.51) B
Laminated wood	Longitudinal Swelling (%)		
	24 h	48 h	72 h
MFM	−0.04 (±0.005) A	−0.02 (±0.004) A	0.00 (±0.004) A
FMF	0.00 (±0.003) A	0.01 (±0.003) A	0.03 (±0.004) A
Laminated wood	Width Swelling (%)		
	24 h	48 h	72 h
MFM	1.42 (±0.48) A	2.07 (±0.35) A	2.62 (±0.31) A
FMF	1.85 (±0.21) A	2.67 (±0.26) A	2.95 (±0.28) A
Laminated wood	Thickness Swelling (%)		
	24 h	48 h	72 h
MFM	0.06 (±0.063) A	0.86 (±0.090) A	2.43 (±0.75) A
FMF	2.61 (±1.14) B	4.79 (±1.04) B	5.41 (±0.83) B

[a] Each value is the mean of six replicates (The directions of the solid wood samples used for the manufacturing of the laminated wood were not predefined); [b] Standard deviation; [c] Letters on each column represent Duncan groupings at 95% level of confidence.

Table 4. Bending properties of the two types of laminated wood.

Laminated Wood	Bending Properties [a]	
	MOR (N/mm^2)	**MOE (N/mm^2)**
MFM	14.47 (±1.56) [b] A [c]	5659.51 (±697.41) A
FMF	13.81 (±1.51) A	4571.99 (±798.47) A

[a] Each value is the mean of four replicates. The directions of the solid wood samples used for the manufacturing of the laminated wood were not predefined; [b] Standard deviation; [c] Letters on each column represent Duncan groupings at 95% level of confidence.

In Table 5, the two window's thermal transmittance values are displayed.

Table 5. Thermal transmittance of the windows (U_W).

Window Type	Window's Thermal Transmittance U_W (W/m^2·K)
FMF	1.687
MFM	1.946

Contour plots of the four treatments based on water absorption after 24 h immersion in water and other physical and mechanical properties demonstrated a general positive relationship between physical and mechanical properties (Figure 4A,B). Based on water absorption and thickness swelling, as well as MOR and MOE values, reported in Tables 1 and 2, it was clear that fir wood had higher mechanical properties in comparison to meranti, though meranti had a higher density. Moreover, it was fir wood that demonstrated higher thickness swelling and water absorption. Thickness swelling in solid wood species occurs in woody mass. Moreover, mechanical properties are also positively related to woody mass in each solid wood species. Although the above mentioned contour plots demonstrated a positive relationship, some inconsistency was also obvious in the graphs.

The inconsistency indicated that there were other factors affecting the outcome of the relationship between physical and mechanical properties. Factors like vessel properties (vessel diameter, frequency, and specific area), type and size of pit openings, and extractive contents are just a few examples of the factors that significantly affect permeability in solid wood species [30–32], which in turn determine thickness swelling and water absorption behaviors. Moreover, the way microfibrils are oriented in wood cell wall (primary and secondary walls) would also significantly affect mechanical properties, regardless of the woody mass. The above mentioned inconsistency was primarily attributed to the interactions of these factors that influence physical and mechanical properties. The inconsistency was also partially attributed to an interference caused by addition of resin in the laminated wood. The effect of the addition was also apparent in the thermal transmittance of boards, resulting in a significant increase in the thermal coefficient values of the laminated wood in comparison to the solid wood specimens (Table 5) [33,34].

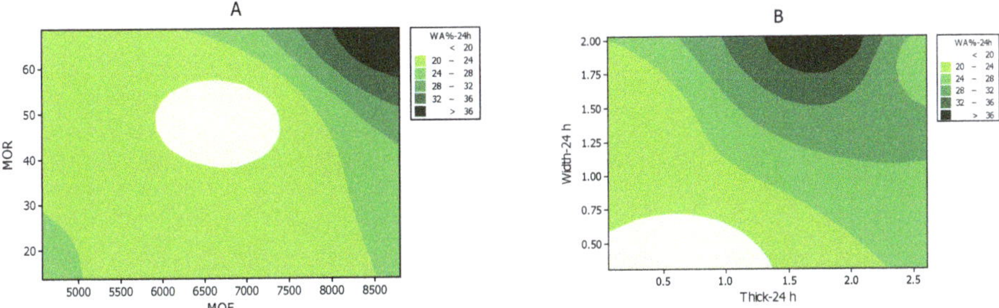

Figure 4. Contour plots between physical and mechanical properties of three-layer laminated wood made from fir and meranti wood (**A**: MOR = modulus of rupture; MOE = modulus of elasticity; and **B**: WA = water absorption; Width and Thick-24 h = width and thickness swelling after 24 h immersion in water).

Cluster analysis was carried out for four treatments, and based on all the physical and mechanical properties studied here. The properties included water absorption, longitudinal, width, and thickness swelling, as well as modulus of rupture, modulus of elasticity, and thermal transmittance coefficient. The four treatments consisted of the two laminated wood (MFM and FMF), along with the values of the two solid wood species of "fir" and "meranti". This statistical analysis revealed that two treatments of meranti solid wood and MFM laminated wood were connected by a vertical line at about number "one" on the scale bar (on top of the graph) (Figure 5). This indicates that these two treatments are very similar based on all the properties studied here. Similar observations were reported by Hartono and Sucipto [13] who made three layer laminated wood from oil palm trunk in the core, a low quality raw material, and applied high density woods in the surface layers, namely sengon and meranti. Based on the similarity demonstrated between meranti solid wood and MFM laminated wood, it can be concluded that the potential applications of this engineered laminated wood (MFM) can be considered nearly the same as can be expected from the physical and mechanical qualities meranti solid wood. Using this engineered laminated wood, forests and woody resources can more efficiently be preserved and more practically be used for different applications. However, the cluster analysis illustrated that "fir" was connected at scale No. 25 with the other three treatments. This indicated that fir, as asolid wood, reacted quite differently, and it was to be considered as a material with completely different properties. Table 2 demonstrated that the mechanical properties of fir solid wood were significantly higher than those of meranti solid wood. The reason for the substantial decrease in the mechanical properties of FMF laminated wood can be elaborated from different perspectives. It was previously reported that shear strength and failure patterns in glue line were significantly dependent on the mechanical properties of the wood [35,36]. The cited authors explained that failure occurred in the glue line when shear strength of the wood used in the experiment was higher than that of the glue itself. However, if the mechanical properties of the wood were low, or thermal modification had made the woody parts weaker, failures occurred in the wooden parts [32,36,37]. In the present study, meranti wood had significantly lower mechanical properties in comparison to fir solid wood (Table 2). Therefore, MFM laminated wood were expected to be very similar to those of meranti solid wood, because fir was used in the core; that is, the neutral zone which is less important. However, fir solid wood had significantly higher mechanical properties; therefore, in FMF laminated wood, the failure occurred in the glue line and before the two fir wood layers on the top and bottom surfaces failed, it was the glue line that failed, resulting in a substantial decrease in the overall mechanical properties. The eventual outcome of the above mentioned facts resulted in close clustering of meranti solid wood with MFM laminated wood, as well as remote clustering of fir solid wood with FMF laminated wood. It was concluded that laminated wood with fir on the surface layers are not recommended with regard to their substantial decrease in mechanical properties, and significant increase in physical properties, when compared with values of fir solid wood specimens.

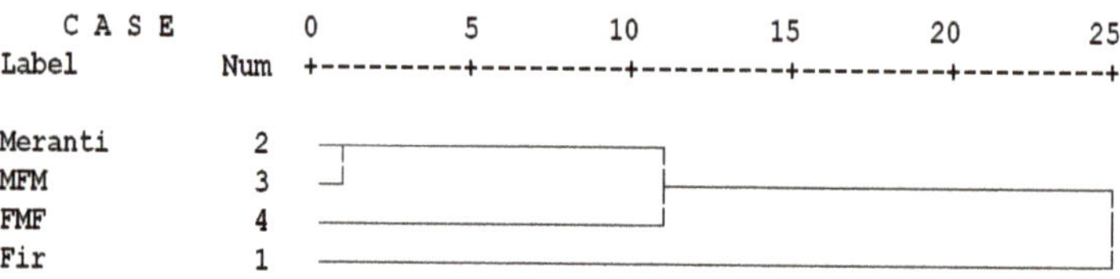

Figure 5. Cluster analysis of the four treatments based on the physical and mechanical properties.

4. Discussion

From the data presented in Table 3, it was revealed that those laminated specimens with fir wood on the surface layers had higher water absorption and thickness swelling values than those with meranti on the core layer. This was quite in agreement with the values of fir and meranti as solid wood specimens (Table 2). That is, meranti showed preferably lower water absorption and thickness swelling values in comparison to fir, both in form of solid wood and in form of laminated wood when the surface layers were made of fir wood. MOR measurement, as a vitally important mechanical property, demonstrated higher values for MFM boards in comparison to FMF ones.

Closer inspection of the data presented in Tables 1 and 3, reveals that the physical properties of the laminated wood made from meranti in surface layers and fir in core (MFM) were nearly the same with those of meranti solid wood. From this observation, it can be concluded that the presence of fir in the core did not affected the physical properties of the laminated wood. On the other hand, the laminated wood made from fir in surface layers and meranti in core (FMF), showed substantially lower values compared to meranti solid wood. It is of interest to mention the high thickness swelling values of FMF laminated wood as depicted in Table 3, which can be attributed to the slight delamination occurred; this was more pronounced in FMF laminated wood.

By referring to the data presented in Tables 2 and 4, it is clear that both types of laminated wood had substantially lower bending properties than solid wood. This in line with the observations made by Puluhulawa et al. [14] who manufactured three layer laminated wood using a light wood in the core (mahang) and meranti in the surface layers.It is interesting to notice that the fir solid wood has MOR values nearly five times more than MFM laminated wood, and the MOE value of MFM laminated wood is nearly 55% lower that the MOE of fir solid wood. These low values in bending properties of both types of laminated wood produced in this study can be attributed to the delamination occurred, and as expected MOR was affected to a greater extent than MOE. A decrease of about 33% in bending properties was also reported by Corigliano et al. [11] for Iroko wood laminates. Papadopoulos reported that the MOR and MOE values of finger jointed beech wood are

76.4% and 10% lower that the corresponding values of beech solid wood [38]. Similar observations were also reported in mixed glued laminated timber of poplar and Eucalyptus grandis clones [39]. On the other hand, Mohamad et al. [40] evaluated the bending strength behavior of laminated wood manufactured using two Malaysian hardwood timber species from different strength grouping, namely keruing (*Dipterocarpus* spp.) and resak (*Vatica* spp. and *Cotylelibium* spp.). They concluded that the strength grouping does not correlate well with the strength of timbers in structural size. They further stated that bending strength values of glulam was mainly affected by the ability of timber to bond well between laminates in relation of the density of the timber. Hayashi and Miyatake [8] reported that bending strength of three Japanese wood species—namely sugi, todomatsu and ezomatsu composite glulam—is decreased by the use of other species of laminae having high MOE for outer layers and this is in line with the observations made in this study. The use of fir, a wood species that has higher MOE values than meranti as depicted in Table 2, in the surface layers, affected the bending properties of the laminated wood since the bending properties of FMF were found to be lower than those of MFM.

Overall, MFM laminated wood presented better bending properties than FMF, as depicted in Table 4. This can be explained by the fact that the core layer experiences lower compressive and tensile stresses compared to the face and back layers. This wood species combination was effective because wood with the lower density can safely be used as a core layer [41].

A clear limitation of this study is the fact that bondability tests were not carried out. It is well known that the bending properties of laminated wood were strongly affected by the ability of wood to bond well between laminates and by the combination of two different wood species, especially without any type of joint. Since the laminating effect in laminated wood stems from the fact that laminations are bonded, the integrity of the cross-section is a crucial factor in the overall product performance [42,43]. As it is mentioned by Frihart et al. [44] and Selbo [45], the resistance to moisture and the bonding quality remains issues that are usually neglected in studies where tropical woods are used for manufacturing of laminated wood. It is reported that the small lumens and the thick cell walls of tropical woods usually lead to a limited penetration of the adhesive used which in turn results inweakened bondlines [44,45]. According to Konnerth et al. [46] and Knorz et al. [47], the bond strength of several European and tropical wood may greatly vary depending on the adhesive system and wood species. Therefore, and taking into consideration the inherent difficulties of bonding tropical woods, the adhesive system and the bondline strength of a given species should be carefully assessed in order to confirm their potential use as a engineered wood product such as laminated wood, especially if there are for structural applications. It has to be mentioned at this stage, that in Canada no laminated wood products made from hardwood and tropical wood species, designed for structural applications, are available on the market. The CSAO122 [48] standard, governing the manufacturing and quality control testing of structural glued-laminated timber, does not include any provision regarding the use of hardwood and tropical wood species.

Table 5, displays the two window's thermal transmittance values. From this it can be seen that the FMF window shows slightly better (lower) thermal transmittance value than the MFM window. This can be attributed to the slightly lower density of the fir wood compared to meranti wood (0.43 and 0.47 $\text{Kg}\cdot\text{m}^{-3}$ respectively), to its lighter color and to the presence of resin canals in the fir wood [49,50]. It is known that thermal conductivity of a material depends on its temperature, moisture and density. Generally, light materials are better insulators than heavy materials because light materials often contain enclosures. When it comes to insulation, the lower the thermal conductivity the better insulation capacity; if a material conducts heat well, a lot of heat will be lost through that material. This difference is translated into a 13.3% better thermal behavior of the FMF window. A smaller U-value means smaller heat transfer rate, so less energy loss.

It was concluded that based on the experiments carried out in the present study, laminated wood with meranti wood on the surface layers are recommended for the industry. Wood-frame windows can be successfully made from laminated wood with meranti wood on the surface layers, therefore employing easily renewable materials, with low environmental impact, that are recyclable and manageable in the medium term.

Future work may involve bondability tests, since it is well known that the bending properties of laminated wood are strongly affected by the ability of wood to bond well between laminates and by the combination of two different wood species, especially without any type of joint. In addition, the application of other adhesive systems may be an avenue for exploitation.

5. Conclusions

The aim of this paper was to investigate the physical (thermal transmittance and dimensional stability) and mechanical properties of two types of three layer laminated wood made from fir and meranti; fir in surface layers and meranti in core (FMF) and vice versa (MFM) and to examine its potential application for wood-frame windows.MFM laminated wood performed better than the FMF as far as the physical and mechanical properties are concerned. The thermal transmittance (rate of the heat transferred) of the FMF window is 13.3% better (lower) compared to the MFM window. The main reason for this is believed to be the lower overall density of the FMF window, which also makes it more competitive as a result of the reduced manufacturing cost since fir is less expensive compared to meranti. It was concluded that wood-frame windows can be successfully made from these types of laminated wood.

Author Contributions: Methodology, K.N., A.M., D.K. and G.N.; Validation, K.N., A.M., D.K. and G.N.; Investigation, K.N., A.M., D.K., G.N., A.V. and N.M.; Writing—original draft preparation, K.N., G.N. and A.N.P.; Writing—review and editing, G.N., H.R.T. and A.N.P.; Visualization, K.N., A.M., D.K. and G.N.; Supervision, G.N., H.R.T. and A.N.P. All authors have read and agreed to the published version of the manuscript.

Funding: This research received no external funding.

Institutional Review Board Statement: Not applicable.

Informed Consent Statement: Not applicable.

Data Availability Statement: The data presented in this study are available on request from the corresponding author.

Conflicts of Interest: The authors declare no conflict of interest.

References

1. Morin-Bernard, A.; Blanchet, P.; Dagenais, C.; Achim, A. Use of northern hardwoods in glued-laminated timber: A study of bondline shear strength and resistance to moisture. *Eur. J. Wood Prod.* **2020**, *78*, 891–903. [CrossRef]
2. Papadopoulos, A.N.; Taghiyari, H.R. Innovative Wood Surface Treatments Based on Nanotechnology. *Coatings* **2019**, *9*, 866. [CrossRef]
3. Moody, R.C.; Hernandez, R.; Liu, J.Y. Glued structural timbers. In *WoodHandbook: Wood as an Engineering Material*; Forest Products Laboratory: Madison, WI, USA, 1999.
4. Moody, R.C.; Hernandez, R.; Liu, J.Y. Glued-Laminated Timber. In *Engineered Wood Products-A Guide for Specifiers, Designs and Users*; Smulski, S., Ed.; PFS Research Foundation: Madison, WI, USA, 1997; pp. 1–39.
5. Janowiak, J.; Manbeck, H.B.; Hernandez, R.; Moody, R.C. Red maple lumber resources for glue-laminated timber beams. *J. For. Prod.* **1997**, *47*, 55–64.
6. Komariah, R.N.; Hadi, Y.S.; Massijaya, M.Y.; Suryana, J. Physical-mechanical properties of glued laminated timber made from tropical small-diameter logs grown in indonesia. *J. Korean Wood Sci.Technol.* **2015**, *43*, 156–167. [CrossRef]
7. Japan's Ministry of Agriculture, Forestry and Fisheries (MAFF). *JAS for Glued Laminated Timber*; Japanese Agricultural Standard: Tokyo, Japan, 2008; pp. 1–42. (In Japanese)
8. Hayashi, T.; Miyatake, A. Recent research and development on sugi (Japanese cedar) structural glued laminated timber. *J. Wood Sci.* **2015**, *61*, 337–342. [CrossRef]
9. Wong, W.C.; Wong, C.N. Grouping species in plywood manufacture. In *Proceedings of the 11th Commenwealth Forestry Conference*; Ministry of Agriculture, Lands and Fisheries: Trinidad Tobago; BMJ Publishing Group: Port of Spain, Trinidad Tobago, 1980; pp. 523–535.

10. SáRibeiro, R.A.; SáRibeiro, M.G. Mechanical properties of amazonian lumber for the development of design stresses. In *Proceedings of the International Timber Engineering Conference*; The University of Tokyo Forests (UTF): Tokyo, Japan, 1990; pp. 189–202.
11. Corigliano, P.; Crupi, V.; Epasto, G.; Guglielmino, E.; Maugeri, N.; Marino, A. Experimental and theoretical analyses of iroko wood laminates. *Compos. B Eng.* **2017**, *112*, 251–254. [CrossRef]
12. Correal, J.F.; Echeverry, J.S.; Ramirez, F.; Yamin, L.E. Experimental evaluation of physical and mechanical properties of glued laminated guaduaangustifoliakunth. *Constr. Building Mater.* **2014**, *73*, 105–112. [CrossRef]
13. Hartono, R.; Sucipto, T. Quality improvement of laminated board made from oil palm trunk at various outer layer using phenol formaldehyde adhesive. *IOP Conf. Ser. Mater. Sci. Eng.* **2018**, *309*, 012049. [CrossRef]
14. Puluhulawa, I.; Alamsyah, A.; Pribadi, J.A. The experimental study on laminated beam by mahang and meranti wood with variations connectors. *Conf. Proc.* **2018**, *3*, 109–115.
15. Ong, C.B. Performance of Glue-Laminated Beams from Malaysian Dark Red Meranti Timber. Ph.D. Thesis, University of Bath, Department of Architecture and Civil Engineering, Bath, UK, 2018.
16. ASTM International. *ASTM D143—94. Standard Test Methods for Small Clear Specimens of Timber*; ASTM International: West Conshohocken, PA, USA, 2007.
17. *EN 385 Finger Jointed Structural Timber-Performance Requirements and Minimum Production Requirements*; European Committee for Standardization: Brussels, Belgium, 2001.
18. BSI 2014. *ISO 9869-1:2014: Thermal Insulation-Building Elements-Insitu Measurement of Thermal Resistance and Thermal Transmittance; Part 1: Heat Flow Meter Method*; BSI: London, UK, 2014.
19. Cuce, E. Role of airtightness in energy loss from windows: Experimental results from in-situ tests. *Energy Build.* **2017**, *139*, 449–455. [CrossRef]
20. Cuce, E. Accurate and reliable U-value assessment of argon-filled double glazed windows: A numerical and experimental investigation. *Energy Build.* **2018**, *171*, 100–106. [CrossRef]
21. Bergman, T.; Lavine, A.; Incropera, F.; De Witt, D. *Fundamentals of Heat and Mass Transfer*, 7th ed.; John Wiley & Sons: New York, NY, USA, 2011.
22. Taghyari, H.R.; Ghorbanali, M.; Tahir, P.M.D. Effects of the improvement in thermal conductivity coefficient bynano-wollastonite on physical and mechanical properties in medium-density fiber board (MDF). *BioResources* **2014**, *9*, 4138–4149.
23. Taghiyari, H.R.; Norton, J.; Tajvidi, M. Effects of nano-materials. In *Bio-Based Wood Adhesives: Preparation, Characterization, and Testing*; CRC Press/Taylor & Francis Group: Boca Raton, FI, USA, 2017; pp. 310–339.
24. Taghiyari, H.R.; Soltani, A.; Esmailpour, A.; Hassani, V.; Gholipour, H.; Papadopoulos, A.N. Improving thermal conductivity coefficient in oriented strand lumber (OSL) using sepiolite. *Nanomaterials* **2020**, *10*, 599. [CrossRef] [PubMed]
25. TESTO 635-2 Temperature & Moisture Meter. Available online: https://www.testo.com/en-US/testo-635-2/p/0563-6352 (accessed on 1 February 2021).
26. BSI Standards. *BS EN ISO 10077-2: 2017 BSI Standards Publication Thermal Performance of Windows, Doors and Shutters—Calculation of Thermal Transmittance*; BSI: London, UK, 2017.
27. Ada, R. Cluster analysis and adaptation study for safflower genotypes. *J. Bulg. Agric. Sci.* **2013**, *19*, 40–47.
28. Taghiyari, H.R.; Layeghi, M.; AminzadehLiyafooee, F. Effects of dry ice on gas permeability of nano-silver-impregnated Populusnigra and Fagusorientalis. *IET Nanobiotechnol.* **2012**, *6*, 40–44. [CrossRef] [PubMed]
29. Taghiyari, H.R.; Majidinajafabadi, R.; Vahidzadeh, R. Wollastonite to hinder growth of aspergillusniger on cotton textile. *An. Acad. Bras. Cienc.* **2018**, *90*, 2797–2804. [CrossRef] [PubMed]
30. Efhami Sisi, D.; Karimi, A.; Pourtahmasi, K.; Taghiyari, H.R.; Asadi, F. The effects of agroforestry practices on vessel properties in *Populus nigra* Var. *betulifolia*. *IAWA J.* **2010**, *31*, 481–487. [CrossRef]
31. Taghiyari, H.R.; Efhami, D.; Karimi, A.N.; Pourtahmasi, K. Effects of initial spacing on gas permeability of *Populus nigra* Var. *betulifolia*. *J. Trop. For. Sci.* **2011**, *23*, 305–310.
32. Taghiyari, H.R.; Efhami, D. Diameter increment response of *Populus nigra* Var. *betulifolia* induced by alfalfa. *Austrian J. For. Sci.* **2011**, *2*, 113–127.
33. Jankowska, A.; Kozakeiwicz, P. Comparison of thermal properties of selected wood species intended to woodwork windows production. *For. Wood Technol.* **2014**, *85*, 101–105.
34. Šilinskas, B.; Varnagirytė-Kabašinskienė, I.; Aleinikovas, M.; Beniušienė, L.; Aleinikovienė, J.; Škėma, M. Scots pine and norway spruce wood properties at sites with different stand densities. *Forests* **2020**, *11*, 587. [CrossRef]
35. Taghiyari, H.R.; Esmailpour, A.; Adamopoulos, S.; Zereshki, K.; Hosseinpourpia, R. Shear strength of heat-treated solid wood bonded with polyvinyl-acetate reinforced by nanowollastonite. *Wood Res.* **2020**, *65*, 183–194. [CrossRef]
36. Taghiyari, H.R.; Majidi, R.; Mohseni Armaki, S.M.; Haghighatparast, M.R. Graphene as reinforcing filler in polyvinyl acetate resin. *J. Int. Adhes. Adhes.* **2020**. Accepted.
37. Pizzi, A.; Papadopoulos, A.; Policardi, F. Wood Composites and Their Polymer Binders. *Polymers* **2020**, *12*, 1115. [CrossRef]
38. Papadopoulos, A.N. The effect of acetylation on bending strength of finger jointed beech wood (*Fagus sylvatica* L.). *Holzalsroh Werkst.* **2008**, *66*, 309–310. [CrossRef]
39. Castro, G.; Paganini, F. Mixed glued laminated timber of poplar and eucalyptus grandis clones. *Holzalsroh Werkst.* **2003**, *61*, 291–298. [CrossRef]

40. Mohamad, W.H.W.; Razian, M.A.; Ahmad, Z. Bending strength properties of glued laminated timber from selected Malaysian hardwood timber. *Int. J. Civ. Environ. Eng.* **2011**, *11*, 4.
41. Forest Products Laboratory. *Wood Handbook-Wood as an Engineering Material*; USDA Forest Service Forest Products: Madison, WI, USA, 2010.
42. Falk, R.; Colling, F. Laminating effects in glued-laminated timber beams. *J. Struct. Eng.* **1995**, *121*, 1857–1863. [CrossRef]
43. Dietsch, P.; Tannert, T. Assessing the integrity of glued-laminated timber elements. *Constr. Build. Mater.* **2015**, *101*, 1259–1270. [CrossRef]
44. Frihart, C.R.; Hunt, C.G. Adhesives with wood materials: Bond formation and performance. In *Wood Handbook: Wood as an Engineering Material*; Forest Products Laboratory: Madison, WI, USA, 2010; pp. 1–10.
45. Selbo, M.L. *Adhesive Bonding of Wood*; Forest Service: Washington, DC, USA, 1975.
46. Konnerth, J.; Kluge, M.; Schweizer, G.; Miljković, M.; Gindl-Altmutter, W. Survey of selected adhesive bonding properties of nine European softwood and hardwood species. *J. Eur. Wood Prod.* **2016**, *74*, 809–819. [CrossRef]
47. Knorz, M.; Schmidt, M.; Torno, S.; van de Kuilen, J.W. Structural bonding of ash (*Fraxinus excelsior* L.): Resistance to delamination and performance in shearing tests. *J. Eur. Wood Prod.* **2014**, *72*, 297–309. [CrossRef]
48. CAN/CSA 0122-16. *CSA Structural Glued-Laminated Timber*; Canadian Standards Association: Toronto, ON, Canada, 2016.
49. Dinwoodie, J.M. *Timber: Its Nature and Behavior*; Van Nostrand Reinhold Company Ltd.: Wokingham, Berks, UK, 1981.
50. Walker, J.C.F.; Butterfield, B.G.; Langrish, T.A.G.; Harris, J.M.; Uprichard, J.M. *Primary Wood Processing: Principles and Practise*; Chapman & Hall Ltd.: London, UK, 1993.

Article

Environmental Impact of Wood Modification

Callum Hill [1,2,*], Mark Hughes [3] and Daniel Gudsell [4]

1 JCH Industrial Ecology Ltd., Bangor, Gwynedd LL57 1LJ, UK
2 Norwegian Institute of Bioeconomy Research (NIBIO), P.O. Box 115, NO-1431 Ås, Norway
3 Department of Bioproducts and Biosystems, Aalto University, Vuorimiehentie 1, FI-002150 Espoo, Finland; mark.hughes@aalto.fi
4 Abodo Wood Ltd., Mangere, Auckland 2022, New Zealand; Daniel.Gudsell@abodo.co.uk
* Correspondence: enquiries@jchindustrial.co.uk

Abstract: The modification of wood involves extra processing over and above what is associated with un-modified material and this will involve an associated environmental impact. There is now a body of information on this due to the presence in the public domain of a number of environmental product declarations (EPDs). Using these data, it is possible to determine what the extra impact associated with the modification is. The process of modification results in a life extension of the product, which has implications regarding the storage of sequestered atmospheric carbon in the harvested wood products (HWP) materials' pool and also extended maintenance cycles (e.g., longer periods between applying coatings). Furthermore, the life extension benefits imparted by wood modification need to be compared with the use of other technologies, such as conventional wood preservatives. This paper analysed the published data from a number of sources (peer-reviewed literature, published EPDs, databases) to compare the impacts associated with different modification technologies. The effect of life extension was examined by modelling the carbon flow dynamics of the HWP pool and determining the effect of different life extension scenarios. Finally, the paper examined the impact of different coating periods, and the extensions thereof, imparted by the use of different modified wood substrates.

Keywords: wood modification; life cycle assessment; carbon storage

Citation: Hill, C.; Hughes, M.; Gudsell, D. Environmental Impact of Wood Modification. *Coatings* **2021**, *11*, 366. https://doi.org/10.3390/coatings11030366

Academic Editor: Yong X. Gan

Received: 15 February 2021
Accepted: 20 March 2021
Published: 23 March 2021

1. Introduction

The production and use of materials in construction has an associated environmental burden and requires the use of energy in order to transform them from the state in which they exist in nature, transport them and process them to create products which have the desired functionality. Life cycle assessment is the methodology that is used to determine the environmental impacts, but it must be used appropriately, especially when making comparative assertions.

The use of wood in construction can potentially provide environmental benefits compared with the use of non-renewable materials. These potential benefits arise from:

- The reduced environmental footprint associated with the production of timber-based building materials when compared with non-renewable alternatives
- The use of timber in construction as a store of atmospheric carbon during the lifetime of the timber product
- The recovery of the inherent solar energy stored in the timber material, which can be recovered at the end of life and be substituted for the use of fossil-derived energy sources

Greater environmental benefits are realised by increasing the longevity of timber products in construction. This is because over the lifetime of a building there is a reduced requirement to replace components, and there may also be an associated reduction in maintenance requirements. In order to make appropriate choices for using materials in construction, it is necessary to evaluate the whole life of a product from creation to final disposal, which requires knowledge not only of the environmental impacts associated with the manufacture of a product but of all parts of the lifecycle, including installation, maintenance, potential replacement and disposal, or re-use at the end of the product lifetime. Although the cradle to factory gate part of the lifecycle can be determined with some confidence, complete lifecycle analyses can be extremely difficult to make accurately and often rely on assumptions regarding service life of the product and the building in which it is installed.

One approach to examining the potential benefits provided by wood modification is to employ the concept of "environmental payback period" [1]. In economics, the payback period is the amount of time required in order to recover the cost of an investment. The same idea can be used when making an investment in a new energy technology, where the initial investment in embodied energy to create the new infrastructure can be recovered over a period of time by the energy savings which are achieved by employing the new technology. It is also possible to apply this idea to environmental burdens, such as global warming potential (GWP). For example, the impacts associated with the production of a parquet flooring made from a siloxane-based modified wood were compared to an unmodified wood-based equivalent. However, it was noted that the outcome of such an analysis was highly sensitive to the assumptions that were made regarding in-service performance [1]. A carbon payback time approach was adopted by Hill and Norton [2] in order to determine the benefits arising from the use of a modified wood in cladding applications. They used published data of the environmental burdens of the modified wood and calculated a carbon payback time based upon the methods recommended in the International Life Cycle Data System (ILCD) Handbook for determining the temporal storage of atmospheric carbon in wood products, using a 100-year discounting period. This showed that longer payback times to "carbon neutrality" were required for modified, compared with unmodified wood. However, the analysis did not include any effects of lifetime extension or reduced maintenance upon the whole life performance of the different products. Moreover, there has not been any internationally accepted method for determining the effects of different product lifetimes upon the climate change impacts in life cycle assessment [3], making the use of the ILCD (or any discounting) approach to payback time unsatisfactory.

An important aspect to be considered when looking at product lifetimes is the issue of durability. Durability is defined as the length of time that a product is able to fulfil its intended function. Wood in service is subjected to different stressors, and how it resists these will be determined by its durability. These stresses can be environmental (e.g., fungal, rain, sunlight, insect attack) or mechanical (e.g., fatigue, loading, wear). The purpose of wood modification is to improve the durability properties of wood with respect to one or more of these stressors, with the expectation that the modified wood will have a longer service lifetime. One way to determine this is to place the modified wood samples with unmodified comparisons in exterior environments and observe the behaviour over many years (probably decades), but this is hardly a practical proposition. A more appropriate approach is to use accelerated tests in order to determine likely service lives. The challenge is how to predict what the extension in service life will be due to a modification, usually based upon laboratory or accelerated tests. Regarding biological durability, this has been discussed comprehensively by Meyer-Veltrup et al. [4], where the issue of service life prediction was considered by using a dose–response model.

Differences in life cycle costs or environmental impacts associated with a system based upon modified wood compared with unmodified wood are determined by:

- Material costs/impacts—related to material composition of the different systems and any differences in the processing requirements
- Inspection intervals—longer times between inspections will result in lower lifetime costs

- Changes in limit states for maintenance—either a longer time taken to reach a specific limit state or a more relaxed limit state will result in longer times between maintenance intervals or longer times between replacement, leading to lower lifetime costs
- Changes in limit states for replacement—a longer time to reach or a more relaxed limit state will result in lower lifetime costs

Failure models can be divided into three main types:

- Failure is modelled using a probability distribution with no assumptions made about the deterioration process that leads to failure. In this model, the condition of the component can only be in one of two states—functioning or non-functioning.
- The deterioration process leading to the failure is modelled. The deterioration should be observable directly or indirectly by the relevant indicative properties. Typically, there are three modelling approaches: (a) three-state models in which the time to potential failure and time to failure are modelled by probability distributions, (b) Markov chain models, where there are several defined degradation states, or (c) continuous stochastic processes, such as Wiener or Gamma processes.
- The physical process causing the deterioration is modelled using a stress input factor (such as load) in order to determine the deterioration of the system over time. This requires the relationship between the likelihood of failure and the stress input factor to be well understood.

The choice of model depends upon the physical processes behind the deterioration and the amount of information that is available on the condition and behaviour of the system. One approach to the problem is to use probabilistic methods and sensitivity analysis to establish the benefits of different inspection and maintenance regimes [5]. Such approaches are useful where there is a clear criterion for expressing a limit state for failure or replacement; the situation for timber in service is often more complex.

With coatings on wood, the failure is more likely to be progressive, with properties such as cracking or gloss changing gradually and with time to failure being highly variable, depending upon exposure conditions and weather. In addition, the definition of failure may be more associated with perception or aesthetic criteria, even though the functionality of the component has not yet been compromised. An example of this would be mould growth on claddings, where the underlying wood has not been degraded but is replaced because the appearance is unsatisfactory. The lifetimes of coated timber products are determined by the performance of the wood coating system and the risk factors related to the exposure conditions.

Tellnes et al. [6] reported on the effect of wood modification upon the service life of decking products by comparing carbon footprints and concluded that this had a significant influence on the results. At the present time, there has not been a comparison of the potential environmental benefits that might be achieved through the use of modified wood as a cladding material in the built environment, nor has the potential benefit of the extended lifetime of coatings on modified wood been examined. The purpose of this study was to determine the environmental consequences of the use of timber as a cladding material and in particular to determine whether the use of modified wood provides any additional environmental benefits compared with the use of unmodified timber equivalents. This paper examines the relevant published data in this area and considers what information is required in order to make a proper evaluation of the research question.

2. Materials and Methods

The first step of this study was to examine the published environmental impact data on modified wood. This was largely based upon environmental product declarations which are currently published. It was decided to examine two parameters—embodied energy (EE) and global warming potential (GWP). Data for the lifecycle stages A1–A3 were collected, as defined in EN15804 (Table 1)(EN15805+A2:2019, Sustainability of construction works. Environmental product declarations. Core rules for the product category of construction products).

Table 1. Lifecycle stages divided into modules according to EN15804.

Module	Lifecycle Stage	Description
A1	Production	Raw material supply
A2	Production	Transport
A3	Production	Manufacturing
A4	Construction	Transport
A5	Construction	Installation
B1	Use	Use
B2	Use	Maintenance
B3	Use	Repair
B4	Use	Replacement
B5	Use	Refurbishment
B6	Use	Operational energy use
B7	Use	Operational water use
C1	End of life	Deconstruction/demolition
C2	End of life	Transport
C3	End of life	Waste processing
C4	End of life	Disposal
D	Beyond the life cycle	Reuse/recovery/recycling

The results for the GWP data and sequestered atmospheric carbon are reproduced in Table 2. In many cases, the GWP impact data were not reported directly and had to be calculated, based upon the reported total GWP and the reported sequestered atmospheric carbon (in kg carbon dioxide equivalents per cubic meter of wood product). The embodied energy data are reproduced in Table 3.

Table 2. Global warming potential (GWP) data for modules A1–A3 (forest to factory gate) (declared unit 1 m^3) (GWP in kg CO_2 eq.). Abbreviations: EPD (environmental product declaration), MC (moisture content), EPD (environmental product declaration), TMT (thermally modified timber).

EPD Registration Number	Date	Country	Description	Density (kg/m^3)	MC (%)	TOTAL (Reported)	Sequestered	GWP (Calculated)	GWP (Reported)
Wood for Good [1]	2014	GBR	Fresh sawn softwood	672	60	−713	−770 [2]	+57	
S-P-00561	2017	AUS	Fresh sawn hardwood	768	26	−851	−1118 [2]	+267	
EPD-EGG-20140246-IBA2-EN	2018	AUT	Sawn timber green	740	70	−779	−798 [2]	+19	
Wood for Good [1]	2014	GBR	Sawn dried softwood	483	15	−679	−770 [2]	+91	
Wood for Good [1]	2014	GBR	Sawn dried hardwood	698	12	−878	−902	+24	
NEPD 307 179 EN	2015	NOR	Sawn dried softwood	450	15	−672	−715	+43	
S-P-00560	2017	AUS	Sawn dried softwood	551	12	−760	−902 [2]	+142	
S-P-00561	2017	AUS	Sawn dried hardwood	735	10	−888	−1225 [2]	+337	
EPD-EGG-20140247-IBA2-EN	2017	AUT	Sawn timber dried softwood	507	15	−784	−808	+24	
S-P-01325	2018	SWE	Sawn dried softwood	455	16	−577	−719 [2]		+138
S-P-00997	2019	NZD	Sawn dried radiata pine	488	11.6	−747	−798		+51

Table 2. *Cont.*

EPD Registration Number	Date	Country	Description	Density (kg/m^3)	MC (%)	TOTAL (Reported)	Sequestered	GWP (Calculated)	GWP (Reported)
13CA24184.102.1	2013	USA	Dried planed softwood lumber	434	0		−795		+73
NEPD 00247N	2014	DNK	Sawn dried planed Siberian larch	650	18	−624	−1010	+386	
(BRE) 000124	2017	GBR	Sawn dried planed softwood	479	15	−712	−764	+52	+107
S-P-00560	2017	AUS	Sawn + dressed dried softwood	551	12	−699	−902 [2]	+203	
SP-00561	2017	AUS	Sawn + dressed dried hardwood	735	10	−731	−1225	+494	
Wood for Good [1]	2014	GBR	Sawn dried planed softwood	482	15	−646	−768	+122	
NEPD 308 179 EN	2015	NOR	Sawn dried planed softwood	420	17	−607	−660	+53	
S-P-00997	2019	NZD	Sawn dried planed radiata	486	11.6	−728	−795		+69
S-P-00997	2019	NZD	Sawn dried planed jointed radiata	475	10.5	−697	−784		+87
S-P-02153	2020	CZE	Sawn dried planed jointed softwood	450	15	−685	−717		+32
4788424634.102.1	2020	USA	Dried planed softwood lumber	460	15	+63 [3]	−733 [2]		+63
NEPD 00259N	2014	EST	TMT spruce	350	5	−97	−611	+514	
NEPD 00259N	2014	EST	TMT pine	450	5	−258	−786	+528	
NEPD 00260N	2014	EST	TMT ash	670	6	−430	−1159	+729	
S-P-01718	2019	GBR	TMT (Brimstone) poplar	409	5	−453	−719	+266	
S-P-01718	2019	GBR	TMT (Brimstone) sycamore	571	5	−639	−1010	+371	
S-P-01718	2019	GBR	TMT (Brimstone) ash	631	5	−704	−1110	+406	
RTS_44_19	2019	FIN	TMT Thermo-D Lunawood rough	430	5	−426	−724	+298	
RTS_44_19	2019	FIN	TMT Thermo-D Lunawood planed	390	5	−342	−657	+315	
RTS_44_19	2019	FIN	TMT Thermo-S Lunawood rough	430	5	−516	−724	+208	
RTS_44_19	2019	FIN	TMT Thermo-S Lunawood planed	390	5	−409	−657	+248	
S-P-01543	2020	NZL	TMT Vulcan radiata sawn	420	7	−535	−758	+224	
S-P-01543	2020	NZL	TMT Vulcan radiata surfaced	420	7	−516	−758	+243	
S-P-01543	2020	NZL	TMT Vulcan radiata finger-jointed	420	7	−469	−758	+290	
NEPD-376-262-EN	2015	NLD	Accoya (radiata)	510	4	−433	−944	+511	
NEPD-376-262-EN	2015	NLD	Accoya (Scots pine)	540	4	−741	−999	+258	
NEPD-376-262-EN	2015	NLD	Accoya (beech)	755	4	−1010	−1397	+387	
NEPD-407-287-	2016	NOR	Kebony Clear (radiata)	480	12	−549	−1435 [4]	+886	
NEPD-408-287-EN	2016	NOR	Kebony Clear (SYP)			−646	−1532 [4]	+886	
NEPD-410-288-EN	2016	NOR	Kebony character (Scots pine)			−738	−1097 [4]	+359	
NEPD-411-288-EN	2016	NOR	Kebony character (Scots pine) roofing			−738	−1097	+359	

[1] Not registered as an EPD, but follows the EN15804 PCR. [2] data not supplied in the EPD, calculated using EN16449. [3] Not clear how this value is calculated. [4] Includes biogenic carbon in the furfuryl polymer.

Table 3. Embodied energy and inherent energy data for modules A1–A3 (forest to factory gate) (declared unit 1 m^3). Abbreviations: PERE (primary energy renewable energy), PENRE (primary energy non-renewable energy), PERM (primary renewable energy in the materials).

EPD Registration Number	Date	Country	Description	PERE (MJ)	PENRE (MJ)	Embodied Energy (MJ)	PERM (MJ)
Wood for Good [1]	2014	GBR	Fresh sawn softwood	34	1040	1074	8090
S-P-00561	2017	AUS	Fresh sawn hardwood	111	1810	1921	11,300
EPD-EGG-20140246-IBA2-EN	2018	AUT	Sawn timber green	97	250	347	8050
Wood for Good [1]	2014	GBR	Sawn dried softwood	853	1650	2503	8120
Wood for Good [1]	2014	GBR	Sawn dried hardwood	328	2840	3168	11,300
NEPD 307 179 EN	2015	NOR	Sawn dried softwood	2270	685	2955	7410
S-P-00560	2017	AUS	Sawn dried softwood	2480	1610	4090	9290
S-P-00561	2017	AUS	Sawn dried hardwood	879	2510	3389	12,600
EPD-EGG-20140247-IBA2-EN	2017	AUT	Sawn timber dried softwood	1330	330	1660	8160
S-P-01325	2018	SWE	Sawn dried softwood	3170	748	3918	6750
S-P-00997	2019	NZD	Sawn dried radiata pine	4200	552	4752	8260
13CA24184.102.1	2013	USA	Dried planed softwood lumber	1640	1228	2868	
NEPD 00247N	2014	DNK	Sawn dried planed Siberian larch	3724	6842	10,566	9180
(BRE) 000124	2017	GBR	Sawn dried planed softwood	2270	1570	3840	8440
S-P-00560	2017	AUS	Sawn + dressed dried softwood	3050	2260	5310	9290
S-P-00561	2017	AUS	Sawn + dressed dried hardwood	1190	3840	5030	12,600
Wood for Good [1]	2014	GBR	Sawn dried planed softwood	1060	2130	3190	8080
NEPD 308 179 EN	2015	NOR	Sawn dried planed softwood	2930	902	3832	6840
S-P-00997	2019	NZD	Sawn dried planed radiata	5330	720	6050	8240
S-P-00997	2019	NZD	Sawn dried planed jointed radiata	6530	991	7521	8140
S-P-02153	2020	CZE	Sawn dried planed jointed softwood	1050	472	1522	7500
4788424634.102.1	2020	USA	Dried planed softwood lumber	2381	1000	3381	10,959
NEPD 00259N	2014	EST	TMT spruce	2184	7426	9610	9180
NEPD 00259N	2014	EST	TMT pine	2761	7697	10,458	9180
NEPD 00260N	2014	EST	TMT ash	6678	10,302	16,980	11,990
S-P-01718	2019	GBR	TMT (Brimstone) poplar	13,000	4180	17,180	7460
S-P-01718	2019	GBR	TMT (Brimstone) sycamore	18,100	5810	23,910	10,400
S-P-01718	2019	GBR	TMT (Brimstone) ash	22,200	6480	28,680	9250
RTS_44_19	2019	FIN	TMT Thermo-D Lunawood rough	30782	5270	36,052	8353
RTS_44_19	2019	FIN	TMT Thermo-D Lunawood planed	31,163	6565	37,728	7604
RTS_44_19	2019	FIN	TMT Thermo-S Lunawood rough	27,924	4177	32,101	8354
RTS_44_19	2019	FIN	TMT Thermo-S Lunawood planed	28,483	5174	33,657	7605
S-P-01543	2020	NZL	TMT Vulcan radiata sawn	4200	2970	7170	7560
S-P-01543	2020	NZL	TMT Vulcan radiata surfaced	4740	3230	7970	7560
S-P-01543	2020	NZL	TMT Vulcan radiata finger-jointed	5680	3850	9530	7560
NEPD-376-262-EN	2015	NLD	Accoya (radiata)	847	14,559	15,406	6574
NEPD-376-262-EN	2015	NLD	Accoya (Scots pine)	932	13,137	14,069	10,372
NEPD-376-262-EN	2015	NLD	Accoya (beech)(	1256	18,069	19,325	7596
NEPD-407-287-EN	2016	NOR	Kebony Clear (radiata)	5576	15,354	20,930	16,476
NEPD-408-287-EN	2016	NOR	Kebony Clear (SYP)	6407	13,335	19,742	17,473
NEPD-410-288-EN	2016	NOR	Kebony character (Scots pine)	3078	5691	8769	12,302

[1] Not registered as an EPD, but follows the EN15804 PCR.

A comparison of the relationship between the GWP and embodied energy for the modified and unmodified solid wood products is shown in Figure 1.

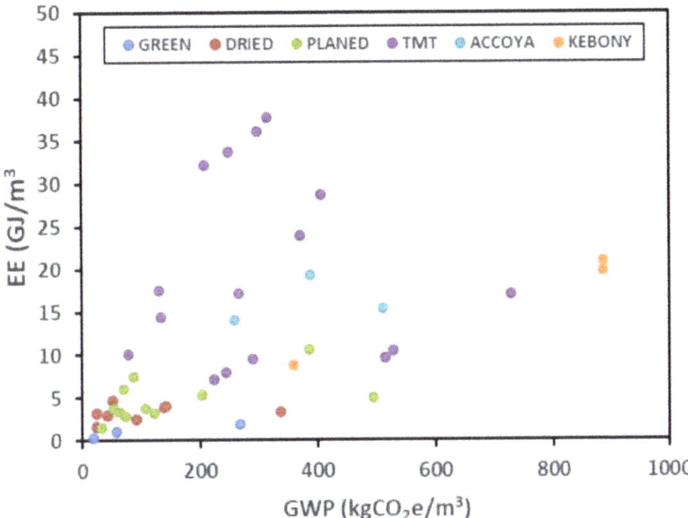

Figure 1. Relationship between embodied energy and global warming potential for unmodified and modified solid wood products, from published EPDs and literature.

In principle, there should be an increase in both embodied energy and GWP emissions as the degree of wood processing increases. This trend is apparent in the data, but there are some obvious outliers which have considerably higher GWP compared with the main group and others which have much higher embodied energy than might be predicted from a trend-line. Those with a higher GWP belong to wood processed in Australia, where the higher carbon footprint of the Australian grid mix is presumably the main cause of the higher GWP values. One data point is associated with the production of sawn and planed Siberian larch, and the higher embodied energy (10.57 GJ/m^3) and GWP (386 kgCO$_2$e/m^3) is due to the long transport distances involved. Since the analysis was performed in a European context, these data were removed from the next part of the study. Some of the data for TMT also exhibit much higher embodied energies than might be predicted from a trend line (data were included in the analysis). The reasons for this are not known but are considered further in the Discussion section.

There has been very little work on LCAs of modified wood published in the literature [7], although a study of the LCA of TMT production has been published [8]. The declared unit was 1m^3 and the embodied energy was 14.38 GJ (production in Portugal) and 17.55 GJ (Spain), with a corresponding GWP of 133 kg CO$_2$e (Portugal) and 131 kg CO$_2$e (Spain). These data are considered typical for production of TMT in a European context and were accordingly included in the analysis and in Figure 1.

Since this was a preliminary study, it was decided to use average values for the different product groups. The average values for GWP are reported in Table 4 for 1 m^3 of unmodified wood (dried, sawn and planed), TMT, Accoya and Kebony, along with the standard deviations. These values were then used in the model.

Table 4. Average values of GWP impact used for the comparison.

Parameter	Unmodified	TMT	Accoya	Kebony
Average GWP	75.6	311.4	385.3	710.33
Standard deviation	29.1	168.0	126.5	304.3

The initial question asked was "what should the lifetime of the modified wood product be in order to compensate for the extra GWP associated with the modification process?" (modules A1–A3). This requires consideration of the initial impact associated with modules A1–A3 and then determining the number of replacements of the unmodified wood that are required before the GWP of the unmodified wood and modified wood are equal (modules A1–A3 plus B4).

Such an analysis only considers the GWP impact associated with materials' production (modules A1–A3) but takes no account of the impacts associated with the transport to construction site (A4) and installation (A5). Typical GWP impact data for these parts of the product lifecycle were obtained from two published EPDs (registration numbers: S-P-01718, BREG EN EPD No.:000124). GWP impacts of 6.9 kg CO_2e/m^3 (A4) and 46.7 kg CO_2e/m^3 (A5) were used as representative, although other values can be used. Therefore, adding the GWP values for modules A1–A5 gives the total impact for production and installation on site. For uncoated products, no maintenance was assumed during the product lifetime (zero impact for modules B1–B7).

The additional effect of applying a coating was also investigated (module B4). For this purpose, a thickness for the cladding boards of 30 mm was assumed, which results in an exposed surface area of approximately 33 m^2; it was assumed that only one side of each board was coated with a water-based alkyd paint but that two coats were applied. A coverage of 3m^2 per litre of paint and a paint density of 1.1 kg/litre were assumed, resulting in a usage of 60 kg of paint per m^3 of wood product. The GWP impact of the wood coating was determined using Ecoinvent 3.1 and using "Alkyd paint, white, without water, in 60% solution state {RER} | alkyd paint production, white, water-based, product in 60% solution state | Alloc Def, S" with the method "IPCC 2013 GWP 100a V1.03".

3. Results and Discussion

3.1. Relationship between Embodied Energy and GWP

The embodied energy of a material or product used in a structure or product is often defined as the primary energy used in the manufacture, which includes all of the energy used in the production, as well as the primary energy used in the transport of materials and goods required for the production process. This definition relates to the initial embodied energy, which is related to the cradle to factory gate stage (modules A1–A3, EN15804) of the product life cycle, as reported in the work herein. This is sometimes referred to as the initial embodied energy. In some definitions, the transport to construction site (A4) and the energy used on site for the erection or installation of the product (A5) are also included. The energy used in the maintenance and refurbishment of the product is referred to as the recurring embodied energy, and where end of life is also considered, there is a contribution from demolition energy. Embodied energy does not include operational energy, which may be associated with the product, e.g., heating a building. The units used are generally MJ (or GJ) per unit mass, or volume, or per defined functional unit, although some workers report this as kWh (=3.6 MJ). Transport of materials to site can have a major impact on the embodied energy and GWP impact of the construction materials.

The embodied energy is invariably reported according to the cumulative energy demand (CED) method, which states that the embodied energy is assessed as the primary energy used for the manufacture, use and disposal of an economic good (product or service), or which may be attributed to it with justification. The method distinguishes between non-renewable and renewable energy use. The cumulative energy demand (CED) represents the primary energy used (both direct and indirect) during the life cycle of a product [9]. This includes the energy consumed during the extraction, manufacturing and the disposal of the product and raw and auxiliary materials. Different methods for determining the primary energy demand exist. For example, the lower or higher heating values of primary energy sources may be used; the use of renewable energy resources may not be included, or it may be reported separately. Primary energy is defined as "the energy requiredfrom nature" (e.g., coal) embodied in the energy consumed by the purchaser (for example, electricity) and the energy used by the consumer as "delivered energy". This means that a process using 1 MJ of electricity in one region of the world may have a different embodied energy compared with an identical process using 1 MJ of electrical energy in another part because the grid mix in the two regions is different.

The current standards do not provide complete guidance and do not address important issues regarding embodied energy reporting. For example, EN15804 does not mention embodied energy, although it does require the reporting of energy inputs as primary energy and requires the reporting of the following categories describing resource use:

- Use of renewable primary energy, excluding renewable primary energy resources used as raw materials
- Use of non-renewable primary energy, excluding non-renewable primary energy resources used as raw materials

It is important to distinguish between embodied energy, which is associated with the production of a good or service, and the inherent (or embedded) energy, which is a physical property of the material. The terms embodied and embedded are sometimes confused in the literature. As noted previously, the embodied energy of a material is the primary energy that is associated with the extraction, processing and transportation of that material from the cradle to the factory gate. In contrast, the embedded energy of a material is a property of that material and can be directly measured. For example, the inherent energy in a wood product can be recovered at the end of its life cycle by incineration, whereas the inherent energy of concrete is zero. The inherent (embedded) energy is reported in EN15804 in the following categories:

- Use of renewable primary energy resources used as raw materials
- Use of non-renewable primary energy resources used as raw materials

However, different LCA practitioners report data for these categories in different ways. In addition, the inherent energy is reported as primary energy in these categories, which does not necessarily represent the true value of the recoverable energy, which is usually more accurately reported for wood as the lower heating value (LHV).

There should be a relationship between GWP and embodied energy because energy production does result in carbon emissions, but the greenhouse gas (GHG) emissions per unit of energy vary enormously, depending on the primary energy source used. For example, there is a very large difference in the GWP associated with the generation of 1 kWh of electricity if coal or wind are used as the primary energy sources. Therefore, the relationship between EE and GWP will exhibit scatter. Other reasons for different GWP impacts per unit of energy are related to processes, such as wood drying using different primary energy sources and differences in efficiency.

3.2. Impact of Modified Wood versus Unmodified (Uncoated)

Comparisons of modified versus unmodified wood are shown in Table 5. The data are shown in terms of a payback time for the use of unmodified wood. The time in years refers to the lifetime that the modified wood product would have to achieve before the GWP impacts for the use of modified wood and unmodified became identical. This is based upon the assumption that the unmodified wood has a lifetime in service of 20 years. Any advantage in terms of GHG emissions would only be realised if the lifetimes of the modified wood products exceeded these threshold values. The first row shows the payback times only taking into account the production of the materials. The second row includes the impacts from material production and installation onsite.

Table 5. Calculated payback times based upon average GWP values for each product group.

Scenario	Unmodified	TMT	Accoya	Kebony
A1–A3	20	82.3	101.9	187.9
A1–A5	20	56.5	68.0	118.3

The more realistic scenario is the one including modules A1 through to A5, but the required lifetimes for the modified wood in order to achieve parity in terms of GWP impact is quite challenging. Re-examination of the data presented in Figure 1 shows that there is considerable scatter in the GWP and EE values. The data for Accoya and Kebony are limited (unsurprisingly), and although the average value has been assumed to be representative, other values could be used in the analysis. Closer inspection of the EPDs for Kebony (NEPD-407-287, NEPD-408-287, NEPD-410-288) reveals two distinctly different GWP values (+886 kg CO_2e/m^3 for Kebony Clear produced from southern yellow pine or from radiata pine, or +359 kg CO_2e/m^3 for Kebony Character produced from Scots pine). These absolute values for GWP impact take no account of the sequestered atmospheric carbon that is stored in the wood and furfuryl polymer of the product. In the EPD, these data have been combined to yield an overall negative GWP impact. This approach to dealing with sequestered carbon requires consideration of the end of life of the product and the ultimate fate of the carbon but is not able to deal with the temporal aspects of carbon storage [3]. For example, the climate change mitigation effects associated with atmospheric carbon storage are quite different for a product lifetime of 5 years, compared with 100 years. At this stage of the analysis, the carbon storage benefits have not been included and the impacts due to GHG emissions were either based on those reported in the EPD or calculated using those data. If the GWP impact for Kebony Character produced from Scots pine is used in the analysis, the payback time (considering A1–A5) becomes 64 years. The important issue of sequestered atmospheric carbon is dealt with later. A similar consideration of the Accoya EPD data (NEPD-376-262) also shows that there are large differences in the reported GWP for Accoya (Accys, Arnhem) made from radiata pine (source—New Zealand), Scots pine (Sweden) and beech (Germany). The GWP impact associated with the production of Accoya was not reported directly and had to be calculated based on the information supplied in the EPD (Table 6).

Table 6. Data obtained from the Accoya EPD (NEPD-376-262), the density, sequestered carbon and GWP total were reported in the EPD, whereas the GWP impact was calculated from these data.

Species	Density (kg/m^3)	Seq. C	GWP Total	GWP Impact	GWP Total [1]	GWP Impact
				($kg\ CO_2e/m^3$)		
radiata	510	−944	−433	511	−709	235
Scots	540	−999	−741	258	−1130	−131
beech	755	−1397	−1010	387	−1540	−143

[1] Includes acetic acid credits allocated on an economic basis

If the use of by-product acetic acid is included in the LCA, then the total GWP impact is reduced, in some cases giving an overall negative impact. From the data provided in the EPD, this indicates that the recovery of acetic acid by-product from the production of $1m^3$ of acetylated wood provides a GWP credit of 276 (radiata), 389 (Scots) and 530 kg CO_2e/m^3 (beech). Furthermore, negative GWP values are obtained for the whole process even when the sequestered carbon is removed from the wood in two cases. Acetic acid is produced as a by-product of the reaction of acetic anhydride with wood. Acetic anhydride is produced by the reaction of acetic acid via the carbonylation process.An overall negative GWP value, even when including system expansion and acetic acid credits, does not seem credible. Without access to the details of the process and the underlying LCA, it is not possible to examine this claim further. In addition, there is a considerable difference between the species when credits are included. Due to these apparent discrepancies, the data including acetic acid credits were not considered further.

The sequestered carbon in the Accoya products used in this EPD were estimated based upon the reported density of the Accoya products (replicated in the table). Using the methodology of EN16449 and these densities and an equilibrium moisture content of 0%, and an assumed carbon content of 50%, the calculated sequestered atmospheric carbon content was close to that reported in the EPD (935, 990, 1384 kg CO_2e/m^3 for radiata,

Scots, beech, respectively). However, these values should not be derived directly from the modified wood density since the modified wood has larger dimensions than unmodified and also weighs more due to the presence of bonded acetyl groups (which are derived from a fossil feedstock). The correct procedure to determine the atmospheric carbon content of the Accoya product is to use the density of the wood before acetylation. Table 7 shows the sequestered carbon determined using EN16449 at a moisture content of 10% with densities typical of the unmodified wood species.

Table 7. Sequestered atmospheric carbon content of radiata pine, Scots pine and European beech, based upon typical densities and a moisture content of 10%.

Species	Density (kg/m^3)	Sequestered C ($kgCO_2e/m^3$)
radiata pine	488	−813
Scots pine	455	−758
European beech	710	−1183

These values are lower than those quoted in the EPD by approximately 150–200 $kgCO_2e/m^3$. However, based on the data quoted in the EPD, the payback time varies from 68 to 135 years.

3.3. *Impact of Modified versus Unmodified Wood (Coated)*

In some markets, cladding is painted, and the impact of this and any improvement in properties due to wood modification was studied by assuming that a water-based alkyd paint was applied as two coats to the exposed face of the board. A coating was applied before installation and then at pre-determined intervals during an assumed 60-year lifetime. As with the previous model, a service life of 20 years was assumed for the unmodified wood. For a total lifetime of 60 years for the building, three replacements of the unmodified wood were accordingly assumed, whereas the modified wood was assumed to have a lifetime of 60 years. For the modified wood cladding, a coating interval of 10 years was assumed, due to the improved dimensional stability imparted by wood modification. The results from this analysis are shown in Table 8. The benefits of extended coating lifetimes are clear from this analysis.

Table 8. The total GWP impact associated a reference service life of 60 years. For the unmodified wood, a lifetime of 20 years was assumed, and coatings were applied every five years. For the modified wood, a lifetime of 60 years was assumed, and coatings were applied every 10 years.

Material	Total GWP ($kgCO_2e/m^3$) over 60 years
Unmodified	4343
TMT	2342
Accoya	2416
Kebony	2741

The assumptions made regarding coating intervals may not be realistic and the analysis was improved by assuming three different coating intervals for unmodified wood (2, 4, 6 years) and examining the effect of changing the coating intervals on the modified wood (from 2 to 10 years). The results from this analysis are shown in Figure 2.

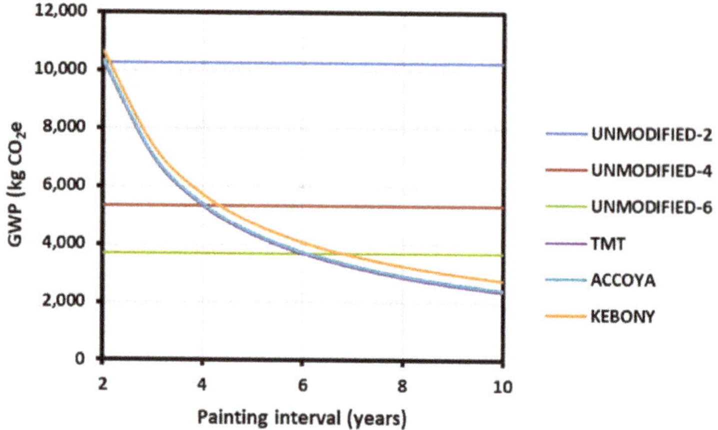

Figure 2. Effect of different coating intervals upon the GWP impact for unmodified and modified wood for a total area of 33.3 m^2. For unmodified wood, the horizontal lines represent the impact for a single coating interval of 2, 4, or 6 years. A total lifetime of 60 years is considered and life-time of 20 years for the unmodified wood, and 60 years for the modified wood was assumed.

This shows that the impacts associated with coating (ignoring travelling to site, scaffolding, disruption, etc.) have a very important contribution to the overall impact. The potential for modified wood to provide a more stable substrate for coating systems and consequently for there to be an extension in maintenance intervals is significant and further studies are warranted.

3.4. Dealing with Sequestered Carbon

There is no agreed method for incorporating sequestered carbon into life cycle assessment [3]. The problem is that the period of time that the sequestered carbon is held in storage is not properly accounted for, and there is considerable debate as to the best way to do this. The most recent version of EN15804 incorporating A2:2019 includes a specific category for reporting GWP-biogenic, in which sequestered atmospheric carbon can be included and reported separately from the GWP-fossil, as well as a table in which the biogenic carbon in the product due to conversion of atmospheric CO_2 into biomass can be reported. The GWP-biogenic indicator accounts not only for removals of atmospheric CO_2 into living biomass (not including native forest) into the product system but also includes any emissions of biogenic carbon into the atmosphere from all sources (except native forests), arising from combustion or disposal. The mandatory information that is provided in Table 9 of the document of the standard EN15804 and it is therefore essential for reporting the biogenic carbon content of the product, so as not to introduce ambiguities. However, other than reporting these values, the standard makes no recommendations regarding the reporting of the time of storage of the atmospheric carbon.

The benefits of carbon storage and extended lifetimes can be illustrated by the use of a simple model which considers the flows of atmospheric carbon into and out of a carbon pool that represents the built environment. Climate change mitigation occurs during the period that the inflows of carbon into the pool exceed the outflows (exactly as would occur with a forest). Although the carbon enters the pool each year as a single pulse as buildings are constructed from wood, the actual residence time or lifetime of the wood products in buildings is much more accurately represented by a distribution, so that the outflow of the carbon now takes place over a number of years. Many different functions can be used to represent this distribution, with an exponential decay often being chosen by default [10,11]. However, an exponential function over-estimates the losses for short time intervals and under-estimates losses over extended time periods [12]. For the purposes of this illustrative model, a normal distribution was chosen to represent the carbon outflow. Two average product lifetimes (+/−standard deviation) were selected: 20 (+/−5 years) and 60 (+/−5 years). The model was based upon an inflow of carbon of one tonne per year, with the outflow being modelled by the standard deviation. Thus, eventually all of the carbon associated with the inflow exits the pool into the atmosphere (assumed to be 100% oxidation of the carbon inflow). The results from such an analysis are illustrated in Figure 3.

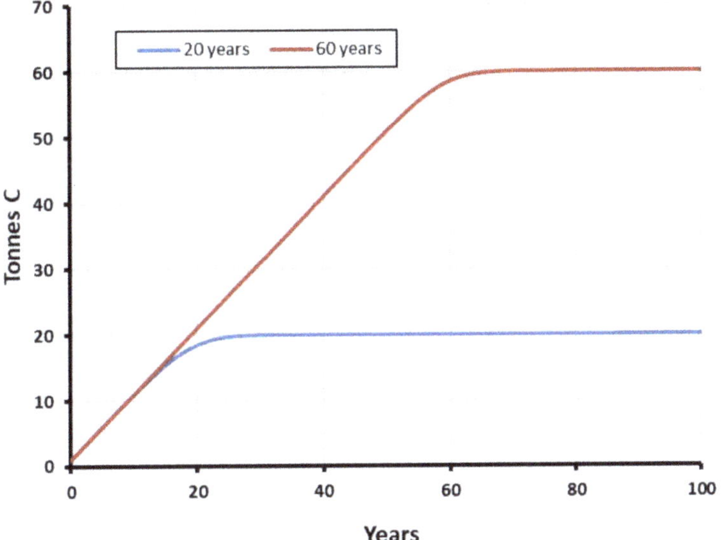

Figure 3. Effect on total carbon stored and time to equilibrium for a carbon pool with an input of 1 tonne C per year and a carbon loss represented by a normal distribution of time.

This shows very clearly that a tripling of the product lifetime leads to a 3-fold increase in the total amount of carbon stored and the time to reach equilibrium. Dealing with sequestered atmospheric carbon using this "stocks and flows" approach shows that eventually the flow of carbon out of the product pool equals the inflow, but there is a crucial period during which there is an increase in carbon stored, resulting in a mitigation effect. Depending on the longevity of the timber products, this mitigation period can be extended. The point here is that this buys time while a low carbon economics is created. Attempts to include the carbon storage benefits into a conventional LCA approach have many difficulties, and a consensus has not been reached as to the best way to do this. A stock and flow approach shows the benefits very clearly. The choice of function to represent the outflow may affect the time to equilibrium and total amount stored to a limited extent but does not change the overall findings. Applying this approach to a real-life situation requires accurate data regarding the lifetimes of timber products in the built environment. To some extent, this can be based on models that use laboratory and accelerated data, but examination of timber stocks and flows in the built environment receives very little attention. This analysis has concentrated on an analysis of average values and compared modified wood with unmodified in a European context. The impact of the treatment of wood with biocides has not been considered. For low hazard applications, the use of wood without coatings and biocides is sufficient, but for more demanding situations the use of biocides would be essential. At the present time, there are no satisfactory published data which could be used to examine the impacts of biocide use.

This study has only examined the GWP impacts for the purposes of comparison of modified with unmodified wood. Other impact categories should not be neglected when performing an analysis of the sustainability of different technologies. For example, in cases where timber is extracted from native forests, impacts upon land use and biodiversity may be a serious issue. At this time EN15804 does not include consideration of such impacts, apart from possible GWP implications of land use change. There is no impact category listed for impacts on biodiversity, apart from potential comparative toxic unit for ecosystems and potential soil quality index (SQP) as a land use related impact. However, these are listed with the disclaimer "The results of this environmental impact indicator shall be used with care as the uncertainties on these results are high or as there is limited experienced with the indicator". Until the scientific data and methodology supporting such impact categories improves, other impacts concerning land use and land use change should not be included in a comparative analysis. Other points that should be included in the comparative analysis are a more comprehensive consideration of the environmental impacts associated with installation, including potential disruption of other activities.

4. Conclusions

The results from this preliminary analysis have shown that any potential environmental benefit that might be realized from the use of modified wood is very clear when there is an extended time between coatings. The benefits of life extension of the wooden products is less clear if the sequestered carbon is not considered. The modification of wood has an associated environmental burden associated with the process over and above that for the production of unmodified wood. There is a considerable range of GWP values reported in the currently valid published environmental product declarations for unmodified and modified wood. In this preliminary analysis average values were taken for different product groups, and these averages were used to examine what the lifetime of the modified wood has to be so that the extra impact is equal to replacing unmodified wood over several lifetimes (herein called a payback time). When the environmental burdens associated with installation are not included, these payback times vary from 80 to 190 years. These payback times are reduced somewhat when the GWP impacts associated with installation are included in the analysis. Closer analysis of the EPDs for the chemical modification processes (Kebony, Accoya) shows that there are reductions possible, particularly when European-sourced wood is used (the study is in a European context). However, this more detailed analysis has also revealed inconsistencies in the Accoya EPD. Further work is needed to clarify the potential benefits (if any). When coated products are included in the analysis, a doubling of the time between re-painting (5 to 10 years) produces clear benefits in terms of a reduced GWP impact. The potential benefits of extending lifetime due to the use of a modified wood results in a much higher total storage of atmospheric carbon in the built environment carbon pool, and a longer time until equilibrium is reached. There is currently no agreed methodology for combining the LCA with the carbon storage benefits. It is essential that the difference in impacts due to material substitution and the effect on carbon storage dynamics are both considered when undertaking an analysis of the total impact on global warming potential. One further point to be considered here is that end of life, or multiple lives due to cascading of timber products, has not yet been considered in the analysis. This will be included in due course as the sophistication of the modelling improves. A particular problem with using the stock and flow approach is the choice of an appropriate model that accurately describes the outflow (oxidation) of carbon from the built environment pool. Current recommendations regarding building lifetimes, maintenance and replacement of components are not based upon scientific studies but upon assumptions. Thus, although qualitatively demonstrating the benefits of carbon sequestration are straightforward, measuring the benefits of carbon storage in timber products in the built environment are much more challenging. Further analysis is required to determine the effects of biocide use and to investigate other environmental impacts, as well as modelling the impacts associated with installation and maintenance in more detail.

Author Contributions: The individual contributions by the authors were as follows: conceptualization, C.H. and D.G.; methodology, C.H.; validation, C.H., D.G. and M.H.; formal analysis, C.H.; writing—original draft preparation, C.H.; writing—review and editing, D.G. and M.H. All authors have read and agreed to the published version of the manuscript.

Funding: Callum Hill wishes to thank Abodo Wood Ltd. for funding part of this study. Part of the work reported in this paper was carried out under the CircWood (Increasing the climate change mitigation potential of wood used in construction) project supported by the Growth and Development from Wood Program of the Finnish Ministry of the Environment under grant decision number VN/5272/2018. Mark Hughes and Callum Hill gratefully acknowledge financial support from this source.

Institutional Review Board Statement: Not applicable.

Informed Consent Statement: Not applicable.

Data Availability Statement: Data was sourced from publicly-available environmental product declarations published by program operators.

Conflicts of Interest: The authors declare no conflict of interest.

References

1. Hesser, F.; Wohner, B.; Meints, T.; Stern, T.; Windsperger, A. Integration of LCA in R&D by Applying the Concept of Payback Period: Case Study of a Modified Multilayer Wood Parquet. *Int. J. Life Cycle Assess.* **2017**, *22*, 307–316. [CrossRef]
2. Hill, C.; Norton, A.J. The Environmental Impacts Associated with Wood Modification Balanced by the Benefits of Life Extension. In Proceedings of the Seventh European Conference on Wood Modification, Lisbon, Portugal, 10–12 March 2014.
3. Tellnes, L.; Ganne-Chedeville, C.; Dias, A.; Dolezal, F.; Hill, C.; Zea Escamilla, E. Comparative Assessment for Biogenic Carbon Accounting Methods in Carbon Footprint of Products: A Review Study for Construction Materials Based on Forest Products. *iForest* **2017**, *10*, 815–823. [CrossRef]
4. Meyer-Veltrup, L.; Brischke, C.; Alfredsen, G.; Humar, M.; Flæte, P.-O.; Isaksson, T.; Brelid, P.L.; Westin, M.; Jermer, J. The Combined Effect of Wetting Ability and Durability on Outdoor Performance of Wood: Development and Verification of a New Prediction Approach. *Wood Sci. Technol.* **2017**, *51*, 615–637. [CrossRef]
5. Besnard, F.; Nilsson, J.; Bertling, L. On the Economic Benefits of Using Condition Monitoring Systems for Maintenance Management of Wind Power Systems. In Proceedings of the 2010 IEEE 11th International Conference on Probabilistic Methods Applied to Power Systems, Singapore, 14–17 June 2010; IEEE: Singapore, 2010; pp. 160–165.
6. Tellnes, L.G.F.; Alfredsen, G.; Flæte, P.O.; Gobakken, L.R. Effect of Service Life Aspects on Carbon Footprint: A Comparison of Wood Decking Products. *Holzforschung* **2020**, *74*, 426–433. [CrossRef]
7. Candelier, K.; Dibdiakova, J. A Review on Life Cycle Assessments of Thermally Modified Wood. *Holzforschung* **2020**. [CrossRef]
8. Ferreira, J.; Herrera, R.; Labidi, J.; Esteves, B.; Domingos, I. Energy and Environmental Profile Comparison of TMT Production from Two Different Companies—A Spanish/Portuguese Case Study. *iForest* **2018**, *11*, 155–161. [CrossRef]
9. Huijbregts, M.A.J.; Rombouts, L.J.A.; Hellweg, S.; Frischknecht, R.; Hendriks, A.J.; van de Meent, D.; Ragas, A.M.J.; Reijnders, L.; Struijs, J. Is Cumulative Fossil Energy Demand a Useful Indicator for the Environmental Performance of Products? *Environ. Sci. Technol.* **2006**, *40*, 641–648. [CrossRef] [PubMed]
10. Marland, E.; Marland, G. The Treatment of Long-Lived, Carbon-Containing Products in Inventories of Carbon Dioxide Emissions to the Atmosphere. *Environ. Sci. Policy* **2003**, *6*, 139–152. [CrossRef]
11. Pingoud, K.; Ekholm, T.; Soimakallio, S.; Helin, T. Carbon Balance Indicator for Forest Bioenergy Scenarios. *GCB Bioenergy* **2016**, *8*, 171–182. [CrossRef]
12. Marland, E.S.; Stellar, K.; Marland, G.H. A Distributed Approach to Accounting for Carbon in Wood Products. *Mitig. Adapt. Strateg. Glob Chang.* **2010**, *15*, 71–91. [CrossRef]

MDPI
St. Alban-Anlage 66
4052 Basel
Switzerland
Tel. +41 61 683 77 34
Fax +41 61 302 89 18
www.mdpi.com

Coatings Editorial Office
E-mail: coatings@mdpi.com
www.mdpi.com/journal/coatings

www.ingramcontent.com/pod-product-compliance
Lightning Source LLC
LaVergne TN
LVHW071028170726
843515LV00006B/1232

* 9 7 8 3 0 3 6 5 2 3 8 2 8 *